The Forc
New Vehicle Engineering and Technical History of the GT-40

PT-113

Published by
Society of Automotive Engineers, Inc.
400 Commonwealth Drive
Warrendale, PA 15096-0001
U.S.A.
Phone: (724) 776-4841
Fax: (724) 776-5760

For permission and licensing requests contact:

SAE Permissions
400 Commonwealth Drive
Warrendale, PA 15096-0001-USA
Email: permissions@sae.org
Fax: 724-772-4891
Tel: 724-772-4028

Front and Back Cover Photo Credits
GT photos courtesy of Ford Motor Co.
GT-40 photos courtesy of Pete Lyons

All SAE papers, standards, and selected books are abstracted and indexed in the Global Mobility Database.

For multiple print copies contact:

SAE Customer Service
Tel: 877-606-7323 (inside USA and Canada)
Tel: 724-776-4970 (outside USA)
Fax: 724-776-1615
Email: CustomerService@sae.org

ISBN 0-7680-1421-2
Library of Congress Catalog Card Number: 200311590
SAE/PT-113

SAE Order No. PT-113

Printed in USA

Preface

When I entered the Massachusetts Institute of Technology in the fall of 1969, I scarcely knew anything about the Society of Automotive Engineers. SAE papers were occasionally referred to in the car magazines that I had been devouring for years, but I had never laid eyes on one.

During my freshman year, I discovered the *SAE Journal* in MIT's Barker Engineering Library. Today, this magazine is called *Automotive Engineering,* and then, as now, the publication covered automotive technology in far greater detail than the mass-market, popular car magazines did, or do. As someone with a powerful interest in cars, and with an increasing affinity toward mechanical engineering, I became a regular reader. With stacks of back issues on hand, I spent many a spare hour in Barker Library.

One day, I discovered that the library also had a vast collection of the SAE papers themselves. This being decades before the computer age, the collection was kept on microfiche. To read a paper, you placed one of the four-by-six-inch photographic slides into a hefty machine that projected each thumbnail-size image of a page onto a rather large screen. The quality wasn't flawless, but it was certainly legible, and I now had even more motivation to spend time in the library.

As I perused the list of the papers in the catalog, I found seven published in 1967 that pertained to the Ford GT Mark II that had, after three years of trying, won the Le Mans 24-hour race for Ford by finishing first, second, and third in 1966. If I had discovered the Holy Grail, I wouldn't have been any more thrilled. Those seven papers are reproduced in this volume.

Paper no. 670066 covered the development of the 427-cubic-inch V-8 that powered the winning car. In its 21 pages, authors Joe Macura (who I would later meet when I joined Ford's Advanced Engine Engineering office in 1968) and Jon Bowers explained in great detail how Ford engineers extracted 485 horsepower from the pushrod engine while preserving decent fuel consumption and maintaining the robust reliability demanded by a race in which the winning car covered 3009 miles in 24 hours.

Paper no. 670067 detailed the development of the car's intake manifold, which was so critical to producing the broad power band needed to succeed on a circuit where speeds ranged from 35 to 215 mph; 670068 dealt with the engine's ignition and electrical systems; 670071 explained the computer simulation system that had been built in Ford's Dearborn engineering center to simulate the accelerations and gear changes of each Le Mans lap. This simulator was used to test the entire powertrain until it could reliably survive for 48 hours—twice the duration of the race.

The Ford GT's engine wasn't the only focus of these papers. No. 670069 dealt with the purpose-built four-speed transaxle, which had to be developed to deal with the torque of the 7.0-liter engine. The 32-page no. 670070 covered the brakes, which had to dissipate 12,597,900 foot-pounds of energy during each lap of the 8.36-mile Le Mans circuit. And 670065 covered the overall development of the car and its various systems.

These papers provided a wonderful insight into the development of this successful car, during what many consider to be the technically most interesting period of sports car racing. In the Sixties, the rules at Le Mans were remarkably wide open, whereas today, every class of racing is restricted by a thick tangle of regulations limiting engine size, number of cylinders, type of aspiration, minimum weight, wing sizes, tires sizes, and virtually everything else.

The Ford GT Mark IIs, with their enormous low-revving pushrod V-8s, competed with Ferrari P3 prototypes powered by 4.0-liter double-overhead-cam V-12s that revved to 8000 rpm. Meanwhile, Porsche campaigned Carrera 6s with 2.0-liter engines, ultra-lightweight chassis, and highly aerodynamic long-tail bodywork. And American innovator Jim Hall showed up with his Chaparrals, bristling with advanced features such as fiberglass chassis construction, automatic transmissions, and wings.

One key reason for this variety was that in the Sixties very little science and engineering had been applied to the art of motor racing. As a result, there was no general agreement about the best technical approach to generating speed on a road racing track.

The various carmakers each viewed the problem through the lenses of their own histories and capabilities. The cars on the starting grid demonstrated how varied these histories were.

When Ford first assaulted Le Mans in 1964, the company followed a similarly casual approach by initially purchasing a race car design from the English firm Lola. This car's numerous shortcomings soon led Ford to apply its considerable engineering and developmental resources to the project, and the result was the one-two-three finish in 1966.

As the SAE papers in this collection illustrate, this excellent result was anything but random. Instead, it was the outcome of a highly organized, well-financed, technically rigorous process that was applied to every aspect of the Ford GT's design.

This process produced an even more sophisticated Ford GT Mark IV that finished first and fourth in 1967. After the two Ford victories, the French organizers of the Le Mans race imposed a 3.0-liter displacement limit on the one-off prototype racers and a 5.0-liter limit on GT cars that were produced in quantities of at least 25. Even this change couldn't stop the latest versions of the earlier Ford GTs, which had been produced in quantity and were by then thoroughly sorted and well understood, from winning Le Mans in 1968 and 1969.

It is these four fabulous victories by Ford in the Sixties that inspired the new Ford GT that is just going on sale. Based on a concept car that debuted at the 2002 North American International Auto Show in Detroit, the new production car embodies the characteristic proportions and styling elements of the original.

Under its skin, however, it has little in common with the original other than its mid-engine layout. In truth, the new car must satisfy a much broader set of demands than did the racing car form the Sixties. In addition to delivering an extremely high level of performance, the 2005 Ford GT must also function as a street car, with a climate control system, moderate interior noise levels, a reasonable ride, and the ability to operate in extremes of hot and cold.

Moreover, modern production cars must meet a complex and wide-ranging set of regulatory requirements that were completely unknown when the original Ford GT was designed. I would guess that 200 of the new GTs produce less smog-producing exhaust emissions than a single original GT. And the crash protection provided by the modern version would have been inconceivable in the mid-Sixties.

The SAE papers about the new GT, included in this volume, explain how Ford engineers managed to meet these numerous requirements while staying true to the sprit of the original. As you read them, compare their solutions to those crafted by the engineers in the Sixties. And keep in mind that those Ford engineers, who 40 years ago developed the GT, were among the first to apply modern scientific and organizational principles to the creation of a successful racing car.

Csaba Csere
December 2003
Ann Arbor, Michigan

Table of Contents

New Vehicle Engineering--The Ford GT

Historical SAE Papers on the Ford GT-40

NEW VEHICLE ENGINEERING - THE FORD GT

2004-01-1251

2005 Ford GT – Melding the Past and the Future

Fred Goodnow and Matthew Zaluzec
Ford Motor Co.

ABSTRACT

The 2005 Ford GT high performance sports car was designed and built in keeping with the heritage of the 1960's LeMans winning GT40 while maintaining the image of the 2002 GT40 concept vehicle. This paper reviews the technical challenges in designing and building a super car in 12 months while meeting customer expectations in performance, styling, quality and regulatory requirements. A team of dedicated and performance inspired engineers and technical specialists from Ford Motor Company Special Vehicle Teams, Research and Advanced Engineering, Mayflower Vehicle Systems, Roush Industries, Lear, and Saleen Special Vehicles was assembled and tasked with designing the production 2005 vehicle in record time. Key to the success of the program was the establishment of co-located cross functional teams from product development and manufacturing organizations from the various companies responsible for the different aspects of the vehicle including body, trim, electrical, chassis, power train, and final assembly.

INTRODUCTION

In January 2002, the Ford GT40 concept car was put on display at the North American International Autoshow and instantly became a vehicle in demand. Less than 45 days after the concept vehicle was introduced to the public, the decision to design and build a production version was announced. The technical challenge was to build a high performance super car that met today's customers demands for performance, styling and quality. Adding to the technical challenges was the requirement to build the first three production vehicles in time for Ford's 100 Year celebration and, soon thereafter, launch full production with a job 1 date targeting early spring of 2004. Although the production vehicle, Figure 1, and the original racecar would share the mystique of the Ford GT name, the production vehicle required the development of a completely new vehicle architecture, radically different from the original steel and fiberglass LeMans racecars built in the 1960's. In order to meet weight targets and timing, the GT team designed the production vehicle around an all-aluminum space-frame. Extensive use of extrusions and castings allowed for a faster development time while minimizing capital expenditures and program costs. In order to preserve the vintage styling of the original 1964 Mark I GT40 exterior, Figure 2, Mayflower Vehicle systems made extensive use of super plastic forming of aluminum to preserve the unique vehicle contours while providing functional benefits in the form of air flow management. In the tradition of the original Ford GT, 8-inch deep hood vents allow the airflow across the heat exchangers to exit. Similarly, all rear-engine deck scoops are functional for engine cooling and heat management. The front fenders drape over 18-inch Goodyear eagle F1 super car tires, while the rear deck wheel openings are engineered to accommodate 19-inch wheels and tires. Due to the mid-engine design of the 2005 vehicle, weight distribution was critical. In order to keep the front-to-rear weight distribution at a 43:57 ratio, an ultra-light rear deck was designed consisting of a four-piece super-plastic formed outer structure assembled to a one-piece carbon fiber rear deck inner structure. The unique design of the original Ford GT40, featuring a right angle door wing that cuts into the roof, was captured in the production body of the 2005 vehicle.

Figure 1. 2005 Ford GT

Like the body, the powertrain also followed tradition by employing a low-profile, powerful V8. The original 1966 Ford GT40 Mark II was powered with a 427 CID (7.0L) engine delivering 485 HP. Working closely with Roush Industries, the Ford GT uses a modified all-aluminum 5.4 liter production power train matched with a Ricardo Transmission and fed by an Lysholm Eaton screw-type supercharger to meet the performance requirements of the vehicle. It features four valves per cylinder and forged components, including the crankshaft, the H-beam connecting rods and the aluminum pistons. The 5.4L V8 power plant can deliver over 500 HP and 500 foot-pounds of torque. Finally, a completely new interior engineered by Lear captured the essence of the original Ford GT40 race interior, while providing functionality coupled with modern craftsmanship and an atmosphere of performance.

Figure 2. - 1964 Ford GT40 Mark I

TIMING

Since official program approval, the 2005 vehicle has been on the fast track for product-development timing. Developing the Ford GT from approval to drivable production models in less than a year meant the team relied heavily on computer models to compress the typical first nine months of engineering work into three months, figure 3. Unlike a typical vehicle program where a vehicle is designed and built over a 48-52 month time frame, the vehicle was designed and built in record time. The total number of prototypes was reduced to 24 and the first 9 workhorse cars were built less than 100 days after program approval. Starting with a pre-production workhorse space frame, the chassis was built using "make like" production castings and extrusions with the intention of carrying over many prototype extrusions into the production vehicle. The build process of the first three production cars kicked off on March 10, 2003.

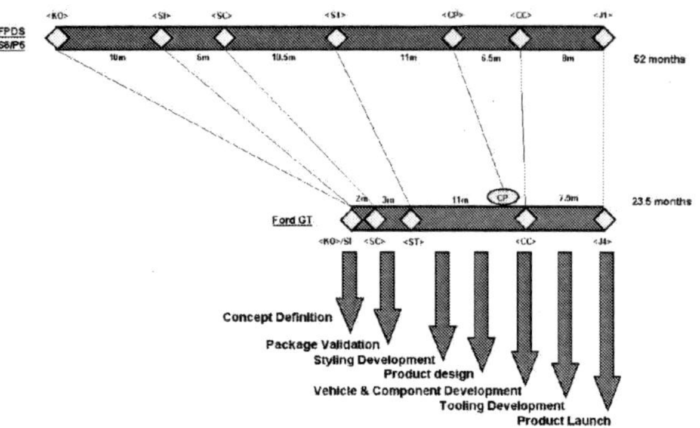

Figure 3 – Ford GT Product Development Timing.

THEME DEVELOPMENT

DESIGN: CONCEPT TO REALITY

The Ford GT is the ultimate Living Legend; a true super car with appeal equal to that of the greatest sports cars in the world but with the addition of a heritage no one can match. The 2005 production car, like the concept, casts the familiar, sleek look of its namesake, yet every dimension, every curve and every line on the car is a unique reinterpretation of the original. The car features a long front overhang reminiscent of 1960s-era racecars (figure 4). But its sweeping cowl, subtle accent lines and high-intensity-discharge (HID) headlamps strike a distinctly contemporary pose. Essential elements of the original – including the stunning low profile and mid-mounted American V-8 engine – continue in this latest interpretation of the classic. The new car is more than 18 inches longer and stands nearly 4 inches taller. Its new lines draw upon and refine the best features of Ford GT history and express the car's identity through modern proportion and surface development.

Figure 4. Ford GT Design Concept

Contrary to typical vehicle development programs, the engineering challenge was to build the supercar foundation within the concept's curvaceous form – and to build it in record time for Ford's centennial. The well-defined project afforded the engineering team early insight: This car required a new way of doing business since the concept car was only 5 percent production-feasible. Body design and manufacturing engineers employed new techniques to shape the car's crafted lines because normal stamping techniques couldn't deliver these curves.

Figure 5 - Ford GT Design Concept

Specification	GT40 Concept	Ford GT Production
Exterior		
Wheelbase	107.0 in.	107.0 in.
Overall Length	179.6 in.	182.3 in.
Overhang		
Front	42.7 in.	42.9 in. 33.23
Rear	30.15 in.	in.
Overall Height	43.7 in.	44.3 in.
Ground Clearance	4.7 in.	5.0 in.
Overall Width	77.3 in.	76.9 in
Tread Width		
Front	32.2 in.	31.4 in. 31.5
Rear	32.5 in.	in.
Tire/Wheel		
Front	245/45R18x8 in.	235/45ZR18x9 in.
Rear	285/45R19x10 in.	315/40ZR19x11.5 in.
Interior	42.	
Legroom	in. . 34.7 ir	43.0 in. 35.3
Headroom		in.

Figure. 6 – GT40 vs Ford GT

The distinctive duck tailed rear clip (figure 5) and the unique door design needed to accommodate the requisite slide-down window. After extensive computer modeling and concessions by designers and package engineers, the window freely moved within the door panel. Aerodynamicists couldn't bend the exterior sheet metal; instead, they came up with unique solutions under the body in the form of an aerodynamic underbody airflow management system. The result: a technological wonder wrapped in the Ford GT40 concept form. Figure 6. provides a comparison of the specifications of the 2002 Auto Show GT40 Concept car, compared to the 2005 Ford GT production car.

AERODYNAMIC AND THERMAL DESIGN

Like the concept car, every air intake and heat extractor on the production vehicle is functional. Preliminary wind tunnel testing showed the concept car had remarkably good internal airflow. A fiberglass replica of the concept car was tested in the wind tunnel. Because the design was so close to that of the Ford GT racecars, the intakes and diffusers were all in the right place, thus only minor changes were needed to improve airflow throughout the car.

Compared with the 1969 GT, the heat extractors in the front cowl area were widened and lowered to pull more heat from the front-mounted radiators. The side intakes under the B-pillar were slightly enlarged, driving more cooling air into the engine bay and transmission cooler. Finally, additional vents on either side of the rear glass, and center of the clamshell (above the engine) help diffuse heat from the engine compartment, during idle conditions.

Improving the aerodynamic stability was not an easy task. The aero team tested an original Ford GT40 racecar in the wind tunnel with computer simulations to measure drag, lift, and down force. The original car exhibited very high frontal lift at speed. Because the new car shared a similar shape, the new aero model exhibited similar lift. So, to preserve the design of the concept car, the aero team had to concentrate on the underside of the vehicle. As a result, a front splitter was designed and installed, which creates a high-pressure area for front down force and limits the volume of air traveling under the vehicle. Also added were side splitters to prevent air from sliding under the rocker panels. A smooth, fully enclosed belly pan reduces underbody turbulence. Finally, venturi tunnels accelerate exiting air, creating a vacuum that literally sucks the car to the pavement. The cumulative result is significant down force at speed and one of the most efficient lift/drag values on a production car.

SUPPLIER/FORD TEAM STRATEGY

The strategy for building the 2005 Ford vehicle relies heavily on a limited number of hand-picked "core system" suppliers to deliver major portions of the vehicle. Logistics for assembling the vehicle were limited to a 250-mile region centered at Dearborn, Michigan. As a result, the vehicle was split into major vehicle systems (figure 7) with representative suppliers delivering their products as fully assembled modules, ready for the next stage of assembly.

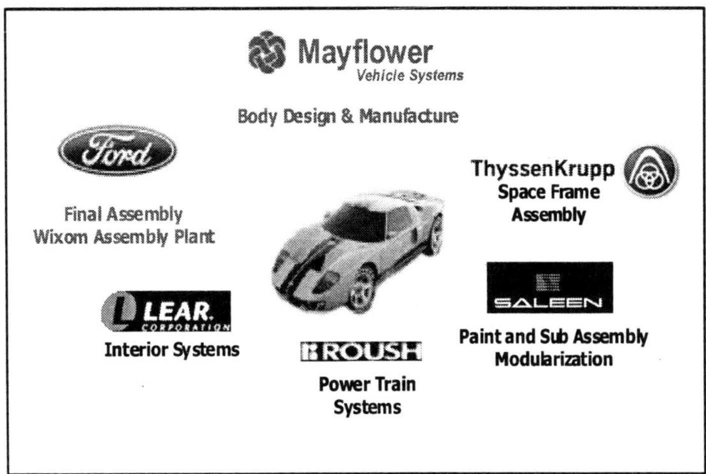

Figure 7. – Ford GT subsystems by supplier

The aluminum space frame, designed by Mayflower Vehicle Systems, is assembled in Detroit Michigan at a Thyssen-Krupp (TK) Manufacturing Facility. Engineered aluminum extrusions, pre-machined aluminum castings and stamped aluminum panels are shipped directly to the TK manufacturing facility where the finished aluminum space frame is MIG welded and assembled from four primary subsystem modules including the front end, rear end, cab and greenhouse subassemblies. Post-assembly computer numerically controlled (CNC) machining is completed to dimensionally set the suspension, powertrain, and body panel attachment points prior to shipment to Mayflower Vehicle Systems Manufacturing facility in Norwalk Ohio. The fully assembled and e-coated space frame, aluminum body-in-white and closures are fitted to the aluminum space frame, dimensionally set for gap and flushness and shipped to the next assembly point for paint and subsystem modularization. Saleen Special Vehicles, located in Troy Michigan, is responsible for painting the space frame and the body in white. They are also responsible for kitting each vehicle into the primary modular powertrain and interior subsystems in preparation for final assembly. Working closely with Lear and Roush Industries, and Ford's Romeo Engine Plant, Saleen Special vehicles is responsible for assembling a rolling space frame from the primary pre-assembled modules. The secondary modules are kitted and shipped with the rolling space frame to Ford Motor Company's Wixom Assembly plant for final assembly where engine decking, windshield installation and final interior installation is completed.

PACKAGE DEVELOPMENT

To maximize passenger comfort the engineering team made extensive use of a virtual-reality computer-modeling device called the digital occupant buck. The device is a revolutionary step in CAD/CAE technology with a virtual re-creation of the interior surfaces translated from the CAD data; a physical mock-up of the seat, steering wheel and pedal assembly; and a test engineer fitted with magnetic sensors, which manipulate a virtual person inserted in the digital interior.

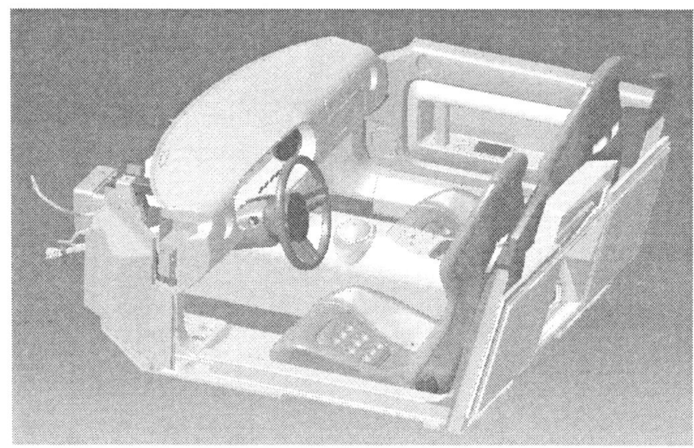

Figure 8 – Digital Interior Buck of Ford GT Interior

The real advantage of digital occupant is that it allows engineers to develop a comfortable interior for a wide range of statures. As a result of this modeling, the occupant package for the maximum range of accommodation is achieved. This included obvious improvements, such as maximizing seat travel and headroom. In addition, it makes use of other subtle improvements, like centering the pedal package relative to the driver's seat, and canting the shift lever toward the driver for optimal driving ergonomics.

Figure 9 – Digital Occupant Package Buck

Figure 10 – Mechanical Package Layout

The benefits of using the CAD/CAE technology for the mechanical, electrical and fuel systems layout for rapid analysis of system configuration and re-configuration were also realized. Crash modeling was used to verify location of the fuel tank under the extruded aluminum tunnel. Using systems modeling, the mechanical components, including the fuel pumps, level sensors, and vapor control valves were designed to be mounted on a steel rail with the fuel tank blow molded around the rail. This not only maximized fuel volume and reduced the number of connections to the fuel systems, but also was engineered in record time with the fewest number of iterations needed before making the final production fuel tank system.

ENGINEERING SOFTWARE

In order to accomplish the monumental task of delivering a production-ready design capable of producing representative parts that could be assembled into a fully functional vehicle in 12 months, it was of paramount importance to employ the latest CAE tools to deliver the 2005 production vehicle. Early in the program, ALIAS was used to generate concept sketches of the vehicle as a visual-aid tool for defining the overall vehicle styling, package and appearance.

Product data management – CATIA design control, bill of material (BOM) & data management was tracked using ENOVIA Virtual Product Modeler (VPM). Full vehicle data control and transmittal processes were accomplished using a proprietary solution installation. Change control system management was tracked through VPM to capture and understand the cost of change throughout the program. The VPM data was made available to all engineers and designers via an intranet user sight and ENOVA portal software.

Product Design – Product design of the production vehicle body and space frame was engineered by Mayflower Vehicle Systems using IBM Dassault CATIA V4.2.2 & V5. Dimensional variation analysis (DVA) was captured using Dimensional Control Systems 3D (DMS) software.
OPTISTRUCT was used to model the vehicles design and mass and was effective in minimizing vehicle weight by design. Class "A" Exterior surfacing was analyzed using ICEM surf. and virtual surface verification was completed using Opticore OPUS Realizer rendering software.

Package Validation - Virtual package validation was made using ENOVIA Digital Mock-UP (DMU), which allowed for visualization of complete 3D vehicle environments by using the latest tessellated models. The DMA software was used for both 3D package & validation of current design as well as for clash detection between modeled subsystems. IDEAS was used for the interior and powertrain systems modeling. Suspension kinematics and compliance analysis was completed using ADAMS and SUSPENSIONGEN, whereas FRAMs was used to model ride and handling.

Manufacturing Simulation – A critical aspect of delivering the program on time was the use of virtual tool validation prior tooling release. Autoform and Pamstamp were used for 1-step & incremental press simulation for both conventionally and superformed aluminum body panels. FLOWCAST and SOLIDCAST modeling was used to optimize and simulate the casting process for the aluminum cast shock towers and midgate rear casting.

Finite Element Analysis – Dynamic crash of the Ford GT was modeled using LS Dyna. Static and transient stiffness analysis and Modal response was modeled using Nastran. Occupant safety analysis and kinematics was evaluated using MADYMO and RAMSIS analysis software was used for anthropometrics (range of human motion) studies.

ALUMINUM SPACE FRAME

The aluminum space frame for the 2005 production Ford vehicle makes extensive use of aluminum extrusions. In all, 35 detailed extrusions are used though out the space frame structure. Four large complex castings are used for the front and rear shock towers and offered the opportunity for part consolidation. The rear shock tower includes the mount brackets and several key attachment points for the stretch bent side rail, rear crash box, and cross car beam stiffening reinforcements.

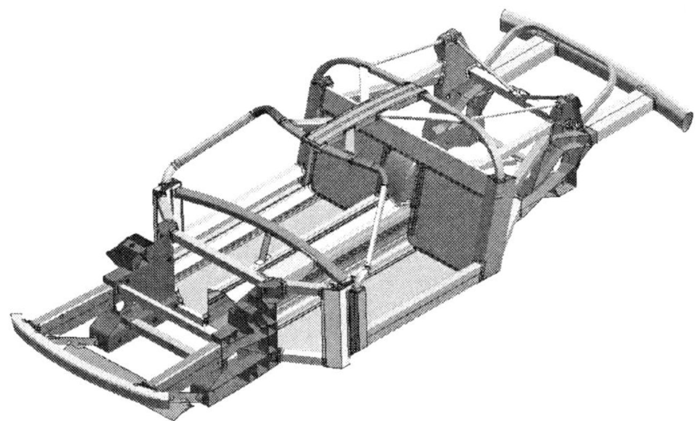

Figure 11. – Aluminum Space Frame

Smaller cast nodes are employed in the upper and lower A-pillars and a cast tunnel housing is used for the rear midgate allowing access to the engine via removable close out panels. Simple brake formed stampings are used for the close out panels with the exception of the floor and tunnel, which use an ultra light weight pressure roll bonded aluminum sandwich panels.

Final assembly of the aluminum space frame uses automated MIG welding with post-weld CNC machining providing the accuracy needed for suspension, power train and body-in-white attachment. A notable contributor to space frame rigidity is the industry's first application of friction-stir welding, used to construct the multi-piece aluminum tunnel (housing the fuel tank). With this technique, a tool rotating at 10,000 rpm applies pressure to a seam and actually blends the metal surfaces, forming a smooth, and a continuous seam. Compared to automated MIG welding, friction-stir welding improves the dimensional accuracy of the assembly, and produces a 30 percent increase in joint strength.

ALUMINUM BODY STRUCTURE

Two types of aluminum products are used to manufacture the body structure. Conventional stampings are used to make the greenhouse inner and outer structures, whereas super-plastic formed aluminum is needed for the complex shapes used to make the doors, front fenders, rear quarter panels and engine deck lid. Unlike conventional stamping, super plastic forming of aluminum involves heating an aluminum blank to 950°F. The heated blank is inserted into a press containing a single-sided steel die representing a positive surface of the body panel. The press is closed, brought up to the super plastic forming temperature and back filled with high-pressure air. The aluminum is then super-plastically formed over the die to create body panels with unprecedented shape compared with conventional aluminum stamping processes.

Most aluminum space frame vehicles use nut inserts paired with shims or washers to tailor the fitment of each body panel. The vehicle team developed a novel new method, called a "plus-nut," to efficiently join the body and frame, as well as locate the body panels in the proper position relative to the space frame.

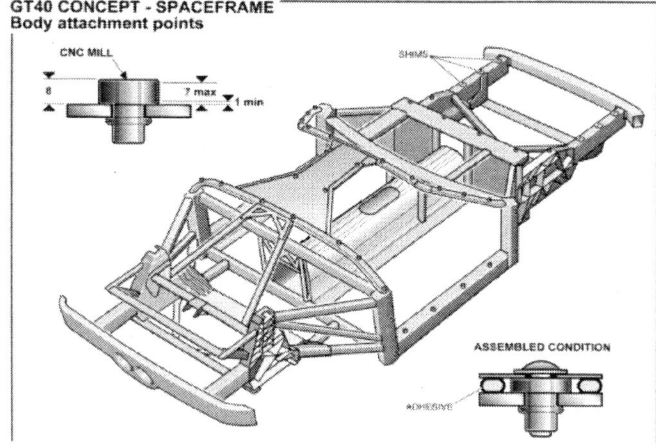

Figure 12. - Plus nut Mechanical Fasteners

These fasteners are essentially aluminum nut inserts, with additional machining stock on the mating surface. While machining the suspension and engine mounts, Computer Numeric Controlled (CNC) milling accurately trims each aluminum plus-nut for precise body positioning. The patent-pending fasteners eliminate the need for shimming the body, reducing assembly costs and improving panel fit.

The assembly sequence starts with the greenhouse inner structure being secured to the space frame using the plusnut mechanical fasteners and adhesive bonding. A two part, high modulus, room-temperature cure, Dow epoxy adhesive is used to supplement the mechanical attachment of the inner panels to the space frame. In additional, the adhesive serves a dual purposed acting as a sealer as well. Once the inner panels are attached the space frame, the outer aluminum panels are bonded and riveted to the inner panel. The combination of a MIG-welded space frame together with a riveted bonded body in white structure provide one of the stiffest body structures in the industry. The estimated torsional stiffness of the 2005 vehicle body in white is approximately 23,000 ft-lbs/degree (30,784 Nm/degree), and the bending stiffness of 54,000 lb/in (9275 N/mm).

The final stages of the body-in-white assembly involve the installation of the two lightweight super-plastic formed doors and the ultra-lightweight aluminum/carbon fiber rear deck. Once the doors and deck are set for gap and flushness, the vehicle is dimensionally checked for margins and flushness, and prepared for shipment to Saleen Special Vehicles for paint and sub assembly modularization.

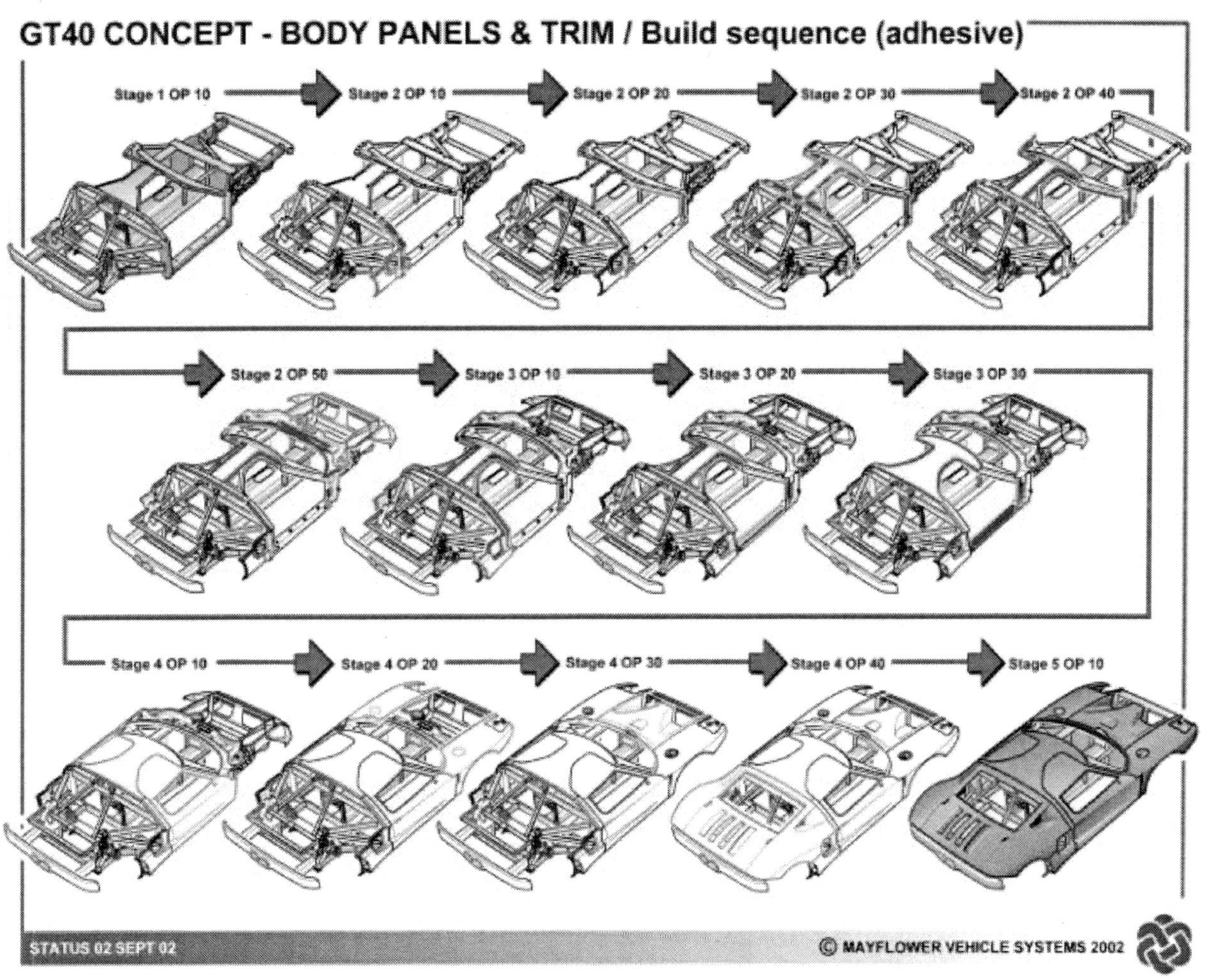

Figure 13. – Body Assembly Sequence

TRIM AND FINAL ASSEMBLY

Saleen Special Vehicles was selected to assemble the GT based on their experience with low volume niche vehicle production. Engineers were present during the development of the car to drive towards modularity for ease of assembly and service. The vehicle is partially assembled at SSV's assembly plant located in Troy, Michigan. Based on a rolling chassis principle, the aluminum space frame is received with body panels and closures attached, set to specified gap and flushness targets. The closures and body panels are removed from the space frame and processed on specially built painting racks that hold each panel in true body position during painting. The space frame is processed separately on rolling carts where electrical and fuel system components are installed. Due to the low-volume nature of the vehicle many traditional hand tools are utilized to minimize cost. Instead of depending on expensive automated equipment for assembly, the selection of highly skilled labor was used insure assembly quality. Once the suspension and wheels are assembled onto the vehicle, the space frame becomes a rolling chassis. The final stages of assembly at SSV include re-attachment of the painted body panels and closures and final preparation for shipment to Wixom for final assembly as shown in figure 13.

The partially assembled vehicle is shipped to the Wixom Special Vehicle Center where a group of hand-selected UAW technicians complete the assembly including glass and interior installation, engine decking and fluid fill. Each GT is subjected to a rigorous and thorough quality inspection and road test. Technicians were chosen based on experience, dedication and enthusiasm about the Ford GT project. Instead of using a "sampling" of vehicles to run through each test, quality checks are done to every vehicle in order to maintain the highest build quality possible.

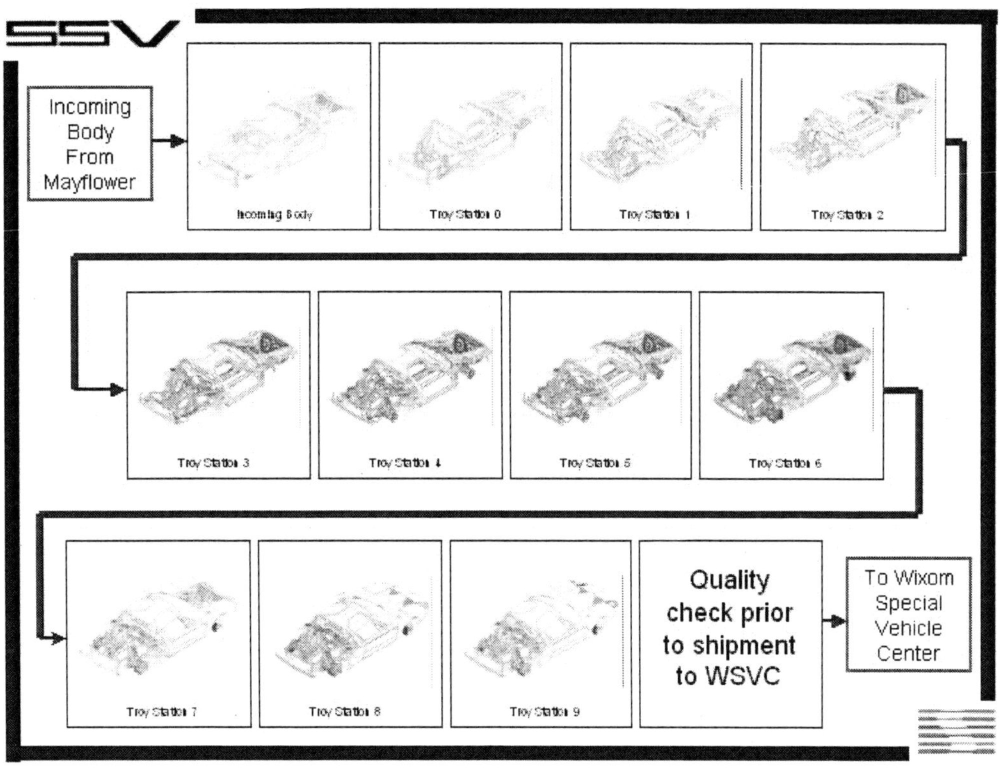

Figure 14. – Paint and Trim Assembly (SSV)

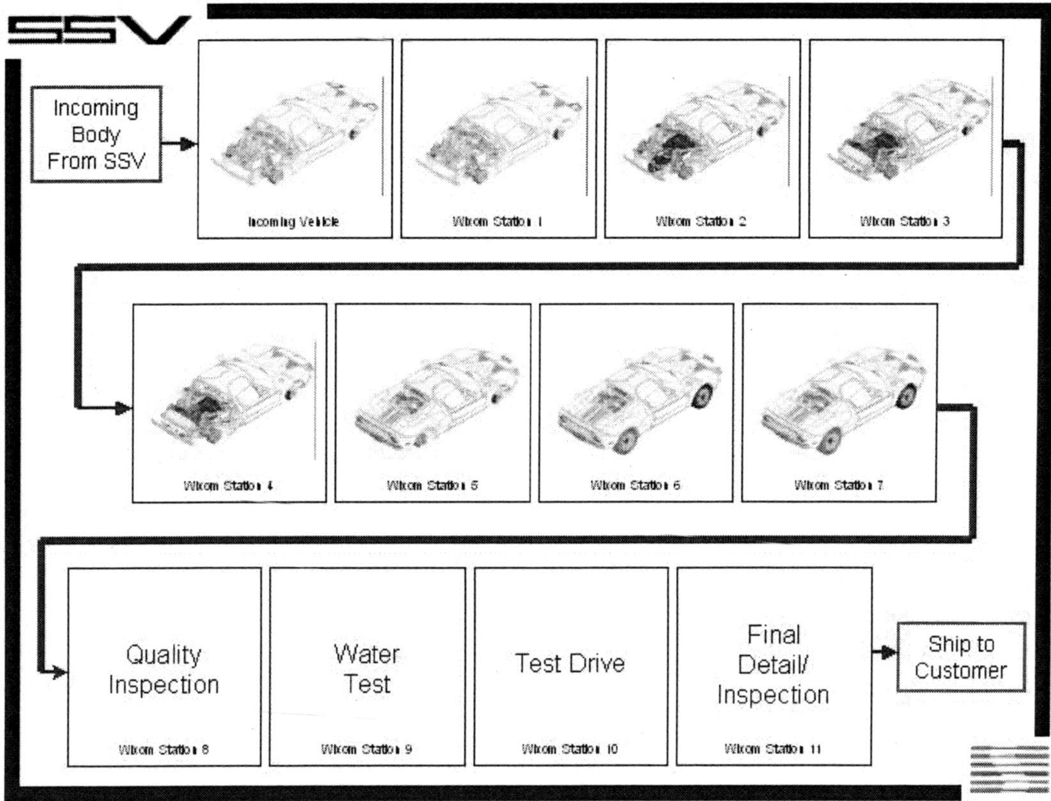

Figure 15. – Final Assembly, Wixom Specialty Vehicle Center.

QUALITY OPERATING SYSTEM (QOS)

The quality operating system adopted for the 2005 production program had to deliver PSW (Part Submission Warrant) and systems to robust manufacturing systems. Additionally, the parts had to be designed and validated for their unique duty cycle. The quality planning stage had to take into account many unique program characteristics:

- The vehicle usage patterns e.g. expected lifetime mileage
- The necessity of using non-traditional suppliers to supply low-volume systems, with associated lack of Quality Operating System (QOS) knowledge and experience.
- Low sample sizes for part and system validation. A normal Production Part Approval Process (PPAP) requirement might expend the first year's production volumes for a low-volume program such as Ford GT.
- Budget and resource restrictions required modifications to normal QOS in order to build a reasonable business case.

THE DESIGN PHASE

The design team had to be handpicked in order to ensure a balance of experience and enthusiasm. Product quality on the 2005 vehicle came as a result of the structured use of experience.

- The definition of the vehicle Design QOS was based on a modified Advanced Product Quality Planning (APQP) gateways and additional unique gateways specific to a low volume specialty vehicle program.
- Development of specific vehicle Design Failure Mode Effect Analyses (DFMEAs) that were clearly linked to the Design Validation Plan (DVP).
- Independent system engineering specialists conducted 'Deep Dive' reviews of DFMEAs and DVPs.
- Independent system engineering specialists conducted 'Fresh Eyes' reviews of the Design Aid buck.
- Early definition of 'High Risk and Non-traditional' suppliers was established.
- Accelerated program timing was linked to quality gateways.

THE VALIDATION PHASE

The design team, working with Ford GT Quality, monitored the DVP for development driven design changes and associated DFMEA updates. Since the vehicle incorporated several new technologies and relied heavily on specialty suppliers for various parts, special attention was given to parts and systems from high risk and non-traditional suppliers.

- Space Frame manufacturing – The aluminum space frame, designed around extruded and cast components, was a Ford First and required the development of new design guidelines for validating the quality and dimensional control of extrusions and castings.

- Aluminum Body-in-White – The application of super plastic formed (SPF) aluminum for the bulk of the class A outer body panels required the development of non-traditional pre-treatment and surface finishing to meet customer quality expectations. Similarly, the development of a SPF QOS was implemented to ensure consistent dimensional and surface quality targets were met.

- Sourcing of non-traditional tooling suppliers of extrusions, castings and super plastic formed aluminum body panels required each to demonstrate a supplier specific QOS suitable to meet Ford's purchasing and engineering requirements.

THE MANUFACTURING PHASE

The limited number of pre-production vehicles required the continual monitoring of 'High Risk and Non-traditional' suppliers. Monitoring started early on in the workhorse phase of the vehicle development and continued through job 1 launch.

The definition of low volume PPAP requirements that will deliver Part Submission Warrant (PSW) parts was verified with low volume sampling and batch build processes. This continued through the Confirmation prototype builds leading up to the production launch.

The development of a manufacturing QOS that incorporates a planning phase (APQP), a launch phase (LQOS) and a manufacturing phase (MQOS) was established for the Ford GT program, not unlike traditional higher volume production vehicles.

The efforts of the co-located teams from both product and manufacturing from Ford and the Ford GT suppliers.

CONCLUSION

The evolution of the 2005 Ford production vehicle project taught the cross-functional program/engineering /manufacturing team a lot of new concepts that could be applied to low-volume, quick-to-market specialty vehicles. The Ford GT program is a clear indication that high-quality, high-performance vehicles do not have to be compromised in processes to accomplish short execution time.

The design and development of the vehicle was accomplished in a fraction of the normal, high-volume mainstream timeframe by: 1) Hand-picking the engineering and supplier team with known expertise and competency, and: 2) Employing the usage of all available current technologies, both electronic and process-wise, to accomplish the end result in a short amount of time, without any compromise to program integrity.

As a result, the Ford GT represents a true high performance American sports car brought to life in the true spirit of winning in competition. Whether the competition is the Le Mans race circuit or the drive to win the hearts of America's driving enthusiasts, the 2005 Ford vehicle is a true winner.

ACKNOWLEDGMENTS

The following team of Ford personnel contributed to the preparation of this paper:

Bill Clarke – Ford GT Body Systems Supervisor

Curt Hill – Ford GT Powertrain Systems Supervisor

Kip Ewing – Ford GT Package & Prototype Supervisor

Kent Harrison – Ford GT Development Vehicle Development Supervisor

Huibert Mees – Ford GT Chassis Systems Supervisor

Camilo Pardo – Ford GT Chief Designer

Bob Brown – Ford GT Quality Manager

Mark Lindgren – Ford GT Process Engineer

Adrian Elliot – Ford GT Space Frame Mfg Supervisor

Rama Koganti - Ford GT Body Mfg Supervisor

2004-01-1252

2005 Ford GT Powertrain – Supercharged Supercar

Curtis M. Hill and Glenn D. Miller
Ford Motor Company

Robert C. Gardner
Roush Industries, Inc.

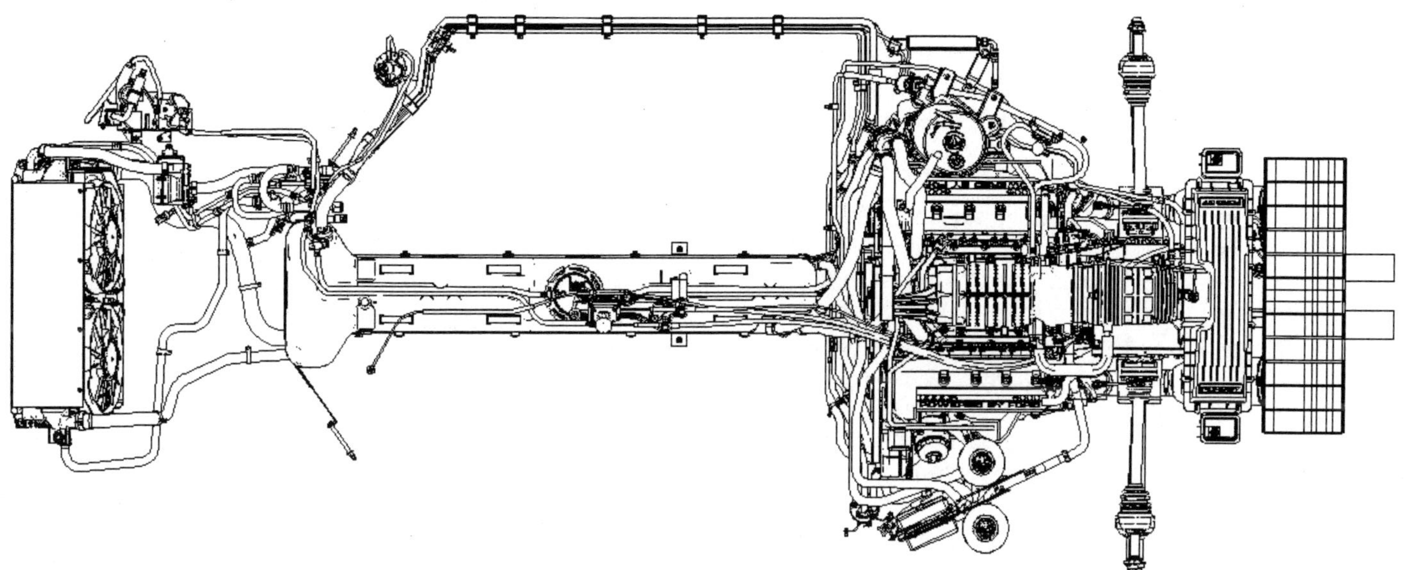

Figure 1. Plan view of Ford GT powertrain

ABSTRACT

The Ford GT powertrain (see Figure 1) is an integrated system developed to preserve the heritage of the LeMans winning car of the past. A team of co-located engineers set out to establish a system that could achieve this result for today's supercar. Multiple variations of engines, transaxles, cooling systems, component locations and innovations were analyzed to meet the project objectives. This paper covers the results and achievements of that team.

INTRODUCTION

Research and analysis indicated that an engine output of 500hp and 500ft-lbs of torque would be required to achieve the vehicle level performance targets for the program. Further it was required that 80% of peak torque be available at 2000rpm. Over thirty engine configurations were investigated, employing Bayesian methodology to capture uncertainties. Constrained by a need to use existing architectures and with less than 24 months to start of production, it was determined that a boosted 5.4L V8 was the most reliable choice to meet the target torque curve. Significant modifications would be necessary to meet durability requirements.

Transaxles of various configurations were evaluated for package, capability, and durability. It was determined that a unique unit would be required, providing some opportunity in gear ratio selection specific to the GT application.

The Powertrain engineering team faced new challenges given the unconventional mid-engine vehicle architecture. Cooling system loads would be very high and difficult to manage in a package that separates the heat source from the heat exchanger by the occupant compartment. Fuel system challenges included not only the technical issues of package and high flow requirements, but also the regulatory requirement of LEV II evaporative emission standards. Induction and exhaust system designs would not only need to support horsepower requirements, they would play an essential role in developing sound quality, an attribute critical to the supercar driving experience.

A small, nimble engineering team was formed, empowered with the authority to deliver the design, development, calibration, and manufacturing/assembly of the entire powertrain system. Roush Industries was selected as the engineering source to work directly with Ford specialists to design, release, and deliver the finished product.

OBJECTIVE

Vehicle performance targets were selected based on the requirement to beat specific competitive supercars for both straight-line acceleration and racetrack lap time at a designated circuit.

The team's objective was to design and integrate all of the powertrain subsystems to provide maximum vehicle performance within the Design Studio's constraints on many vehicle design elements essential to the historic image of the Ford GT. Systems engineering principles would be critical to achieve appropriate balance of each subsystem's performance. The objective was not only performance based, but also required compliance with regulatory emission and safety requirements. Reliability and durability could not be compromised from the high standards that are now common in the marketplace.

TECHNICAL CHALLENGE

With less than 24 months to deliver an all new powertrain, it was necessary to manage technology selection carefully to avoid extended development efforts normally associated with invention or new application. Powertrain Control Module (PCM) selection was driven by development needs essential to calibration delivery inside of a one-year development timeframe. This precluded the team from using some desirable PCM features. Overcoming this disadvantage would require maximizing the capabilities of each individual system while optimizing the integration of the whole powertrain system.

The packaging challenge proved extremely difficult. Hardware structural requirements needed to support race level loads while encumbered by the requirements for a street legal road car. Aggressive program timing negated effective application of a hardware buck, causing all packaging work to be done in a virtual environment.

Durability and reliability metrics had to be completely rethought for an unconventional duty cycle. Standard design verification procedures would need modification, accordingly.

ENGINE PERFORMANCE

Not unlike the original 289 and 427 CID LeMans engines, the new GT engine is derived from production architecture. Unlike the originals, the current engine has to meet considerable regulatory and customer requirements, including emission compliance and driveability.

The average power of the LeMans engines was 380hp for the 1964 289 CID MK 1 and 485hp for the 1966 427 MK II. Whereas displacement provided the necessary power upgrade to win LeMans, boosting provides the modern equivalent. Macura, et al., reviews the 1966 LeMans engine in reference 1. The new Ford GT engine produces over 500hp.

ENGINE DETAILS

A 90° V8 engine configuration was chosen to maximize aftermarket hardware opportunities for GT componentry. 5.4L was the maximum displacement available in a V8 configuration and a compatible high flow cylinder head design had previously been developed for the 2001 Cobra R. A successful previous effort of supercharging the 4.6L Modular Engine validated the approach for a production application. Roback, et al., reviews the modular engine supercharging in reference 2. At 5.4L, the GT minimum power requirement of 96 HP/L significantly exceeds the 4.6L application specific output of 85 HP/L. This increased demand on engine subsystems led to significant upgrades for the GT engine. The cylinder block would have to be strengthened, coolant flow increased, and oil flow increased to support piston cooling. An ADAMS model of the engine was developed and executed to determine maximum dynamic engine loads. The loads calculated by the dynamic analysis were used in the block stress finite element analysis (FEA). Cylinder head combustion sealing would require refinements for increased cylinder pressures. It would also require improved exhaust valve cooling and added clearance for high lift cams. The induction system would need to provide more airflow and pressure from the supercharger. The fuel injection system would have to supply very high fuel flow for peak power demands while maintaining control at idle.

BASE ENGINE DESIGN

Working within the vehicle package and basic Modular Engine architectural constraints, an all-new aluminum cylinder block provided the foundation for a short-block assembly and provides piston cooling via integral oil squirters in the bulkheads. A dry sump lubrication system was utilized to reduce crankshaft centerline to ground distance, to reduce windage, and to provide sufficient flow for piston cooling. The 0.85 bore/stroke ratio drives very high piston speeds, and when coupled with high IMEP, the resultant stress on the piston mandates cooling.

Subsystem level targets cascaded to the component level, driving a set of well balanced component designs.

CYLINDER BLOCK DESIGN

356T6 aluminum was selected as the block material for its superior fatigue and strength properties. A low pressure sand cast process was utilized to produce the block casting. Extensive solidification modeling was conducted to predict casting integrity. Over 30 iterations of the block casting model were completed before final design of the casting tools. First run castings correlated closely to the modeling results, validating the model's effectiveness to predict effects of block design refinements.

The casting process employed chills at locations critical for the 356T6 aluminum to minimize shrink, porosity, and to provide required minimum strength. These locations included bulkheads and cylinder deck faces.

Cylinder block design constraints were primarily package driven. Low-volume manufacturing processing reduced many of the conventional feasibility constraints as did the overriding need to pass durability testing in as few iterations as possible. Durability and performance enhancing design features include:

- Valley webbing for improved torsional rigidity between bores. See Figure 2.
- The oil drain back passages were relocated and lengthened. They have been located on the external block wall and extended to match cast-in returns on the oil pan, eliminating windage from returned oil. See Figure 3.
- Increased cylinder head deck thickness.
- Increased bulkhead thickness
- Elimination of bulkhead windows
- NVH ribbing
- Extended China wall (valley end bulkheads)
- Coolant passages enlarged and designed for optimized coolant flow and balance.

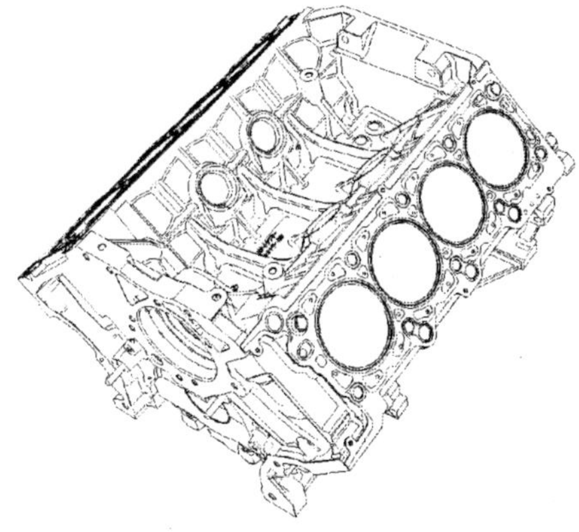

Figure 2. Valley cross webbing

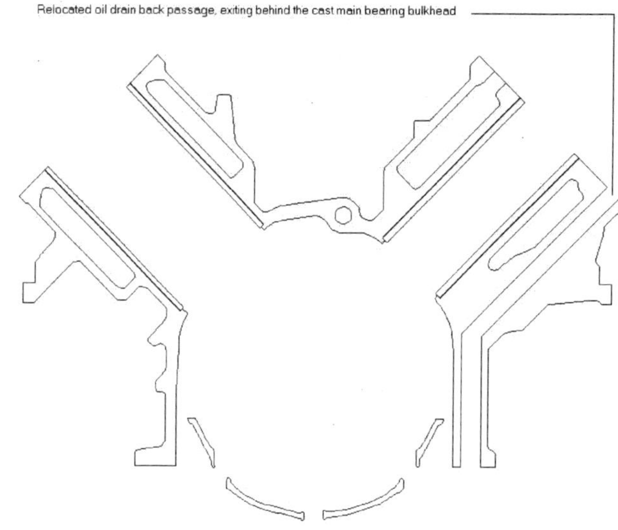

Figure 3. Oil Drain back passage

Engine Block Stress Analysis

Designing a durable aluminum cylinder block, supporting 100hp/L, and with little time for development, relied heavily on computer aided engineering (CAE) tools for design guidance. Large finite element models were created for two design iterations of the cylinder block assembly, the first for development of the structure and a second model to confirm the final design.

FEA was used to help select materials for the main bearing caps, as well as the specific aluminum alloy and heat treat to be used in the engine block casting. Models of the engine block assembly were solved for various engine speeds and angular positions in the crankshaft cycle. FEA stress results were post-processed in fatigue software to predict the expected life of the engine component.

Inputs to the engine block stress finite element models were: loads on the main bearings, piston side loads on the cylinder walls and forces at the cylinder head bolt attachment locations. Forces were calculated using an Adams multi-body dynamic model of the cranktrain; see Figure 4. The inputs to this cranktrain model were the predicted cylinder pressures, which were estimated using a commercially available engine simulation package. Figure 5 shows the estimated gas forces that were input to the multi-body dynamic model of the cranktrain for a 6,000rpm steady-state engine speed simulation. Figure 6 shows a stress contour plot for the engine assembly at a steady-state engine speed of 6,000rpm and at a specific phase in the crankshaft cycle.

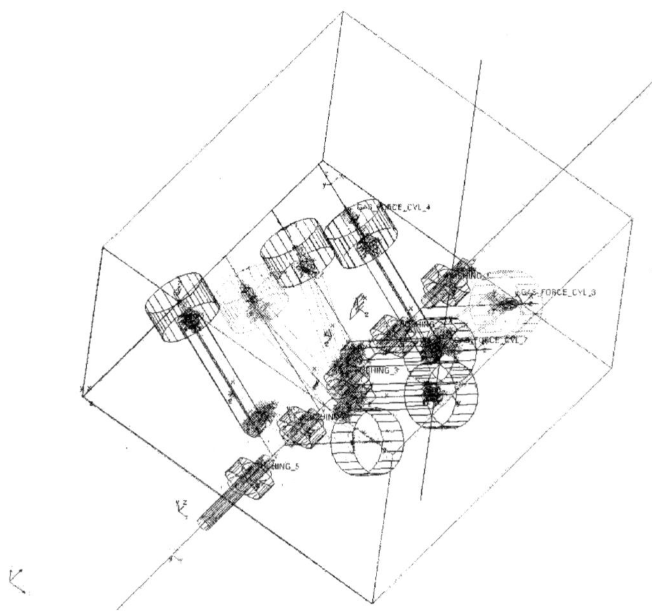

Figure 4. Adams multi-body dynamic model of cranktrain

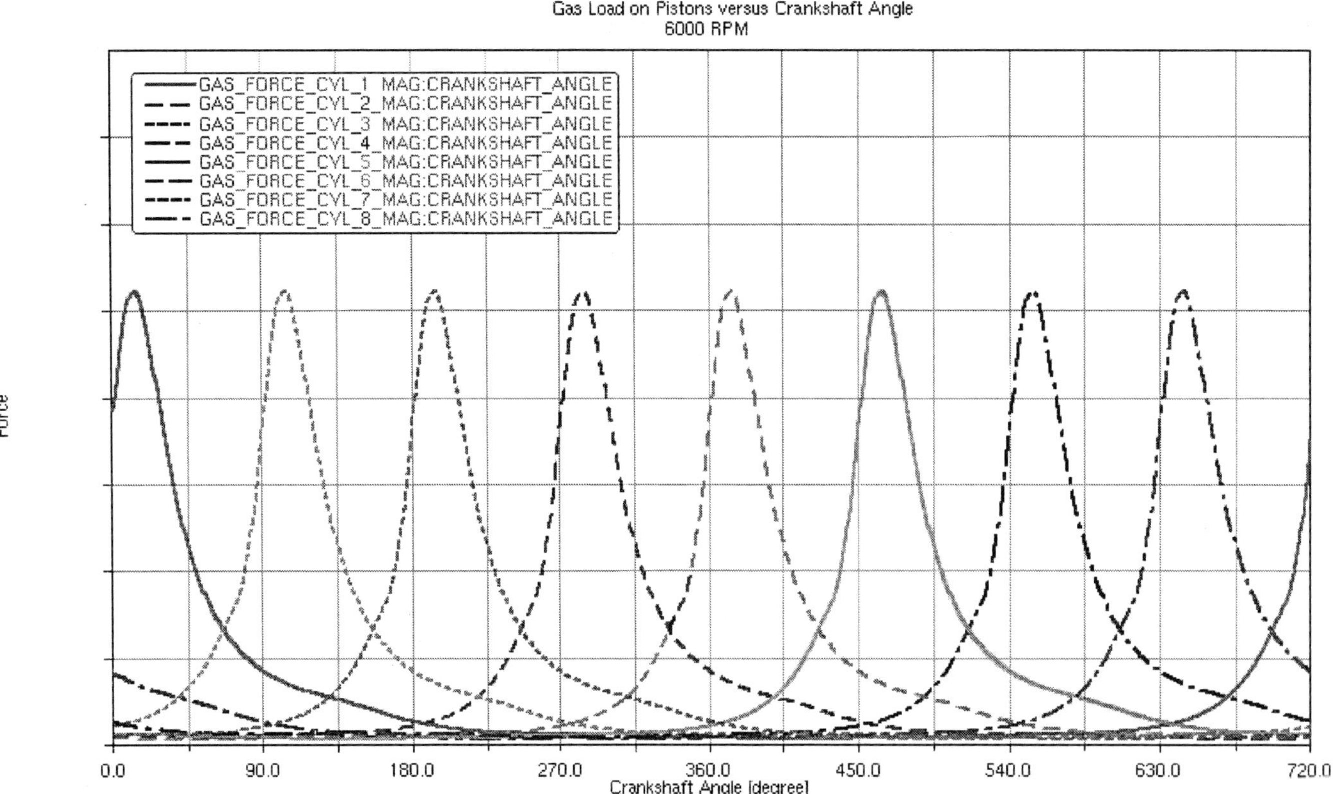

Figure 5. Gas forces input to the multi-body dynamic model of the cranktrain, 6000 rpm steady state engine speed simulation

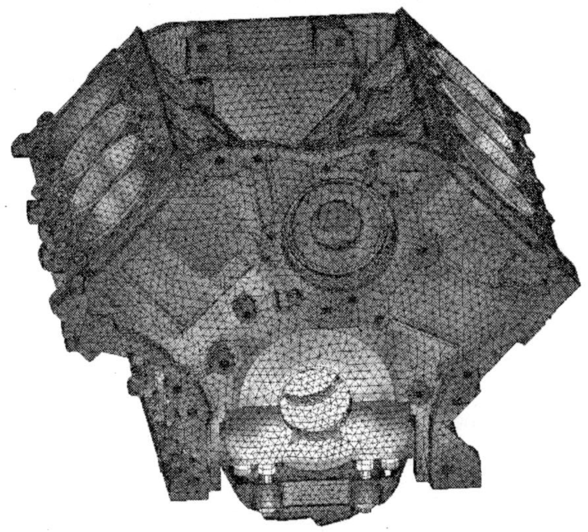

Figure 6. Stress contours in the engine assembly at a steady state engine speed of 6000 rpm and at a certain time in the crankshaft interval.

Optimization of the engine block design was conducted using the stress FEA results; for example, material was added at the sides of the engine block where the engine mount brackets attach for increased strength and stiffness. Other minor revisions were made to the engine block design in this manner.

Stress FEA and fatigue post-processing early in the program showed that the engine block assembly as designed would be sufficiently strong and durable for a specialty/performance vehicle application, and furthermore the analysis work showed that the engine block would meet durability requirements for mainstream vehicle applications

Therefore, the analysis work was used to confirm the design and material choice for the engine block before physical engine prototypes could be obtained for dynamometer cell durability testing.

Engine Block Modal Analysis

Modal finite element analysis of the engine block assembly was performed to confirm sufficient dynamic stiffness. The goal was to exceed the maximum firing frequency of the engine with the first free flexible mode of the engine block assembly. This goal was achieved for the design of the engine block assembly.

Further modal analysis studies were used to help in the selection of a main bearing cap ladder (MBCL) design. In addition to adding strength to the bottom of the engine, it was desirable that the main bearing cap ladder add stiffness to the bottom of the engine and lower the density of the main bearing cap modes in the frequency range of interest. Figure 7 shows a lateral bending mode shape of the engine block assembly with a preliminary main cap ladder design that was later rejected, based on FEA results and other concerns. Nevertheless, the modal finite element analysis was able to quickly calculate the mode shapes and frequencies of the engine block assembly design iterations, and thus the overall program could be completed in a short time.

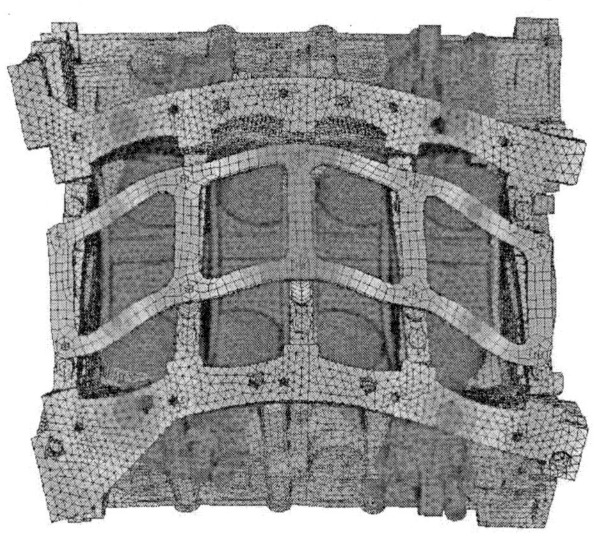

Figure 7. Lateral bending mode of block with preliminary ladder design.

Crankshaft

A twisted-steel crankshaft forging is common with the 5.4L truck engine. The machining is unique on the post end for increased durability to withstand the increased torsional inputs associated with the high IMEP.

Pistons

Mahle high strength, forged, aluminum pistons with anodized top ring land were selected due to the high engine combustion pressures. The outer piston skirt is anti-friction coated.

Connecting Rods

Finite element stress analysis was used to optimize the design and material of the connecting rod assemblies during the Ford GT engine design program. A forged steel "H-beam" connecting rod was chosen early in the project due to the success of a similar design for the 4.6L supercharged engine application. During the connecting rod analysis study, the H-beam connecting rod design was optimized for this engine application. Special attention was paid to the small end of the rod to decrease mass as much as possible while ensuring the strength and durability of the connecting rod.

Many different connecting rod design versions were developed with the help of the designers and design engineers. To analyze many design iterations in a short period of time, second order tetrahedral solid elements were used to represent the connecting rod assemblies. Finite element analysis was of the inertia relief type, with loads applied at the small end and big end of the connecting rod assembly simultaneously. Loads were calculated with a multi-body dynamic model of one of the piston and connecting rod assemblies; see Figure 8. The input to the multi-body dynamic model was the estimated Ford GT cylinder pressure.

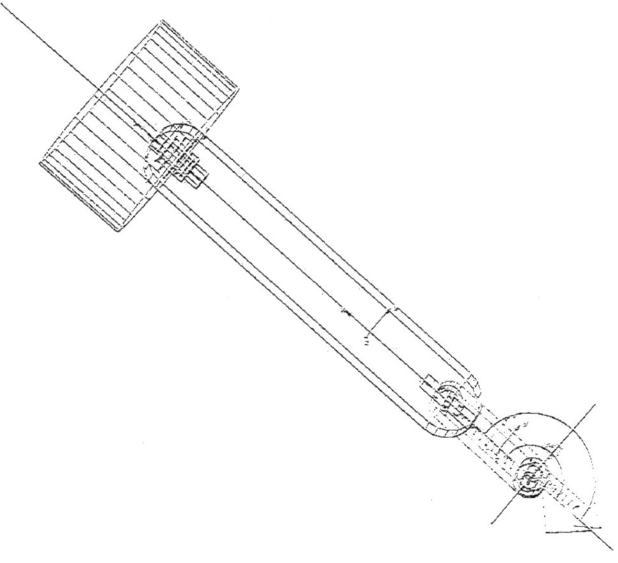

Figure 8. Multi-body dynamic model of one piston and one connecting rod assembly

Forces at the small end and big end of the connecting rod were calculated using a multi-body dynamic model with unique mass and inertia properties for each design iteration under investigation. The multi-body dynamic models were simulated at several steady-state engine speeds in order to determine the engine speeds that produced the highest forces on the connecting rods. At very high engine speeds the gas load was assumed to be an insignificant contributor to the forces on the connecting rods and was ignored. See Figure 9 for a graph that shows example output from one of the multi-body dynamic models at 7,000rpm, with no gas load.

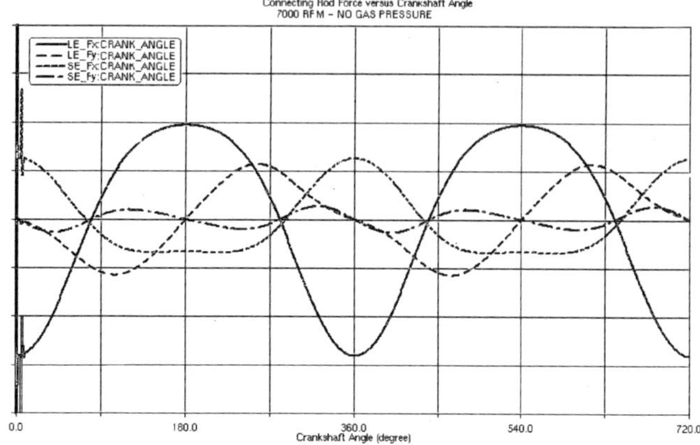

Figure 9. Forces on the connecting rod assembly at 7000 rpm steady state speed, no gas load

In the 7,000 RPM, no gas load steady-state engine speed simulations, the point in time in the crankshaft cycle selected for loads transfer to finite element analysis was at piston TDC in the model. This is where the connecting rod was in pure tension; that is, the connecting rod was being "pulled apart" by the wrist pin and crankshaft pin at this point in time in the crankshaft cycle. Figure 10 shows the stress contours in one of the connecting rod assemblies that was analyzed at 7,000 RPM no gas load, piston TDC.

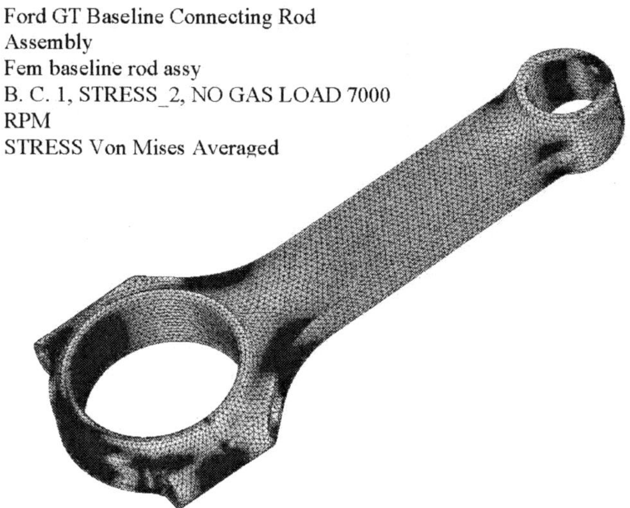

Figure 10. Stress contours in one connecting rod at 7000 rpm, no gas load and piston at TDC

The finite element analysis helped to define the tapered shape of the small end of the connecting rod, and it helped in the selection of 4340 steel as the connecting rod material. Other materials were considered for the connecting rod assembly, including grade 5 titanium alloy. In the case of the titanium alloy, stresses decreased in some components of the connecting rod assembly, and increased in other components of the assembly, mainly due to significant differences in material properties compared to steel such as static stiffness. Finite element analysis was used to show that a significantly more expensive material such as a titanium alloy would not necessarily increase the expected life of the connecting rod assembly as a whole in the Ford GT engine application.

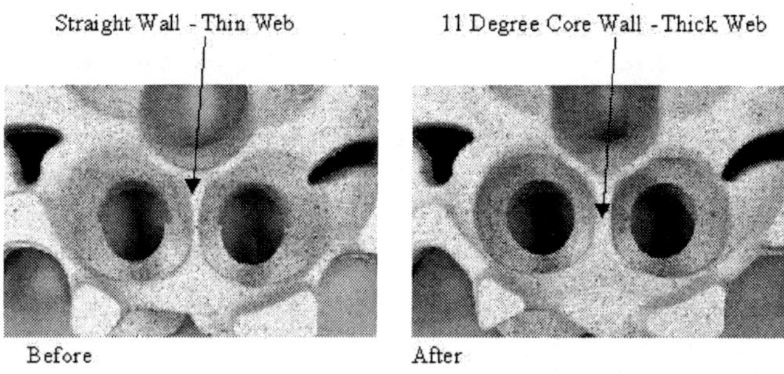

Figure 12. Cylinder head exhaust port core comparison

Main Bearing Cap Ladder

A cast aluminum ladder was selected to strengthen the bottom end of the engine and reduce windage from the crankshaft. See Figure 11.

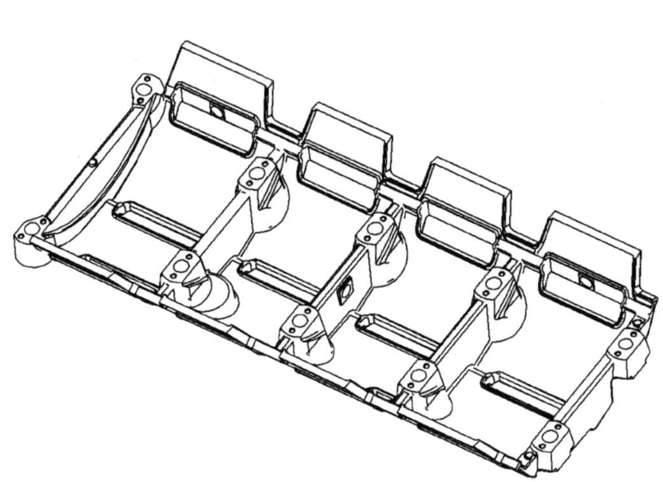

Figure 11. Bearing ladder and windage tray

HEADS

The four-valve cylinder heads derived from a previous 5.4L performance engine are cast in T356 aluminum. They contain detail core revisions for improved casting, machining, and performance. The exhaust port core wall went from straight to 11º, providing thicker sand core walls. See Figure 12. This allows more coolant to flow around the exhaust valve giving improved valve cooling, a necessity for the high temperatures and pressures generated.

Dual-doweled, individual cam caps replaced the ladder cap assembly used on other Modular 4V cylinder heads to accommodate the higher lift cam profiles. See figure 13.

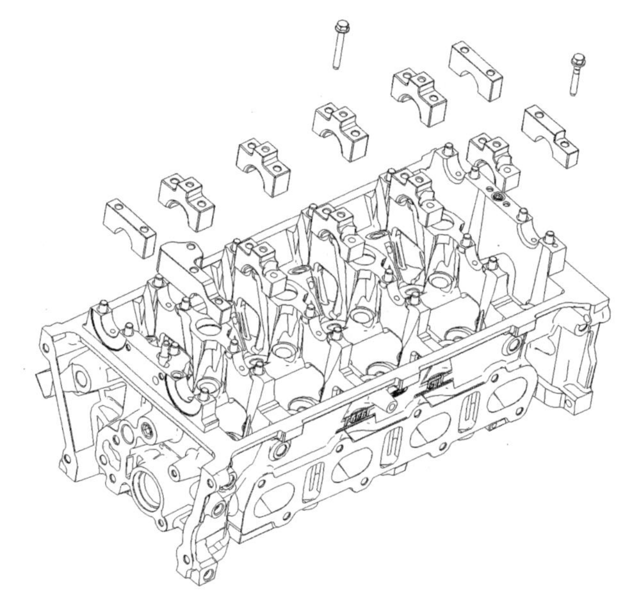

Figure 13. Cylinder head with cam caps and double dowels

INDUCTION SYSTEM

Combustion air is supplied to the air cleaner assembly through "J" ducts molded into the rear vehicle clamshell. The air then passes into two lower chambers of the air cleaner housing then through two separate filters and out the single oval outlet with a drop-in mass airflow sensor. The air then enters the engine through a dual 70mm bore throttle body. See Figure 14. The throttle plates are actuated by a cable connected to a four-bar link to improve low speed driveability. The air then enters the rear of the 2.3L Lysholm screw type supercharger (see Takabe, et al. Ref. 3). The supercharger casting is a unique and package specific design developed to mount the throttle body, EGR system module, idle speed control motor, and boost bypass valve. The boost bypass valve is controlled by manifold vacuum and is designed to reduce parasitic/pumping losses and heating effects at idle and part throttle. The air is compressed by the supercharger and discharged into the upper intake manifold. It then passes through a water-to-air intercooler bolted directly to the upper intake. The intercooler reduces the air charge temperature by up to 130° F improving detonation resistance and increasing high load performance. The air then fills the lower intake plenum and is routed to the cylinder heads by eight short bell-mouthed runners. Each runner contains two 32 lbs/hr fuel injectors. Each injector sprays in a dual conical pattern and is mounted such that there is an even distribution of fuel within each intake runner. To achieve acceptable fuel control during idle, part throttle conditions and fuel vapor purging the engine control system enables only one injector per cylinder during these conditions. The second injector is activated only when higher fuel flow rates are required. Fuel is delivered to the injectors through a unique self-dampened fuel rail. This design is key in reducing pressure pulsations and improving evaporative emissions by eliminating external dampers.

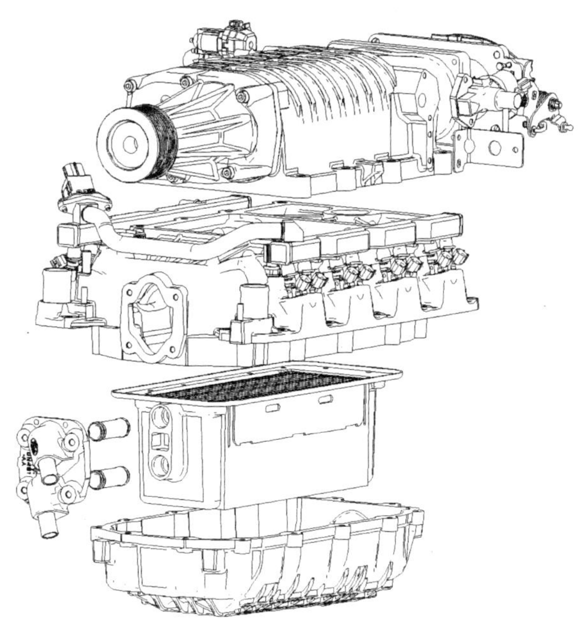

Figure 14. Air intake system

FRONT-END ACCESSORY DRIVE (FEAD)

The Ford GT FEAD system (Figure 15) utilized a unique approach to the overall system design. It was based on previous supercharged systems within the Ford family such as the F-150 Lightning and the Mustang Cobra, but with balanced loads on the crankshaft. Unlike most vehicle platforms, the GT frame layout created a unique challenge to package a FEAD system with less axial clearance but more lateral package space. The minimal clearance to the driver's compartment bulkhead meant that a traditional "spider" style external crankshaft bearing support could not be used to support the secondary sheave as is typical on many supercharged engines. The design team decided to create a "dual sheave" layout that would be shorter than the typical two sheave "spider" designs while still allowing the supercharger to be the lone high load component on its sheave to reduce belt wear. The oil pump drive would then have it's own dedicated third sheave and still fit within the axial package constraints.

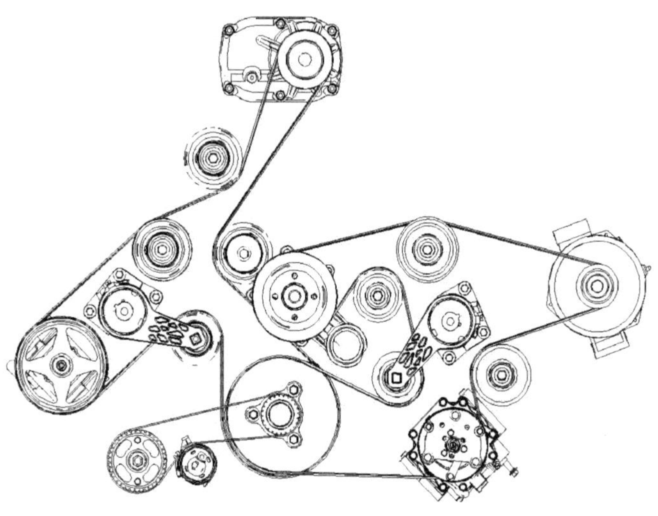

Figure 15. Front view of FEAD

The packaging challenges for this system revolved around the supercharger being moved to the 1st sheave to clear the rear window glass during engine roll. However, the power required to drive the new screw-type supercharger coupled with its drive inertia meant that most other accessories had to be moved from the base sheave out to the second sheave to minimize belt wear and slip concerns. Although this practice is atypical for most FEAD designs, it allowed the GT team to move the water pump drive out to the 2nd sheave thus protecting for the front inlet water pump design which would be critical for delivering enough coolant to the engine to support sustained power (see water pump section for more detail). The key to this new "dual sheave" design was to arrange the accessories such that the forces on the crankshaft from the 1st and 2nd sheaves would be balanced. This reduced the bearing loads and prevented starving the #1 main journal oil feed. (See Figures 16 and 17).

Pulley Hubload Angle for Load Condition # 3

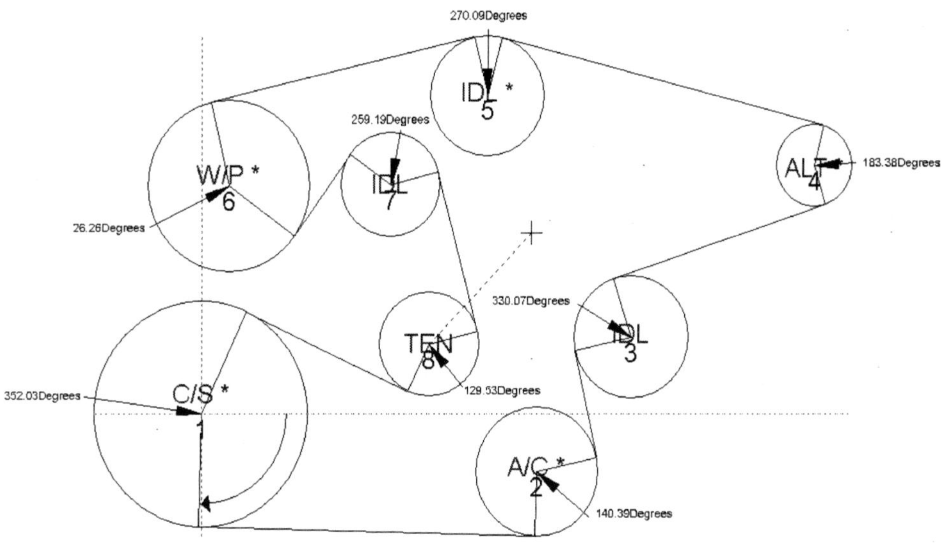

Figure 16. Force diagram of second sheave

After multiple layout iterations, the team was able to create a layout that not only created a favorable loading balance on the crank, but actually reduced the overall load on the crank compared to other benchmark programs. Functionally, the system allowed for standard or better belt wrap over nearly every single component in the system while making sure that neither drive belt would be trapped by the new lower coolant hose connected to the water pump inlet. This maintained acceptable drive belt serviceability despite the 15 total pulleys contained in the system. No coolant hoses or wiring takeouts would need to be disconnected in order to service either drive belt.

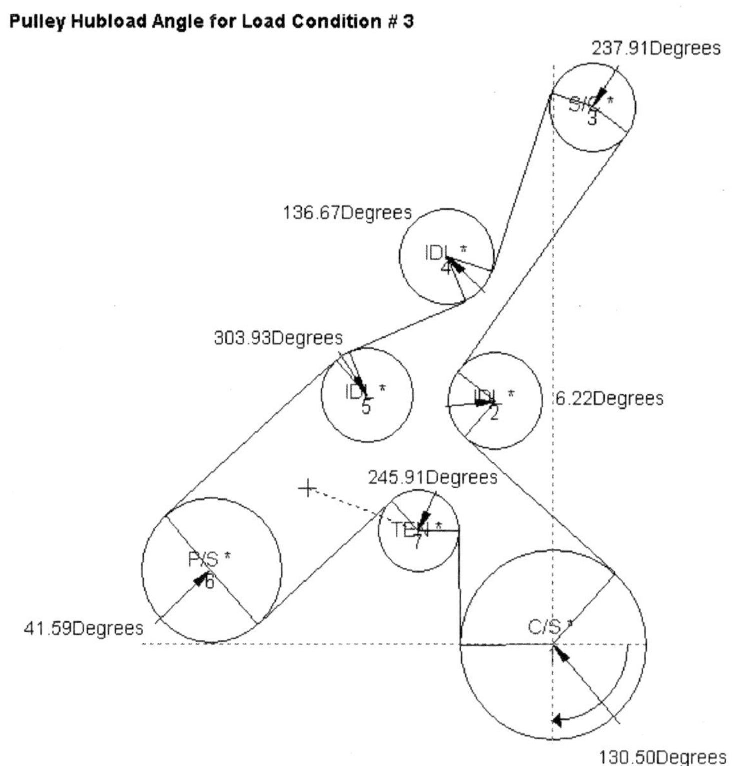

Figure 17. Force diagram of first sheave

The team performed several belt/bearing analyses throughout the design process to make sure that they would not encounter durability concerns in the field. The analyses showed that a 6-rib belt and pulley 2nd sheave system would be sufficient to drive the auxiliary components (A/C, alternator, W/P), but the extreme loads of the supercharger (approx. 80hp alone to drive at full boost) would prematurely wear a typical 8-rib drive belt. Analysis showed a 10-rib system would meet durability requirements but 10-rib components often were only used for heavy truck applications and not for passenger car use. The 10-rib components were not designed for high speed use and proved unacceptable for bearing life or were deemed "overkill" in terms of parasitic loss to the total system. Finding existing parts proved difficult. The final step was to work with suppliers to create unique 10-rib idlers/pulleys that would be feasible for high-speed use.

FEAD Modal Analysis

Since this is a mid-engine car, the front-end accessory drive is located only a few inches behind the driver's right ear. This meant that the driver would likely hear noise generated by the FEAD. To minimize this potential NVH issue, each part of the FEAD and the completed assembly was analyzed using finite element analysis to determine the modal frequencies. The goal was to have none of the components or the completed assembly with a modal frequency below the 4th order firing frequency of the engine. To achieve this any component which had a frequency lower than the target was returned to the designers with suggestions on how to make it stiffer. Through this design iteration process the FEAD was made stiff enough so that it was less likely to be excited by normal engine operation, thus reducing the possibility of any resonant noises coming from the FEAD. See Figure 18.

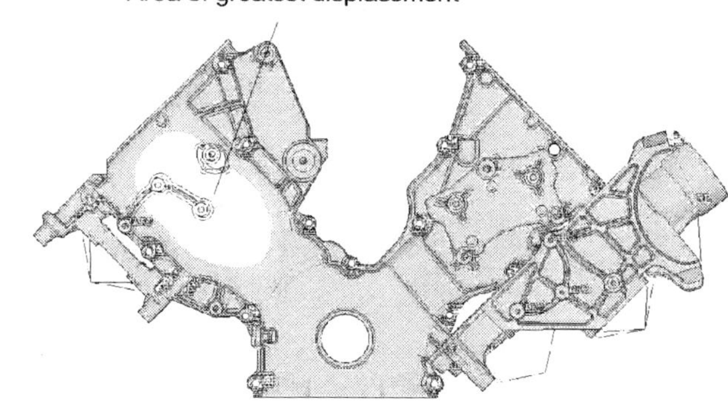

Figure 18. FEA of the FEAD for first Panel mode of the front cover

LUBRICATION

The vehicle design and performance requirements forced changes and upgrades to the lubrication system. The engine would require more lubricant flow to aid in piston cooling and to dissipate the heat generated by the added power. The engine height was limited due to the desire to keep the vehicle center-of-gravity (CG) low. The engine was designed with piston cooling oil jets aimed at the underside of the piston. The oil jets reduced overall piston temperatures to improve durability and had the additional benefit of improving wear on the piston pin and connecting rod bushing. The oil jet however increased the oil flow requirement by approximately 16 l/min. The existing gerotor oil pump could not be modified to produce this flow without significant revisions to other major engine sub-systems and block architecture. A dry sump oiling system mounted inside the oil pan was then investigated but two major issues prevented further investigation. First, the vehicle package required the engine to drop 100mm to maintain the external body shape. Second, the 105.8mm stroke of the engine coupled with the lower engine position eliminated the package space required for an oil pump in the pan. Next, an externally mounted pump driven off the crankshaft was investigated. A three stage dry sump pump (two scavenge sections, one pressure section) is mounted on the structural oil pan (see Figure 19). It is doweled to maintain alignment between the pulley and the crankshaft. An external "helical offset tooth" (HOT) belt (see Fig. 20) drives the oil pump. This belt design smoothes out tooth engagement compared to other tooth profiles. This style belt is self-aligning, eliminating the need for sprocket flanges and reducing drive face width. The "HOT" belt also has a higher power capacity than traditional "toothed" drive belts, which reduces overall drive belt width and improves vehicle package in a critical area.

The oil is scavenged from the pan through two traditional screen and cover assemblies into the two scavenge sections of the pump. It then passes through an external line to the remote mounted oil reservoir where the oil is de-aerated. It next passes though another external line to the pressure section of the oil pump. From there the oil flows through a cast line on the front of the oil pan and into the oil filter adapter bracket. The oil filter adapter bracket (OFA) is a cast 356 Al component which mounts the A/C compressor, alternator, oil filter and oil cooler. A water-to-oil cooler cools flowing oil as it passes to cast passages in the OFA. It is then routed to the high-flow cartridge-style oil filter. The filter is located for easy serviceability and minimum captured used oil. It then exits the OFA to the cylinder block and the main oil gallery.

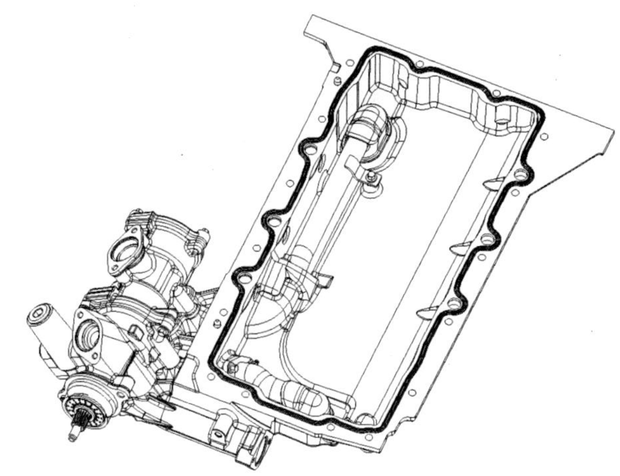

Figure 19. Dry sump pump and oil pan

The engine block has oil squirters aimed at the pistons to aid in lubricating the piston wall while cooling the piston. A cast oil pan mating the block and transmission helped the engine and the vehicle structurally. The twelve-quart (11.4L) oil reservoir is used to supply the vast amounts of oil required to circulate at any given moment.

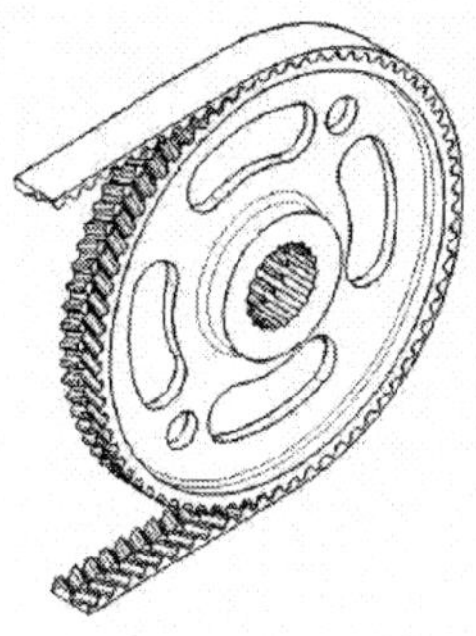

Figure 20. Helical offset tooth oil pump drive

SEALING

Head bolts were upgraded by changing the material from SAE J404 4037 to SAE 4140 for greater yield strength. This revision improved the clamping force as indicated in the following table:

System	4.6L 4V	S/C 4.6L 4V	S/C 5.4L 4V
Clamp Load	54 kN	60 kN	68 kN

The upgraded multi-layer head gasket used on the supercharged 4.6L has proved to be adequate for the cylinder pressures generated.

WATER PUMP

The Ford GT water pump utilizes a front inlet design to achieve the flow rate required to maintain cooling capacity during extended maximum engine power conditions. Preliminary calculations indicated that the cooling system would require a coolant flow rate upwards of 100 gpm to support extended operation at maximum power output. The current Modular design is a rear inlet pump fed by a side inlet at the left front of the engine block (See Figure 21). While this design optimizes packaging, it creates a flow separation through the block prior to the pump as well as water temperature rise as the water passes in front of the #5 water jacket thus pre-heating the water. The front inlet GT design eliminates these restrictions with a gentle transitioning inlet neck that minimizes flow separation as it feeds water into the eye of the pump impeller for distribution to the water jackets (See Figure 22).

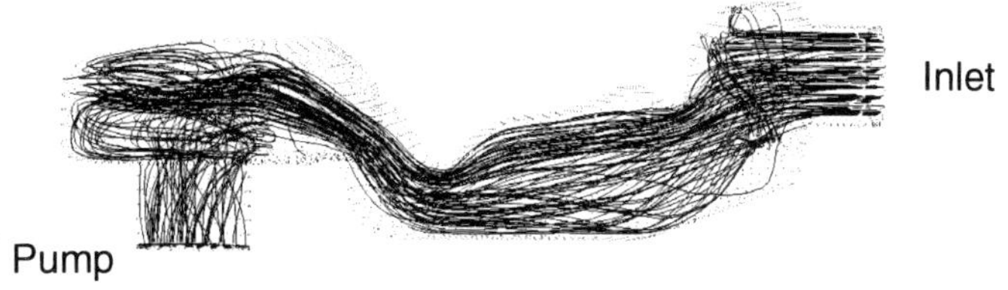

Figure 21. Coolant streamlines with side block entry

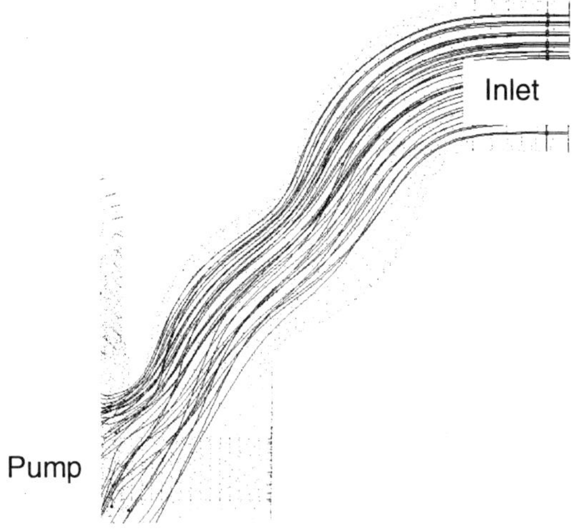

Figure 22. Coolant streamlines of front entry pump

The GT pump also utilizes a full-vane cast pump impeller with vanes designed to generate maximum pressure while minimizing power consumption (see Figure 23). While the existing pump uses a stamped steel "half vane" impeller, the GT team decided to utilize a cast impeller to reduce vane flexing while allowing for a compound vane geometry over a flat, stamped vane. The GT pump generates up to 110 gallons per minute peak flow and approximately a 10 psi pressure increase over current designs at peak operation.

Figure 23. Water pump impeller

A computational fluid dynamics (CFD) study of the side inlet compared to the front inlet eliminated an inlet restriction where coolant passed around a head bolt boss and an oil passage before entering into the pump. See the following table for this comparison.

Design	**% Improvement**
Cast Inlet in Current 5.4L V8 Block	-
Cast Inlet Without Oil Line	10.7%
Front Inlet Water Pump	73.5%

Hill et al. in reference 4 covers an extensive review of the thermal management system.

ENGINE SPECIFICATIONS

Configuration	Longitudinally mounted mid-ship 90° V8
Bore & Stroke	90.2mm X 105.8mm
Displacement	5.409L (330 in^3)
Compression ratio	8.3:1
Horsepower (Tentative)	500+ HP (388 kw) @ 6000 rpm
Torque (Tentative)	500 lb.ft. (628 N-m) @4250 rpm
Specific output	96 bhp/L
Redline	6500 rpm
Valvetrain	Chain drive double overhead camshafts, roller finger followers with hydraulic lash adjustment
Intake valves	Two per cylinder, 37mm head diameter
Exhaust valves	Two per cylinder, 32mm head diameter
Injectors	Two per cylinder, 32lbs/hr
Ignition system	Distributorless coil on plug
Induction system	Eaton supplied Lysholm S2300 screw type supercharger at 12 psi (83kPa) maximum boost with water-cooled intercooler
Crankshaft	Forged steel, internally balanced
Pistons	Forged aluminum, dished, anodized head and anti-friction coated
Connecting rods	Forged high strength steel 'H' beam

TRANSMISSION PERFORMANCE

Miller, et al., reviews transmission performance in reference 5. The transmission is a manual six speed, with triple cone synchronizers on the first four gears and double cone on fifth and sixth. It has an internal oil pump for lubrication and cooling, and the starter mounts to transmission case

The transmission ratios were matched to the final drive ratio to give a progressively smooth drop between gear selections and close to ideal decreasing rpm drop when shifting to higher gears as can be seen in the sawtooth chart. (Figure 24). First gear ratio was also purposely designed for zero to sixty mph and zero to 100 kph tests to be accomplished without a shift to second gear. A GT Power tractive effort curve (Figure 25) indicates that the projected vehicle top speed will be attained in fifth gear.

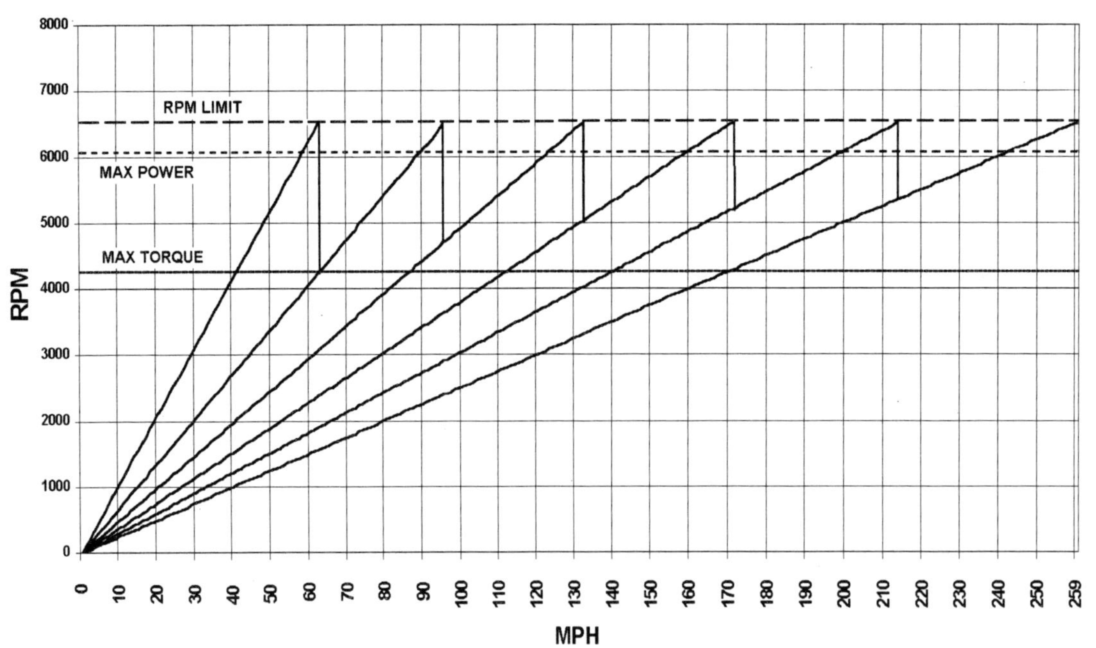

Figure 24. Shift speed drop in gears

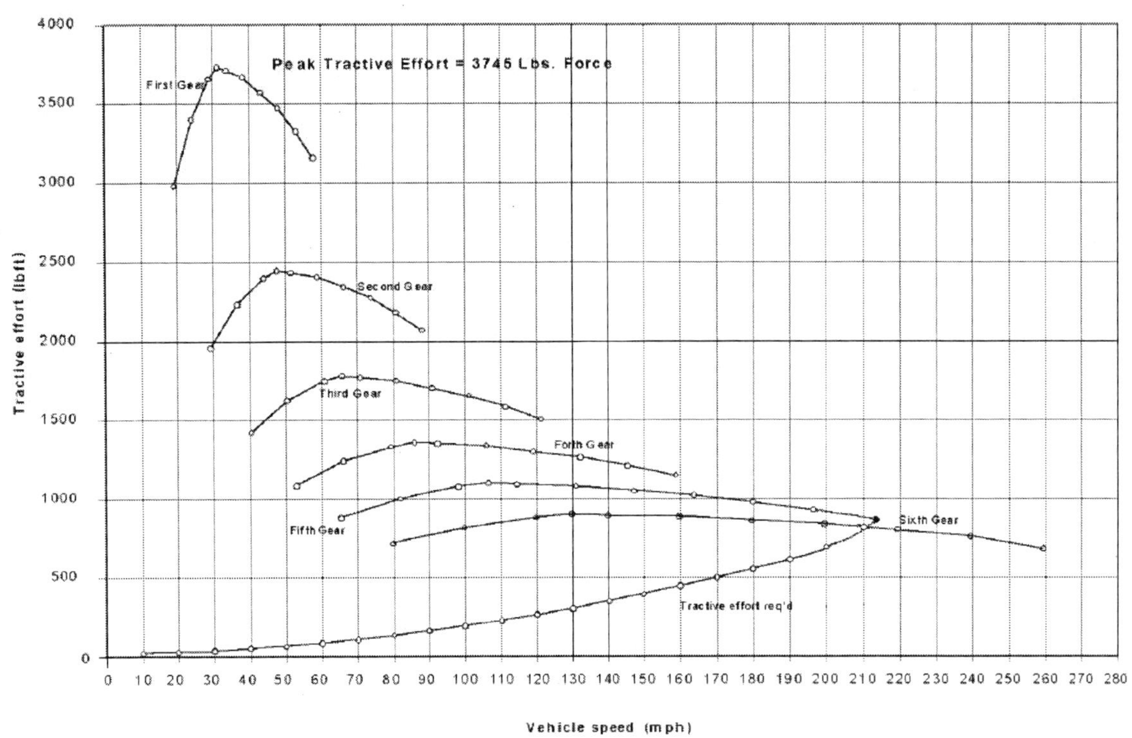

Figure 25. Tractive effort

CLUTCH

A hydraulic controlled twin disc 240mm clutch with an integral starter ring gear mates to a 294.5mm flywheel. The clutch housing is LM25 aluminum and the flywheel is spherical nodular iron EN-GJS-500-7.

DIFFERENTIAL

A torque sensing helical gear driven limited slip differential (LSD) was selected for its characteristics of traction improvement and steering performance. This type of differential has less hysteresis and has quicker response time than disc type LSD's. Its torque bias ratio is 2.3:1 (+/- 0.25) in drive and 2.0:1 (+/- 0.25) during coast.

HALF SHAFTS

Thirty-one spline hollow 35mm diameter half shafts are used to contain the torque generated throughout the drive range of the vehicle. They have a capacity of 5535 Nm (B10 UTS) and yield strength of 4100 Nm.

ENGINE AND TRANSMISSION ASSEMBLY

Modal FEA of the overall powertrain was conducted in order to confirm that the frequencies of the free flexible modes of the powertrain were above the target frequency. Figure 26 shows the first global lateral bending mode shape of the overall powertrain.

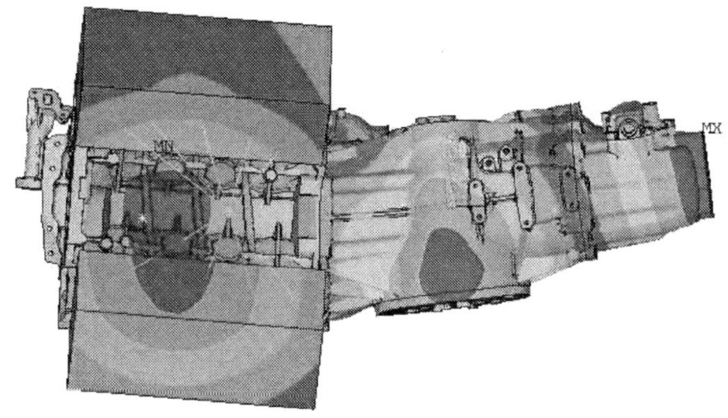

Figure 26. Top view of powertrain finite element model

Figure 27 shows the first global vertical bending mode shape of the overall Ford GT powertrain

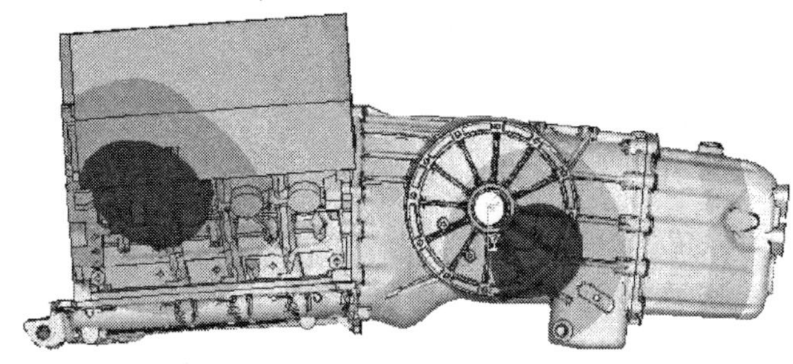

Figure 27. Side view of powertrain finite element model

The powertrain modal FEA was used to validate structural attachment of the engine oil pan to the transmission housing. The first vertical bending mode frequency of the overall powertrain is increased by sixteen percent and the first lateral bending mode frequency of the overall powertrain is increased by five percent.

FUEL SYSTEM

The fuel storage and delivery system features the first usage in the industry of a capless fuel entry system and a "ship in a bottle" (SIB) blow molded tank. These were developed to meet LEV II evaporative emissions requirements.

FUEL INLET

Hidden beneath the Ford GT's race-inspired fuel inlet door is a direct fill capless fuel inlet. (See Figure 28). This is an auto industry first. It has as a spring-loaded self-sealing trapdoor that opens with insertion of the gas pump nozzle. This feature removes operator error from one of the 'check engine' light warning system paths.

With only two openings and the majority of the hose connections internal to the tank, this dramatically reduces potential permeation paths from the fuel system.

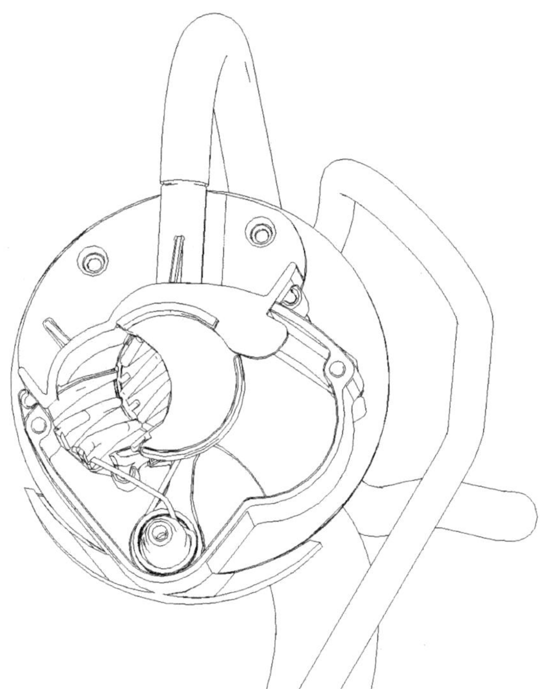

Figure 28. Direct fill inlet

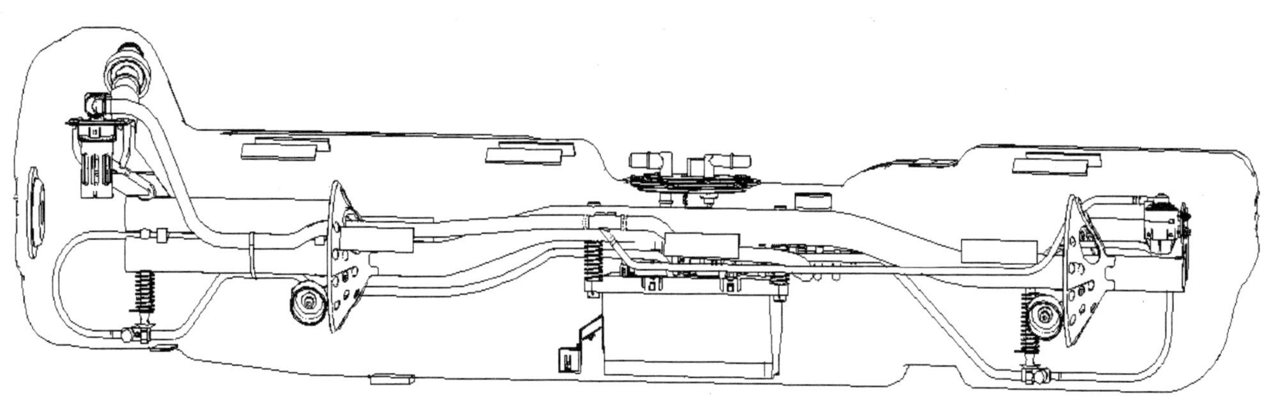

Figure 29. Fuel tank with internals

FUEL TANK

The fuel tank is blow-molded around a fuel module carrier. The tank requires only two openings: one for the inlet pipe, a second for the outlet hose, wiring, and access for service, if required. (See Fig. 29)

The SIB fuel tank system architecture is located in the center tunnel of the vehicle. The tank utilizes a stainless steel fuel module carrier. Mounted on the carrier are: two turbine and three jet fuel pumps, a piezoelectric fuel level sensor, baffles, filters, and internal hoses and wiring for these systems. The jet pumps transfer fuel inside the tank to supply the turbine pumps that feed the engine.

EXHAUST

The midship design of the vehicle has led to a compact exhaust system. This dual exhaust system is completely powertrain mounted and suspended from exhaust manifold outlets on the front end of the system and from the rear of the transmission housing by means of flexible bracketry system at the rear. This mounting method maintains the vertical position of the exhaust systems relative to the powertrain while allowing flexible fore-aft motion for exhaust system longitudinal growth.

MANIFOLDS

A cast iron (high Silicon- Molybdenum) exhaust manifold design was optimized for maximum flow given the limited packaging space. The exhaust manifolds include a new 2 ¾" Ford designed manifold outlet flange required to reduce flow restriction. Performing an FEA analysis and exhaust manifold cracking test validated the durability of the design. The manifolds are completely encapsulated in a multi-layered, manifold-mounted heat shield. The shield is required to protect surrounding components and aid in catalyst light off times.

CATALYSTS

Each bank of the engine has a typical three-way catalyst assembly that carries exhaust gases from the engine to the muffler. The light-off/underbody catalyst assemblies contain two 5.2" diameter round ceramic substrates each, with a mid-bed sensing strategy. Each converter assembly utilizes a tourniquet-style canning process for consistent and even support mat GBD for substrate stability. Converter inlet pipes match the exhaust manifold outlet of 2 ¾" and are thermally insulated and heat-shielded to protect surrounding components and aid in light-off times. The converter cans have 3" cone inlets and 2 ¼"outlets. This provides improved flow across the front face of the catalyst (increasing effectiveness) as well as increased gas flow.

MUFFLER

A single cross-car mounted muffler provides the volume necessary to tune the engine output sound to desired target. It also helps manage exhaust heat introduced to the engine compartment from the midship design (see Figure 30). The entire muffler assembly is manufactured from 304 stainless steel and is heat-insulated on its front, top, back and sides. This insulation helps maintain adequate surface temperatures of surrounding componentry. Internally the exhaust gas is diverted through 2 1/4" piping inside the muffler, passing it through several chambers separated by partitions, baffles and tuning links. Exhaust gases exit the bottom center of the muffler. The exhaust system exits at the center and rear of the vehicle through a bright, 304 stainless steel, oval-formed exhaust tip.

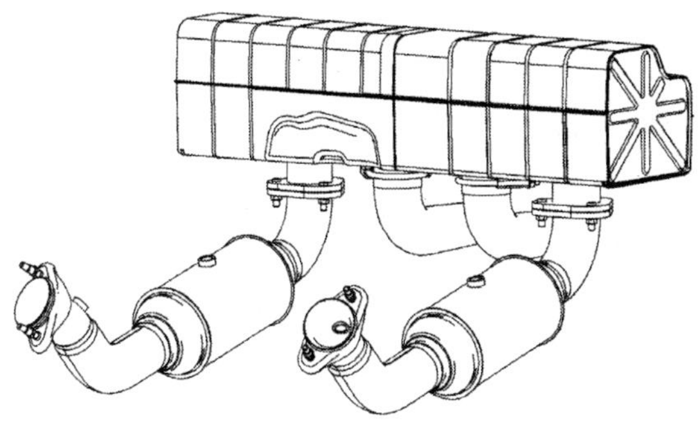

Figure 30 Muffler

COOLING

The cooling module located at the front contains the A/C condenser and the radiators for the intercooler and engine. Cooling the powertrain is significant since the system is designed to meet both extended low and high performance requirements. For more information on cooling see reference 5.

ENGINE RADIATOR

The radiator was based on the available air inlet of the vehicle. The maximum performance requirement that was projected was for airflow of 13,800 cfm, which would be adequate for any extended condition of vehicle performance. However, this projection was for a vehicle operated at maximum power for extended periods of time, long past its total fuel availability time. Therefore, the available airflow of 7,500 cfm would be sufficient for all but the most extreme conditions.

INTERCOOLER

To increase the air charge density, a separate air-to-water intercooler system was developed. It consists of a large radiator in the front of the vehicle, an electric coolant pump and supercharger intercooler in the intake manifold assembly.

Intercooler Radiator

The intercooler radiator was developed with the assumption that this vehicle may be run at extended high performance conditions, requiring prolonged intercooler performance. This requires increased cooling compared to ordinary supercharged muscle cars, where intercooler performance is only required intermittently. The core is 680 mm X 411mm X 32.25 mm.

Intercooler Pump

The intercooler 12V drive pump is capable of flowing 10 gallons (38L) per minute at 13.5 psi (93 kPa).

TRANSMISSION COOLER

An air-to-oil transmission cooler is packaged in the engine bay near the body side air duct.

CONCLUSION

Reliance on Computer Aided Engineering, including virtual packaging tools proved highly effective in delivering the program in unprecedented timing. Application of system engineering to balance and optimize the entire powertrain subsystems delivered vehicle level attributes exceeding targets, providing supercharged, supercar performance that meets the demanding quality requirements of today's market.

ACKNOWLEDGMENTS

The authors would wish to thank those from within and outside of our companies for their invaluable assistance and donations to the success in designing and developing this powertrain. There have been numerous persons that made significant contributions. Those who have helped compile this record are; Mr. R. Grenkowitz, Ford Casting Technical Specialist, Ms. J. Bastian, Roush Industries CAE Senior Engineering Analyst, Mr. T. Wernholm, and Mr. J. Thomson.

REFERENCES

1. Macura, J.F., and Bowers, J. SAE Paper 670066 Mark II-427 GT Engine
2. Roback, B.J., Holl, M.R., Eble, M.L., and Thornton, D. SAE Paper 2003-01-3209 Supercharging Ford's 4.6L for Affordable Performance
3. Takabe, S., Hatamura, K., Kanesaka, H., Kurata, H., Iguchi, Y., and Matsubara, H. SAE Paper 940843 Development of the High Performance Lysholm Compressor for Automotive Use
4. Hill, C.M., Miller, G.D., Evans, M.R., and Pollock, D.M, SAE Paper 2004-01-1257 2005 Ford GT – Maintaining Your Cool at 200MPH
5. Miller, G.D., Cropper, A., Janczak, R.M., and Nesbitt, S, SAE Paper 2004-01-1260 2005 Ford GT Transaxle – Tailor Made in Under Two Years

2004-01-1253

Ford GT Body Engineering – Delivering the Designer's Vision in 24 Months

William Clarke
Ford Motor Co.

Jon Gunner
Mayflower Vehicle Systems, Inc.

Figure 1

ABSTRACT

The objective was to engineer a world-class supercar body that faithfully reproduces the 2002 Concept and pays homage to the 1960's road racer. The car had to be designed, developed and launched in 24 months, while meeting tough requirements for function, weight, occupant package and aerodynamics. Challenging features such as the cantilevered door, "clamshell" engine decklid and a deeply contoured hood were to be included. This paper will discuss how a dedicated team of enthusiasts can have a flexible approach to the engineering process, material selections and manufacturing processes to achieve the designer's vision in 24 months (Figure 1).

INTRODUCTION

In January of 2002, the Ford GT Concept Car was unveiled at the Detroit International Auto Show. The car was well received and Ford decided to proceed with a production version of the car. In May of 2002, the Engineering Team was established to deliver the production version, all new from the ground up, in just 24 months. The teams' mantra was to minimize any alterations to the Concept Car while meeting all requirements for a car to be sold to the public and driven on public roads. In addition, the team was told to deliver 3 fully production representative cars for Ford's Centennial Celebration in June of 2003

THE ASSIGNMENT

Prior to the Detroit unveiling of the Concept Car, a small team had been experimenting with engineering concepts to bring back the Ford GT motor car. The results were the Ford GT90, which was shown at the 1xxx Detroit Auto Show and later a secret project call "Petunia" to further explore the possibility of a remake of the mid-engined sports car. The lessons learned during these exercises laid the road map for the architecture in the Ford GT Concept Car.

The Concept Car shown at the 2002 Detroit Auto Show had been developed beyond what is typical of a "show car". Much of the elements of a functional car were packaged into the Concept Car, raising confidence of its production feasibility. However, many challenges remained, not the least of which were basic durability, closures feasibility with the doors that cut into the roofline and a rearward opening "clamshell", opening side window, bumper standards and other legal requirements. Public reaction to the Concept Car was strongly favorable, leading to the February 2002 Program authorization.

In March of 2002, a Development Plan was established: Provide 3 cars for Ford's Centennial (June 2003), start full production in 24 months to build a small volume of cars. World Class performance targets were set using supercars as benchmarks.

In May of 2002, the Program Team was formed. A unique engineering process and an empowered team were required to develop this unique vehicle in highly compressed timing. The team was lean with each team member having broad responsibilities and the empowerment to make decisions related to their responsibilities. Key suppliers were on board, including Mayflower to handle the package integration for the vehicle, the design and engineering for the body and spaceframe and to handle the manufacturing of the aluminum body assembly. Lear Corporation joined the team to integrate the interior and vehicle electrical systems. Approximately 45 other suppliers were ultimately sourced for design, engineering and manufacturing of the body system components.

CONCEPTUAL DESIGN

During an initial conceptual phase, the team developed basic elements of the design to determine how the theme of the Concept Car could be maintained in a feasible, lightweight design.

Early renderings were used to discuss the engineering concepts, sourcing, manufacturing and assembly processes and to develop a robust cost foundation.

The complete vehicle was designed simultaneously, with a tight liaison among all team members. The meetings were focused on engineering and design matters.

Surface and styling work was performed simultaneously with the engineering/design of the body and interior.

The occupant and ergonomic package was established early, with key focus on achieving superior occupant spaciousness for a mid-engined supercar. The new car significantly improved headroom, legroom, shoulder room and ingress/egress over the 1960's original, while maintaining the theme of the car. The only significant change to the theme for package was a 17mm increase in vehicle height to improve headroom.

An aluminum spaceframe concept was selected for advantages in weight, low volume production and structural performance.

The body concepts were defined, including cantilevered doors, a rear-opening clamshell, tilt-forward hood w/fender release levers.

Materials and processes were selected. The body skin would be unstressed, and would be made primarily from aluminum, with some composites. Machined "plus-nuts" on the spaceframe would be precisely machined to assure body panel fit. Superplastic forming was selected for major aluminum body panels. Carbon fiber was selected to form a 1-piece clamshell inner panel. Ford Research Lab personnel filled key roles and brought aluminum and composites manufacturing experience acquired through Jaguar and Aston Martin programs as well as advanced research projects.

STYLING & SURFACE DEVELOPMENT

The first clay model reflected the Concept Car styling and surfaces. This required a process by which the car was design from the outside in.

A fast track styling and surfacing process was required. Approximately 4 months of styling development was allotted. CAD was the master, while verification and review clays were cut directly from the CAD data.

The exterior and interior were styled and surfaced simultaneously, with styling freeze of the interior being achieved just one week after the exterior styling freeze.

Changes from the original Concept Car were subtle, but necessary to achieve package, aerodynamics and legal requirements (Figure 2). To comfortably package an occupant of generous proportions, the team decided to increase the overall height of the vehicle by 17mm. The height increase was achieved by essentially "morphing" the top surfaces higher, while holding the overall length and width of the car. The occupant package achieved a proper relationship of occupant seating, pedal placement and ergonomic placement of other key controls including the steering wheel and gearshift control. Achieving the ideal man/machine relationship is important in a supercar, but often difficult to achieve. The fuel tank was positioned in the center tunnel, thus allowing a small step-over rocker width for ingress/egress, unlike the 1960's original. This tank position also benefits the polar moment of rotation for the vehicle.

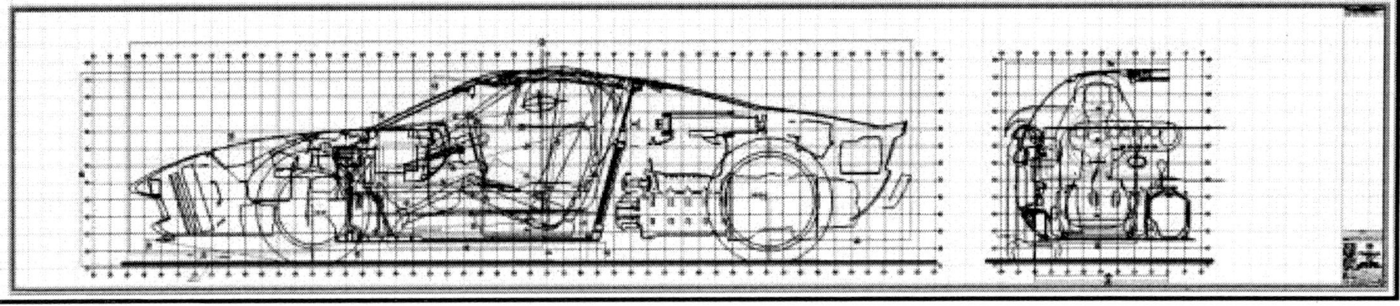

Figure 2.

The Concept Car did not have opening side glass. The surface of the glass would not allow a conventional drop-glass system, since the glass would travel beyond the Class A surface of the door skin during its downward path. Several alternatives were briefly considered by the team, including articulating windows and even gullwing windows and doors. The final solution involved complex surfacing of a new door glass and a helical travel path to achieve a full-drop door window. The window surface was moved inboard, particularly along the beltline. A division bar was added, creating a fixed front quarter window. The drop-glass curvature was tightened and the radius of curvature changes from front to back. This resulted in a surface with a front radii of 688mm at the front and 597mm at the rear.

A black diecast molding was added around the perimeter of the window. This molding maintains the original perimeter of the Concept Car's side window and thus a pleasing appearance (Figure 3 and 4).

Figure 3. Ford GT Concept

Figure 4. Final ICEM surface.

In order to meet the requirements for FMVSS Part 581, a rear bumper was added (Figure 5). The team explored several ways to meet the requirements, ultimately choosing to preserve the "ducktail" shape and character lines of the back of the car. Also, the "floating" rear bumper design helped to visually minimize the "weight" of the rear bumper.

The car went to the wind tunnel to determine the aerodynamic validity of the shape. To aid in downforce at high speeds, several aerodynamic aids were added. Starting at the front of the car, a chin spoiler was added below the front fascia. The entire underside of the car is enclosed with smooth belly pans that form a diffuser at the rear of the car. Side Splitters along the rocker panels and a small rear spoiler at the top of the rear fascia were also added to tune the front-to-rear downforce.

The deep depressions in the hood are key styling features from the 1960's car and function to evacuate air from the back side of the cooling pack. Computer modeling showed that the depressions needed to be deeper from the Concept Car to provide adequate cooling for the 500hp powertrain.

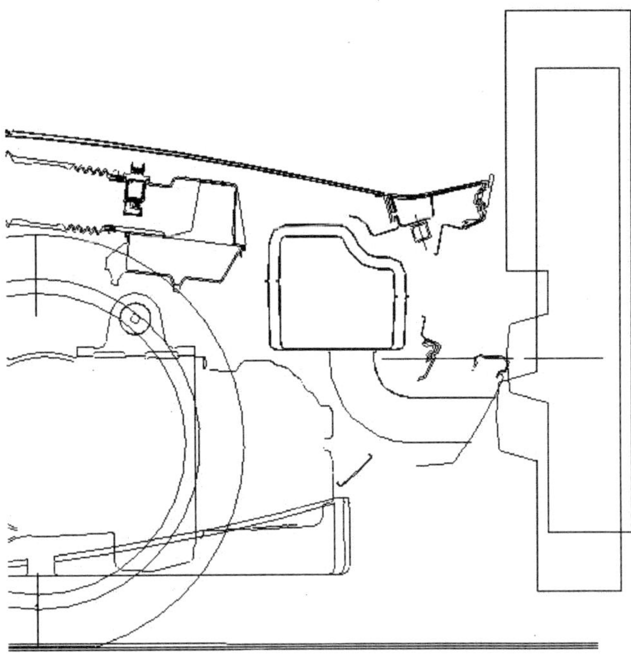

Figure 5. Side view of car with pendulum

All of the vents and grilles in the body are functional. The front grille passes air to the engine cooling pack. The scoop and grille at the base of the windshield was deepened from the Concept Car and provides intake air for the HVAC system. Aero modeling showed that this location would receive ambient air and not heated air off the cooling pack (Figure 6).

Figure 6. Flow streams

The large side scoops behind the doors allow cooling air to enter the engine compartment. The passenger side scoop passes air to the transmission cooler. The intake scoops on the decklid quarter panels serve as fresh air intakes for the powertrain. The design was revised from the Concept Car's "Mark I" type to a shape the directs the air rearward through duct work inside the decklid (Figure 7). Several vents are positioned on the topside of the decklid to allow heat from the engine compartment to escape. Grilles on the rear fascia further aid engine compartment cooling.

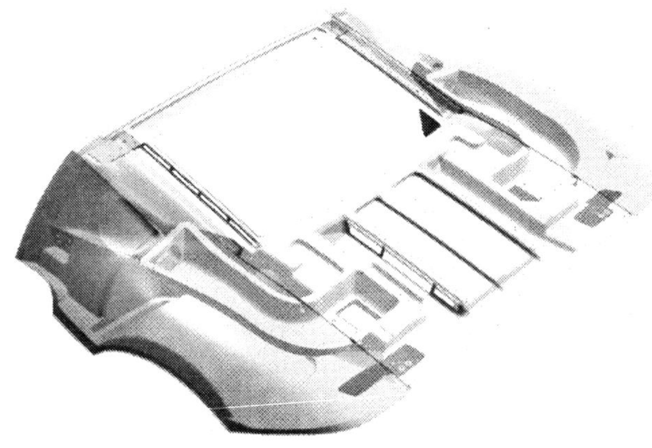

Figure 7. Decklid lid with see-through ducts

Bi-functional HIDs were selected for forward lighting (Figure 8). These lamps provided feasibility for the steeply raked headlamp lenses, maintaining the theme and the aerodynamic benefits. The rake angle is 65 degrees, believed to be among the most rakish in the industry.

Figure 8. Headlamp

Other changes from the Concept Car included the addition of windshield wipers, CHMSL, larger side mirrors, redesigned door shut-faces to package a sealing system. A novel approach to the lower door sealing system was required to maintain a high door cut-line (consistent with the theme) and a low step-over height. The decklid's B-pillar cut-line was revised to provide decklid swing clearance to partially-opened doors. The front fender surface was revised to meet tire clearance & coverage with P235/45ZR18 front tires. World class supercar performance was a top priority.

The interior design maintained much of the "exposed Spaceframe" theme of the Concept Car. The frame is exposed on either side of the occupants' feet and along the rear bulkhead. In keeping with the theme, a stamped aluminum headliner was added, while the console top and I/P center stack are cast magnesium.

Creative solutions were required to achieve a lightweight interior design. Lear and GE provided interior trim components using Azdel ™ as a substrate. This material was applied to the instrument panel, door trim and pillar trim. These are industry-first applications for this material. Azdel ™ reduces weight and tooling investment by up to 30% each compared to conventional injection-molded parts. The seats have carbon fiber seat pans and backs overlaid with foam and leather trim. The carbon fiber weave is exposed on the seatbacks and recliner knobs.

Manufacturing feasibility was conducted simultaneously with the styling and surface development of the body and interior.

ALIAS ™ was used in the studio while CATIA ™ was the primary CAD system used for engineering, development and manufacturing feasibility. ICEMsurf™ was used for Class-A surface development and Opticore ™ was used for Surface Signoff (Figure 9). Minimal hard modeling was used during the Surface Signoff process.

Figure 9. ICEM Highlight analysis

PRODUCT DESIGN

The engineering team chose CATIA™ as the primary CAD system, used on Body, Spaceframe, suspension and vehicle integration. IDEAS was used for Interior and Powertrain. The vehicle integration team used Enovia VPM & DMU to control and validate the vehicle Bill –Of-Material and package.

The CAE team were an integral part of the engineering development team and the FEA modeling was done in parallel with the CAD engineering. LS Dyna™ was used for dynamic and transient analyses, while NASTRAN™ was used for the static work.

MADYMO™ was used for the occupant safety analysis. An important element of the CAE work was to utilize a common model for all FEA. This ensured that common development concepts were developed. ICEM™ CFM was used for thermal analyses while Fluent™ was used for the aerodynamic analyses.

To ensure that the fast track development process was upheld, most engineering decisions were made during formal design reviews or impromptu meetings using master CAD data.

A flat organizational team structure was put in place. System engineering leaders were empowered to make decisions therefore no decision-making hierarchy was required.

With the simultaneous styling & engineering development process, the styling team was involved throughout. This assured that the Designer's Vision was achieved down to each detail. The team was made up of experienced enthusiasts who could make decisions on the spot.

The product uniqueness and low production volume led to selecting many suppliers outside of the normal Ford supply base. This required the integration of several CAD systems and the development of unique processes for engineering and commercial matters.

Enovia VPM was chosen as the Product Data Management system to control the Bill of Materials & all CAD data. To allow the team to control & validate the fast developing package, Enovia DMU was used which reads the data directly from the master BOM in VPM. This enabled Mayflower to feed in any new proposals &

ensure their validity before accepting them into the master package.

Due to the compressed timeframe, and the high percentage of production tooling on first builds, the body system had to be engineered and released only 2 months after Surface Sign-off.

Just one prototype build phase (15 units for the entire program) was utilized, with a high content of production tooled components. This required an intense engineering phase where the initial concept work, panel feasibility, and new surface all had to be combined into production release data.

The "right first time" approach with minimal prototype tooling demanded a high confidence level in the CAD designs, which was gained through the extensive use of virtual verification for manufacturing feasibility, dimensional control, surface quality, package validation, and structural performance.

The initial assumptions calling for non-stressed body panels, which relied on the Spaceframe for structural performance, had to be revised. Several non-skin body panels became structural including the roof inner and cowl. This was effective in optimizing the structure given the large cutouts for the doors.

The cantilevered door development required that the door maintain a tight fit at high speeds. Aero modeling provided the loads and the door was modeled in CAE to develop the structure. Upper guide pins between the inner door and roof cant were developed & optimally positioned (Figures 10 and 11).

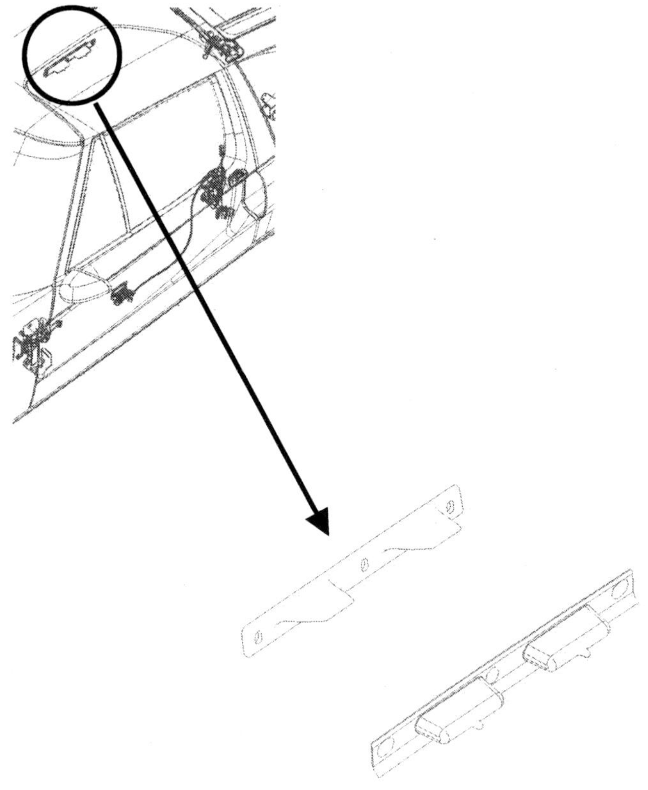

Figures 10 & 11

The large rear hinging Engine Cover was a significant engineering challenge (Figure 12). For optimal structural performance and tight dimensional control, a 1-piece multi-directional 1.2mm thick carbon fiber inner panel was incorporated. Local steel reinforcements are bonded on and J-ducts are added. The inner sub is then married to a superformed multi piece aluminum outer shell that is sub-assembled using Henrob self-piercing fasteners. Robotic roller hemming was chosen as the ideal peripheral joining method for the inner to outer subs. This allows the clinch pressure to be controlled to prevent crush of the carbon fiber inner panel. This is complemented by the use of adhesives to increase rigidity.

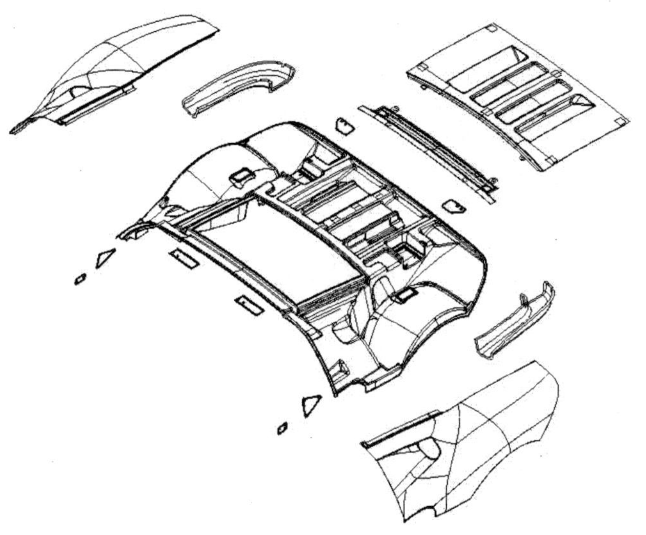

Figure 12

The upper engine intake scoop form is part of the superformed rear quarter panel, with the final finishing form and trim completed by follow-on operations. Ventilation for the engine compartment and water management is integrated into the inner panel. Simple hinges were utilized but required careful location so as to ensure that the sweeping rear cut-line with its tight margin to the fascia would open without fouling. Two gas struts were added to assist opening efforts.

With the support of the Ford Research Lab, an adhesive system was developed & implemented. This had to meet the structural performance and thermal capability requirements of the program.

- Dow 2096 – Slow cure for aluminum-to-aluminum joints.

- Dow 7300/10 – Fast cure for aluminum-to-aluminum and aluminum-to-carbon joints

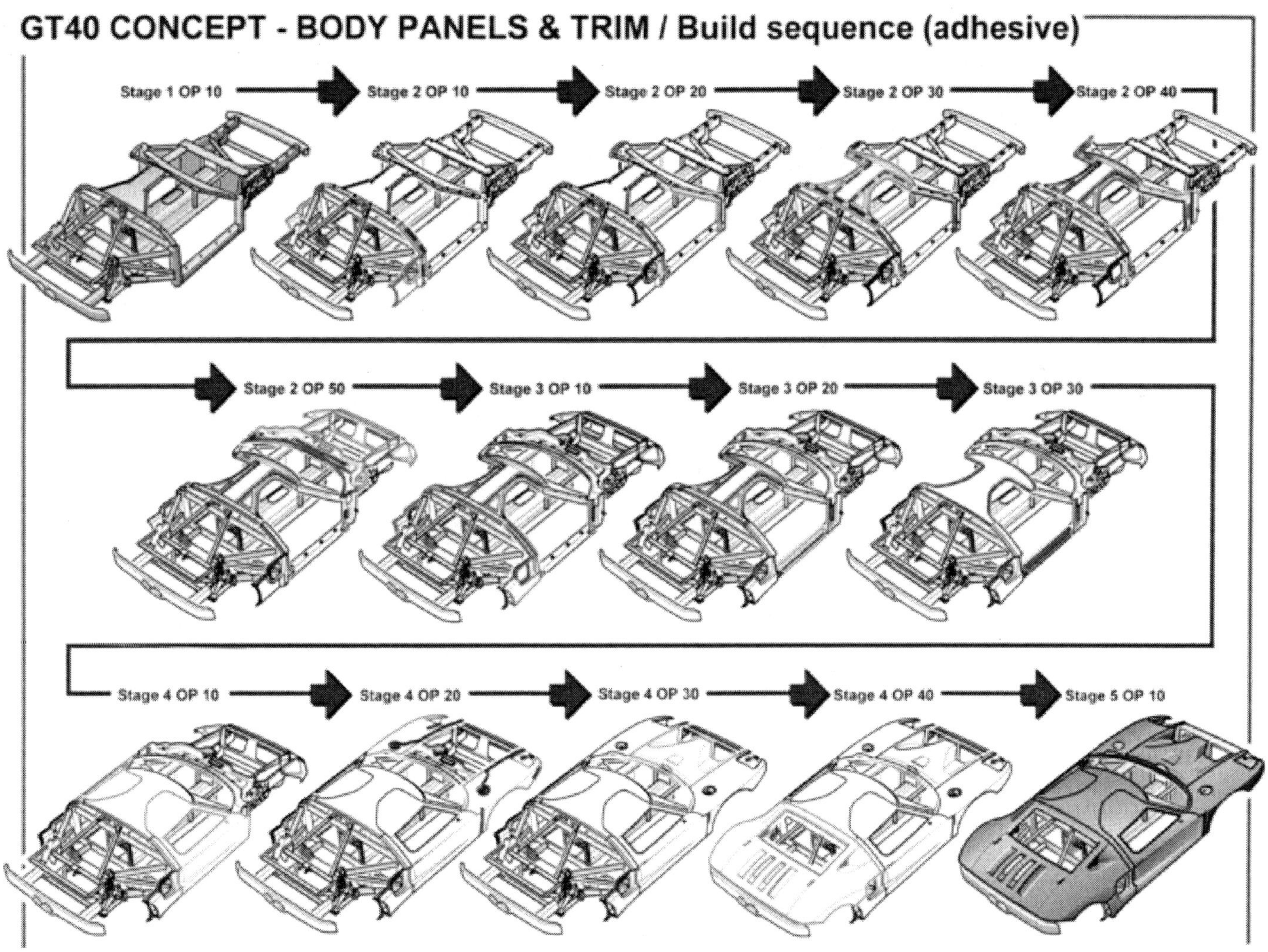

GT40 CONCEPT - BODY PANELS & TRIM / Build sequence (adhesive)

Stage 1 OP 10 → Stage 2 OP 10 → Stage 2 OP 20 → Stage 2 OP 30 → Stage 2 OP 40

Stage 2 OP 50 → Stage 3 OP 10 → Stage 3 OP 20 → Stage 3 OP 30

Stage 4 OP 10 → Stage 4 OP 20 → Stage 4 OP 30 → Stage 4 OP 40 → Stage 5 OP 10

Figure 13.

The Body assembly concept was developed at an early stage (Figure 13). The concept utilizes oversized 'plus-nuts' installed onto the Spaceframe which are then post machined after heat treat and E-coat to give a nominal location of the inner body shell (Figure 14). Body shell components are mechanically fastened & bonded to the Spaceframe. The plus-nuts are sized to accommodate the +/- 3mm tolerance that was defined as part of the DVA study based on 5 Sigma capability. The inner shell defines the door and windscreen apertures as well as the mounting interface for the outer A-Class panels.

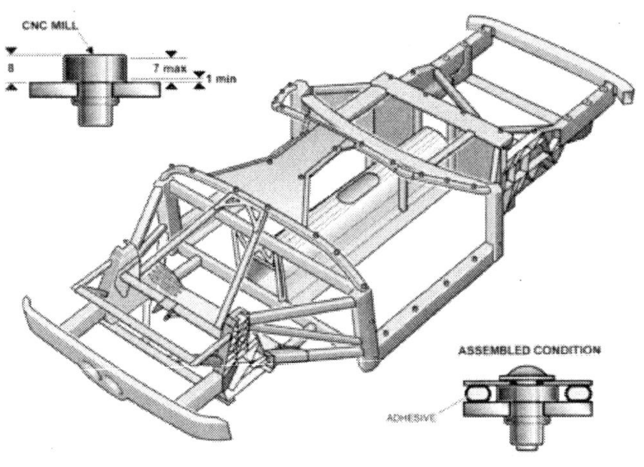

Figure 14.

Full vehicle DVA was done using 'DMS Dimensional Control Systems 3D™' software in parallel to assure that the design, manufacturing and assembly processes would achieve the required fit & finish targets. This required the definition of all Master and Secondary locations for the Body-in-White complete. A strategy of utilizing common control points through each process stage was implemented. This allowed for simpler troubleshooting in try-out and maintains a tighter capability.

One of the most significant steps taken in the development was to use a common CAE model used for all vehicle development, including body performance, closures, modal responses, vehicle structural performance, safety and durability. This ensured that common development paths were followed.

MATERIALS & FEATURES

The body panels are primarily aluminum. Some panels are formed using Superplastic-forming process to achieve the complex curves and surfaces as envisioned by the Designer. Using specialized material, 5083SPF, the Superplastic forming process can achieve up to 200% elongation compared to conventional stamping of 5754 or 6111, which achieves approx 22-26% elongation (Figure 15). The fenders, door inner & outers and the decklid quarter panels are superplastically formed. The door inner and outer panels are each made from a single sheet of aluminum. The roof, cowl panels, pillars, aprons and other components are aluminum stampings (Table 1).

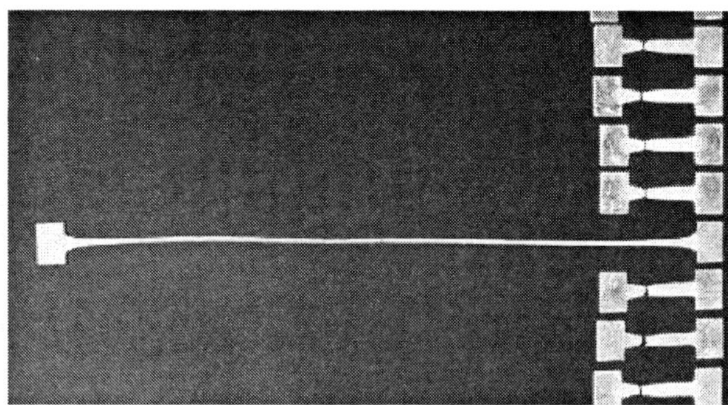

Figure 15. Comparisons of elongation – std alloys vs. SPF alloys (image courtesy of Superform USA)

Body panels are joined to the Spaceframe with machinable, self-locating, aluminum riv-nuts, which are precisely machined to locate the body panels. High strength 2-part structural epoxy adhesives were used to supplement mechanical fasteners and to achieve a durable, heat resistant structure and to improve NVH. The hood and lower quarter panels are made of chopped carbon fiber strand allowing the deep pockets of the hood in conjunction with the tight radius folds. This also results in a lightweight part. The inner panel for the engine decklid is a 1-piece carbon fiber. The 1-piece construction offers solid structure and excellent dimensional integrity lay-up. Bayflex 180 polyurethane was selected for the front and rear fascias, rockers and rear bumper cover for its heat resistance and energy absorption properties.

ALUMINUM GRADE	WHERE USED	PROPERTIES
5083SPF	Fenders front / rear, Door inner, Door Outer, Engine Cover.	Al-Mg—Mn Suitable for Superforming, high Mg content giving good strength & dent resistance.
5754	Inner Body Structure – Non A-Class. Conventional stampings.	Al-Mg-Good conventional stamping grade material.
6111	Roof Outer, Header outer, B-Pillar upper–A-Class. Conventional stampings.	Al-Mg-Si -Highly ductile A-Class conventional stamping grade material.

Table 1. Table of material grades

The fuel door is a forged aluminum piece, as are the fender release levers. The driver's side fender lever opens the front hood (also can be released with the key fob) and the passenger fender lever opens the fuel door. From the outside, the doors are released via electronic push button in the door handle pocket. The inside release is mechanical. The engine decklid utilizes two latches, each with an ajar sensor, and a secondary catch. The mid-lite assembly is made up of two pieces of laminated glass separated by an air gap. This construction was optimized to minimize sound and heat transmission.

The interior utilizes Azdel as a substrate for most of the trim panels. The console top and I/P center stack are magnesium castings that are painted satin silver. The console houses the climate controls that have unique knobs that radiate light from below to illuminate the control graphics on the console. Aluminum stampings make up the trim ring on the door panel as well as the headliner (Figure 16 and 17).

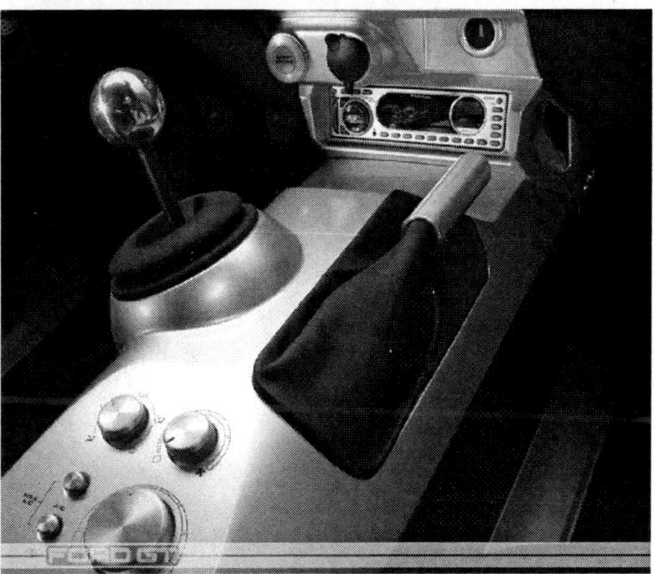

Figure 16 and 17 Body interior aluminum panels

MANUFACTURING

The manufacturing process utilizes "Center of Manufacturing Excellence" whereby the vehicle moves to where the required expertise and infrastructure has already been established. Major assembly starts with the Spaceframe assembly at Thyssen/Budd. The Spaceframe is then sent to Mayflower, where the stamped and superformed aluminum body panels are assembled to the Spaceframe. Next, the Body-in-White is shipped to an SSV facility in Troy, Michigan for paint and some assembly. Finally, the painted body is shipped to Ford's Wixom Assembly Plant for final assembly and finished vehicle inspection.

Many component manufacturing processes are new or nearly new to the auto industry, including Superplastic aluminum forming, INVAR tooling for the carbon fiber decklid inner panel, robotic roll hemming of an aluminum outer panel to a carbon fiber inner, and bonding processes for attaching body panels to the spaceframe.

DEVELOPMENT & TESTING

A unique development and testing plan was developed, including a 24-hour durability test at competition speeds. Also, the body needed to endure many high performance events at race track speeds and resulting engine thermal conditions.

A series of tests were conducted at high speed to assure stability and function of components such as wipers, climate control, door seal integrity, door glass sealing & function, body panel stability and others.

CONCLUSION

The end result is a World Class supercar that carried out the designer's vision from the Concept Car encompassing the quality and performance aspirations of the Ford GT team (Figures 18 and 19).

Figure 18.

Figure 19

ACKNOWLEDGMENTS

Dan Fisher, Ford Motor Company;

Fred Goodnow, Ford Motor Company;

Malcolm Young, Mayflower Vehicle Systems;

Mike Mitchell, Mayflower Vehicle Systems;

Alex Zaguskin, Lear Corporation;

Matt Zaluzec, Ford Motor Company;

Eric Kleven, Ford Motor Company;

Simon Irigbu, Ove Arup

2004-01-1254

2005 Ford GT - Vehicle Aerodynamics - Updating a Legend

Kent E. Harrison, Michael P. Landry and Thomas G. Reichenbach
Ford Motor Company

ABSTRACT

This paper documents the processes and methods used by the Ford GT team to meet aerodynamic targets. Methods included Computational Fluid Dynamics (CFD) analysis, wind tunnel experiments (both full-size and scale model), and on-road experiments and measurements. The goal of the team was to enhance both the high-speed stability and track performance of the GT. As a result of the development process, significant front and rear downforce was achieved while meeting the overall drag target.

INTRODUCTION

Based on overwhelming reaction to the Ford GT showcar introduced at the 2002 Detroit International Auto Show, Ford Motor Company decided to develop a production version. Due to public reaction as well as the need to build three production cars for Ford's centennial celebration, it was agreed that the surface of the showcar would become the surface of the production car, with minor exceptions. For the group charged with meeting aerodynamic targets, this meant that the underbody had to be the main area of focus. The team began work in April 2002, building a full-size aerodynamic test buck, primarily used for cooling airflow studies, and a 0.45-scale model to be used in a rolling-road wind tunnel for drag/lift studies. Additionally, because the new vehicle had similar shape and proportions to the 1968 GT40, an original GT40 was tested in the wind tunnel. The original vehicle was configured like the Gulf GT40, chassis #1075 that won at Le Mans in 1969 for the JWA Racing Team. This early investigation allowed the team to establish, at least directionally, the program's starting point.

AERODYNAMIC DEVELOPMENT

CFD ANALYSIS

Computational Fluid Dynamics (CFD) was used on the show-car surface to generate full vehicle maps of airflow and pressure fields. The CFD analysis helped direct the team to areas of interest.

CFD work was conducted while full and scale model properties were being constructed. This gave a head start on the experimental phase of the work. When the physical properties were constructed, the most efficient path was to begin wind tunnel testing.

Figure 1. Example CFD Image

1968 FORD GT40

The team obtained use of an original GT40 (chassis # 1030), a 1968 model configured like the Gulf GT40, chassis #1075 that won at Le Mans in 1968 and 1969.

Concurrent with the CFD analysis, the car was tested in a fixed-ground plane Wind Tunnel. The purpose of the test was to obtain an early indication of airflow across a similar shape as well as initial drag and lift indicators. The aerodynamic performance of the 1968 GT40 is shown relative to Ford GT team targets in Table 1. Note: all data presented in this paper has been normalized with Ford GT Team targets set to unity as shown in Table 2.

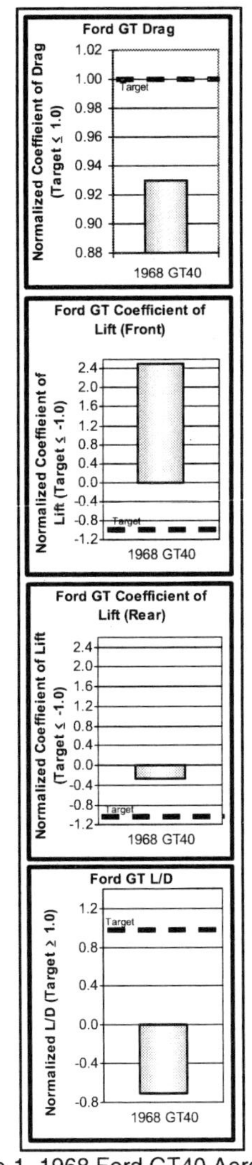

Table 1. 1968 Ford GT40 Aero Data

As can be seen from the data, it was clear why GT40 drivers complained of front-end lift on the Mulsanne straight!

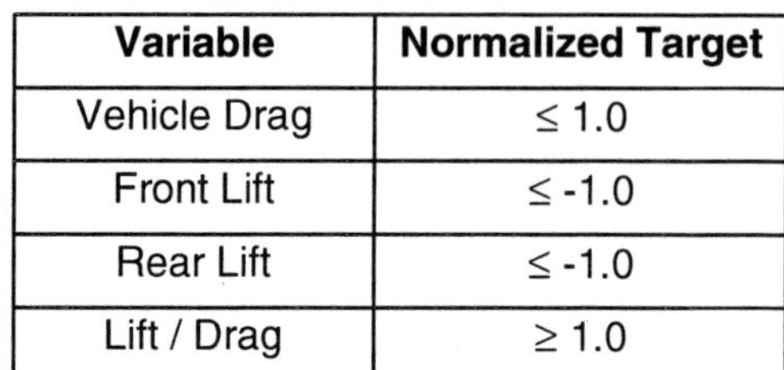

Variable	Normalized Target
Vehicle Drag	≤ 1.0
Front Lift	≤ -1.0
Rear Lift	≤ -1.0
Lift / Drag	≥ 1.0

Table 2. Normalized Ford GT Aerodynamic Targets

Figure 2. 1968 GT40 in Ford Wind Tunnel

FULL-SCALE AERO BUCK

A full-scale aerodynamic buck was constructed to obtain an early indication of airflow across the top surface as well as cooling and engine bay airflow. Significant development time was spent maximizing airflow across the radiator. Several different radiator exit shapes were investigated, including versions similar to the double exit of the Mark II and the single large exit of the Mark IV GT40s. Ford GT cooling development is discussed in detail in SAE 2004-01-1257, "*2005 Ford GT - Maintaining Your Cool at 200 MPH*".

The aerodynamic performance of the full-scale model is highlighted in. While the first full-scale aerodynamic test showed slightly better coefficient of drag than the original GT40, the lift at the front of the car was an area that required significant improvement to meet program targets.

Figure 3. Full-Scale Aero Buck at Ford Wind Tunnel

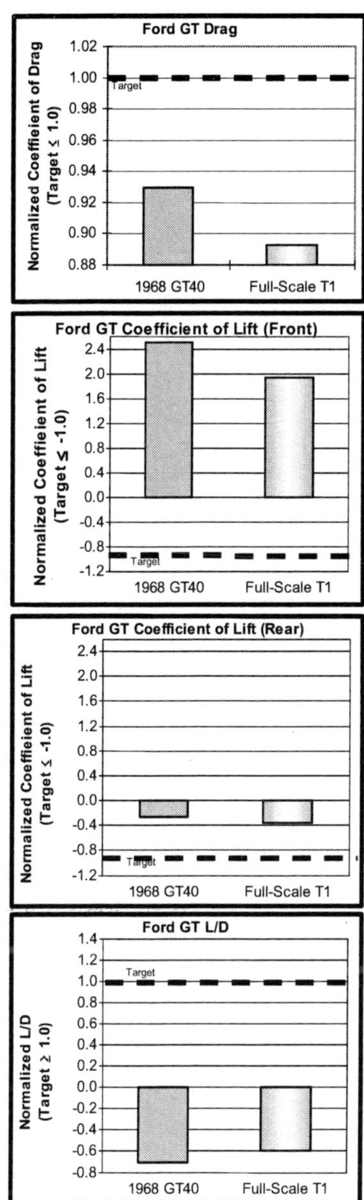

Table 3. Full-Scale GT Aero Data

0.45 SCALE AERO BUCK

A 0.45-scale aero buck was constructed at Swift Engineering in San Clemente, California using precisely scaled measurements from the design studio styled Ford GT surface, as well as correct tire, wheel, and brake sizes and ride-heights.

There were four phases of development with the 0.45 scale aero buck on a moving road surface wind tunnel. Each phase produced an enhanced aero package. The aerodynamic performance from the four phases is presented in Table 4. T1 was the initial testing with the design studio outer surface and basic or simplified aerodynamic devices (splitter and diffuser). T2 was completed with more developed aerodynamic devices (splitter, diffuser, airdams, spoiler extension). T3 was completed with fully developed (for function but not style) aerodynamic devices including additional underbody treatments. T4 was a combination of developed (functional and styling) aerodynamic devices.

Figure 4. 0.45 Aero Buck at Swift Wind Tunnel

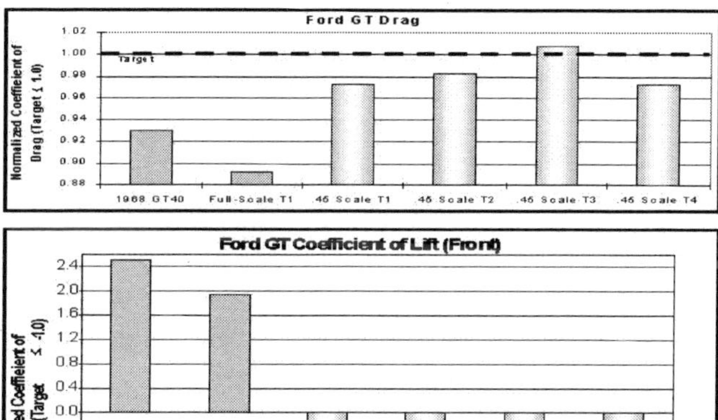

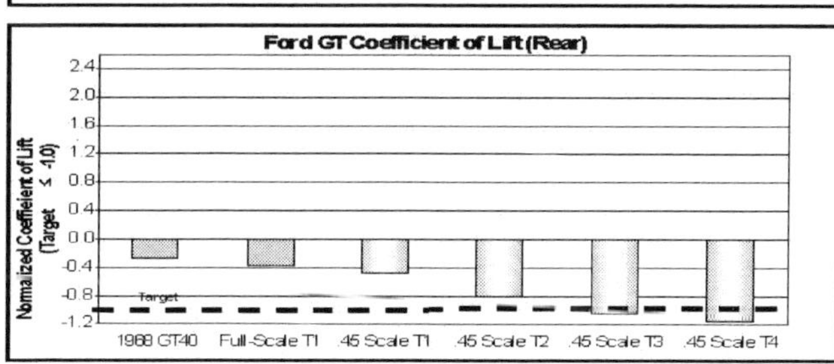

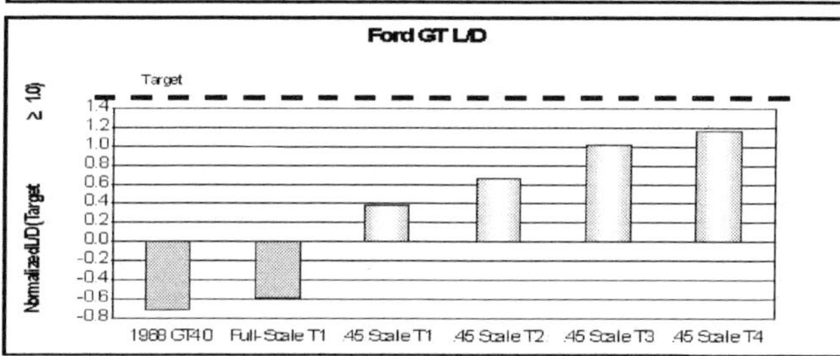

Table 4. 0.45 Scale Aero Buck Data

The final phase (T4) was initiated by the Ford design studio. There was some difficulty in obtaining an integrated appearance at the rear of the vehicle with the proposed tunnels of the T3 level (functional but not styled) rear underbody diffuser. The additional work resulted in the visual impact that the design studio desired and provided the team the additional time required to exceed some of the performance targets. This final phase of scale model development produced an aerodynamic package that minimized the vehicle drag (better than target), produced balanced (40/60) front and rear downforce (both better than target), and a lift over drag (L/D) ratio that was better than target.

DESIGNED AERODYNAMICS PACKAGE

The final aerodynamic package developed with the 0.45 scale model is shown in Figure 5. The package includes a front spoiler with an integrated airdam and splitter, mini underbody diffusers feeding into all four wheel wells, side splitters, a rear underbody diffuser, full under body cladding, and a rear spoiler extension.

The front spoiler affects downforce in a variety of ways. By reducing the gap between the vehicle and the road, the front air dam reduces the flow area and increases the speed of flow under the vehicle. This produces a high-speed, low-pressure region of flow that pulls downward on the nose of the vehicle. The front airdam also creates a high-pressure stagnation zone at the front of the vehicle, which pushes down on the horizontal splitter adding additional downforce. The mini underbody diffusers in front of the front tires accelerate airflow into the wheel wells improving the front spoiler efficiency.

The side airdams direct airflow away from the front tires and wheel wells out to the sides of the vehicle reducing drag.

The rear underbody diffuser produces downforce by accelerating the underbody airflow and reducing underbody pressures. The converging/diverging shape

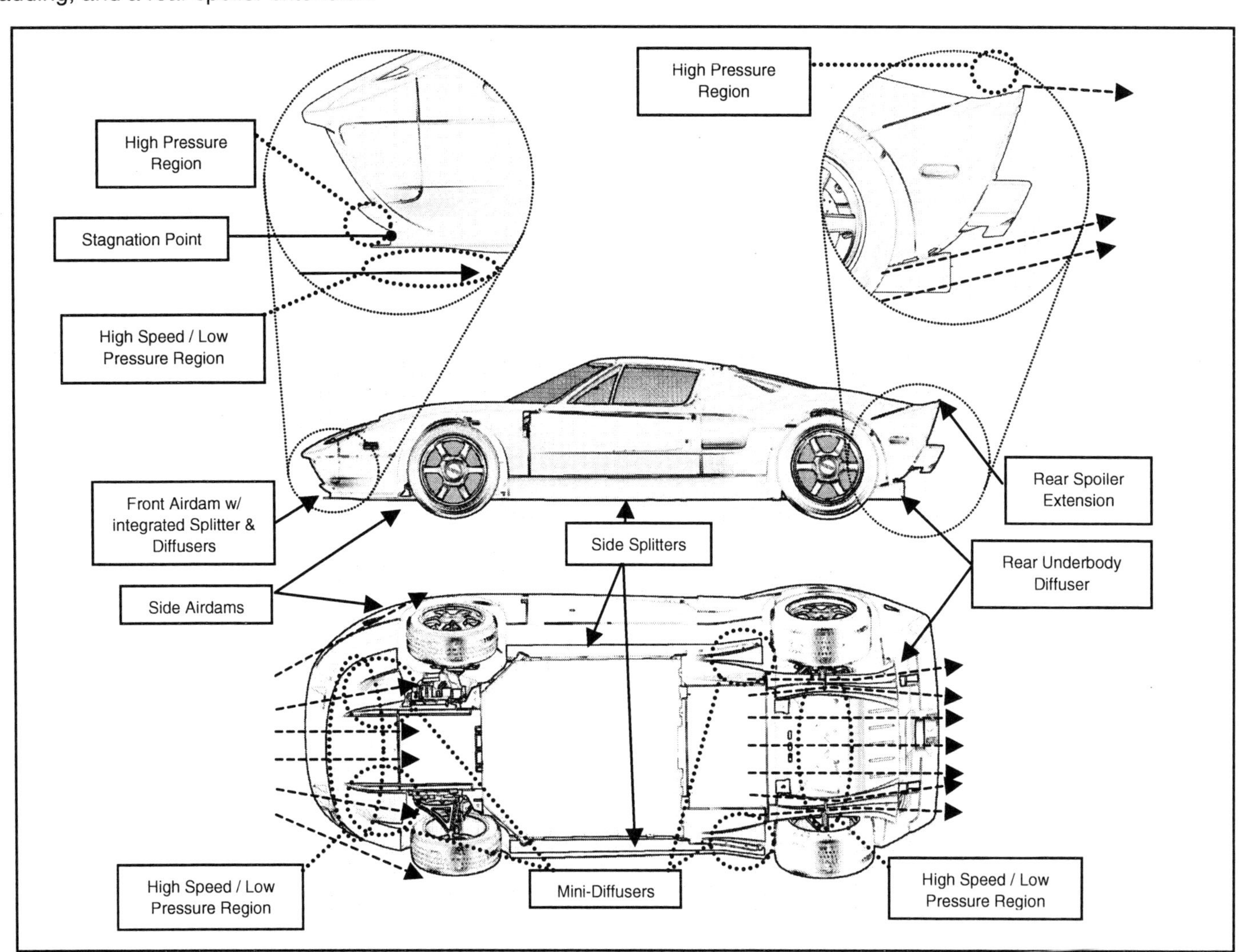

Figure 5. Aerodynamic Package

of the diffuser in both the y and z directions creates a venture-like effect. The side splitters enhance the rear underbody diffuser's efficiency by segregating the side and underbody airflows. The NACA (National Advisory Committee for Aeronautics) style mini-diffusers in front of the rear wheels accelerate airflow into the wheel wells; this also enhances the underbody flow. The rear spoiler extension creates a high-pressure stagnation zone that pushes down on the rear of the vehicle. In addition, the rear underbody diffuser and rear spoiler extension work in conjunction to reduce drag by reducing the size of the low pressure wake at the rear of the vehicle.

ON-ROAD DATA ACQUISITION AND DEVELOPMENT

On road testing was completed to validate and confirm the wind tunnel data and as an assessment of vehicle stability.

On Road Testing Phase I

The first phase of the on road aerodynamic performance assessment was completed on the runway surface at the Oscoda-Wurtsmith Airport in Oscoda Michigan. The runway surface at Oscoda is approximately 11,800 feet long and 220 feet wide. In addition, there is an approximately 3600 foot taxi way leading to the north end of the runway. These surfaces allowed for safe evaluation at speeds in excess of 120 mph.

In this phase, a prototype vehicle (Workhorse #2, WH2) was retrofitted with confirmation prototype (CP, fully functional, design intent vehicle) level aerodynamic parts configured to the final aerodynamic package developed on the 0.45 scale model. For "on road" lift (downforce) measurement, WH2 was equipped with linear potentiometers to measure shock travel, and wheel force sensors, which were developed in the Ford Racing Technology group and currently used for vehicle testing in various racing series (Figure 6 and Figure 7).

Figure 6. Front Wheel Force Sensor

The on road data was collected with a Pi 4+ Electronic Control Module running Pi V6 software. Signals collected included vehicle speed, lateral and longitudinal accelerations, yaw rate, airspeed, wheel forces in the z-direction, wheel moments about the y-axis, and shock travel at all four corners of the vehicle. The specifics of each signal are presented in Table 5.

Figure 7. Rear Wheel Force Sensor

Signal	Sensor Type	Manufacturer / Model #	Sample rate
RR Wheel Speed	Wheel Speed Sensor	Pi Research / 21A-0062-A	5 Hz
Vehicle Speed	Multi-Axis Wheel Force Sensor	Sensor Developments Inc., Model 12001	2 Hz
Lateral Acceleration	Accelerometer	Pi 4+ EMC	5 Hz
Longitudinal Acceleration	Accelerometer	Pi 4+ EMC	2 Hz
Yaw Rate	High Precision Gyro	Pi Research / 01B-033911	20 Hz
Air Speed	Pitot Tube & Sigma/Delta Single Aero sensor	Pi Research / 01B-050245 01B-601219	25 Hz
F_z & M_z (LF, RF, LR, RR)	Multi-Axis Wheel Force Sensor	Sensor Developments Inc. , Model 12001	25 Hz
Shock Travel (LF, RF, LR, RR)	Linear Potentiometer	Penny + Giles / Model HLP190	25 Hz

Table 5. On-Road Equipment Specifications

	Variable	Sensitivity	Direction
Front Axle Lift	Front Under Tray	High	Offset ↑ = Lift ↓
	Front Splitter Width		Width ↑ = Lift ↓
	Vehicle Rake		Rake ↑ = Lift ↓
	Side Splitter Position	Mild	Position ↑ = Lift ↓
	Front Splitter Height		Height ↑ = Lift ↓
	Rear NACA Ducts		Open = Lift ↓
	Rear Spoiler Extension	Negligible	NA
Rear Axle Lift	Front Splitter Width	High	Width ↑ = Lift ↑
	Vehicle Rake		Rake ↑ = Lift ↑
	Front Under Tray	Mild	Offset ↑ = Lift ↑
	Side Splitter Position		Non-Linear
	Front Splitter Height		Height ↑ = Lift ↑
	Rear Spoiler Extension		Height ↑ = Lift ↓
	Rear NACA Ducts	Negligible	NA
Coastdown/Drag	Front Splitter Height	High	Height ↑ = Drag ↓
	Rear NACA Ducts		Open = Drag ↓
	Front Under Tray Position	Mild	Offset ↑ = Drag ↑
	Side Splitter Position		Position ↑ = Drag ↑
	Front Splitter Width		Width ↑ = Drag ↓
	Rear Spoiler Extension		Height ↑ = Drag ↓
	Vehicle Rake	Negligible	NA

Table 6. Oscoda Phase I DOE Interactions

A 36-run, Latin Hypercube design of experiments (DOE) was conducted using 7 variables. The DOE was developed to confirm the wind tunnel investigations as well as understand the impact of minor tuning to the aerodynamic devices on the vehicle's high-speed aerodynamic performance. In addition to the linear potentiometer and wheel force transducer data, an experienced driver also completed subjective vehicle performance evaluations. A summary of the interactions highlighted by the DOE is shown in Table 6. In addition to highlighting interactions, the on road data directionally confirmed the wind tunnel data particularly in regards to vehicle balance.

On Road Testing Phase II

The second phase of on road testing was a sign-off of the design intent parts for high-speed performance. This phase was also completed at Oscoda-Wurtsmith airport. In the second phase, a confirmation prototype CP (fully functional, design intent vehicle) with the full designed aerodynamic package was driven at high speeds for initial aerodynamic sign-off. Linear potentiometer, wheel force sensor, and subjective data were taken. Again, the on road data directionally confirmed the wind tunnel data.

On Road Testing Phase III

The third and final phase of on road testing is scheduled for the first quarter of 2004. The final high speed aerodynamic performance sign-off will be completed on the "Speed Bowl" at the proving grounds in Nardo Italy. The Speed Bowl is a 7.8 mile circular track allowing for safe high speed testing at speeds in excess of 150 mph.

CONCLUSION

Based on the investigations and test results presented in this paper, the following conclusions can be drawn:

1. Current day vehicle aerodynamic technology and knowledge can be applied to a basic vehicle shape from the 1960s to enhance the vehicle aerodynamic performance (lift and drag).

A vehicle's aerodynamic performance (lift and drag) can be greatly enhanced through design changes limited primarily to underbody treatments. A modern vehicle should no longer be performance limited by poor aerodynamics.

3. On road aerodynamic performance can be effectively assessed with the use of linear
2. potentiometers to measure shock travel, and multi-axis wheel force sensors.

   Linear potentiometers and wheel force transducers may also be used to directionally validate or confirm wind tunnel investigations.

5. Analytical and experimental tools can be effectively combined depending on problem constraints. Many more tools are available today than were in the 1960's when the original GT40 was designed.

4.

ACKNOWLEDGMENTS

The authors would like to thank the following people/companies for their support with the work/investigations discussed in this paper:

Wayne Koester and William Pien, Aerodynamics, Ford Motor Company

Frank Hsu, Thermal/Aero Brake Cooling, Ford Motor Company

Pat DiMarco, Ford Racing Team/Performance Vehicle & Equipment Options, Ford Motor Company

Jay Novak, Vehicle Dynamics, Ford Motor Company

David Buche, Quality, Reliability, Robustness, Design for Six Sigma, Ford Motor Company

Dennis Breitenbach, Jamie Cullen, Mark McGowan, Alex Szilagyi, Brian P. Tone, Jeffery Walsh, and Chris White, Ford GT Vehicle Dynamics, Ford Motor Company

Swift Engineering Inc., San Clemente, CA

Sverdrup Technology, Inc., Drivability Test Facility, Allen Park, MI

Tom Salter and Team, Oscoda-Wurtsmith Airport Authority, Oscoda, MI

Bob Nowakowski, Bill Green, and team, Technosports

Rich Reichenbach and Jim Ikach, Roush Racing

REFERENCES

1. Aird, Forbes, Aerodynamics for Racing and Performance Cars, The Berkley Publishing Group, 1997.

2. Katz, Joseph, New Directions in Race Car Aerodynamics: Designing for Speed, Robert Bently Automotive Publishers, 1995.

3. McBeath, Simon, Competition Car Downforce: A practical Guide, G.T. Foulis and Company, 1998.

4. Scibor~Rylski, A.J., Road Vehicle Aerodynamics, Second Edition, Pentech Press Limited, 1984.

CONTACT

Author Information:

1. Kent E. Harrison, Performance Development Supervisor, Ford GT, SVT Engineering, Ford Motor Company, 333 Republic Drive, Suite B56, Allen Park, MI, 48101, phone: (313) 20-68385, fax: (313) 390-4908, e-mail address: kharriso@ford.com

2. Michael P. Landry, Technical Standards Engineer, Surveillance & Compliance, Vehicle Environmental Engineering, Ford Motor Company, Henry Ford II World Center, Suite 224-E6, One American Road, Dearborn, MI 48126-2798, phone: (313) 594-0564, e-mail address: mlandry@ford.com

3. Thomas G. Reichenbach, Vehicle Engineering Manager, Ford GT, SVT Engineering, Ford Motor Company, 333 Republic Drive, Suite B56, Allen Park, MI, 48101, phone: (313) 323-6503, fax: (313) 390-4908, e-mail address: treichen@ford.com

ADDITIONAL SOURCES

FLUENT, computational fluid dynamics software.

2004-01-1255

Design and Analysis of the Ford GT Spaceframe

Huibert Mees
Ford Motor Co.

Celyn Evans
Mayflower Vehicle Systems Inc.

Simon Iregbu and Tim Keer
Ove Arup & Partners Detroit Ltd.

ABSTRACT

The Ford GT is a high performance sports car designed to compete with the best that the global automotive industry has to offer. A critical enabler for the performance that a vehicle in this class must achieve is the stiffness and response of the frame structure to the numerous load inputs from the suspension, powertrain and occupants.

The process of designing the Ford GT spaceframe started with a number of constraints and performance targets derived through vehicle dynamics CAE modeling, crash performance requirements, competitive benchmarking and the requirement to maintain the unique styling of the GT40 concept car.

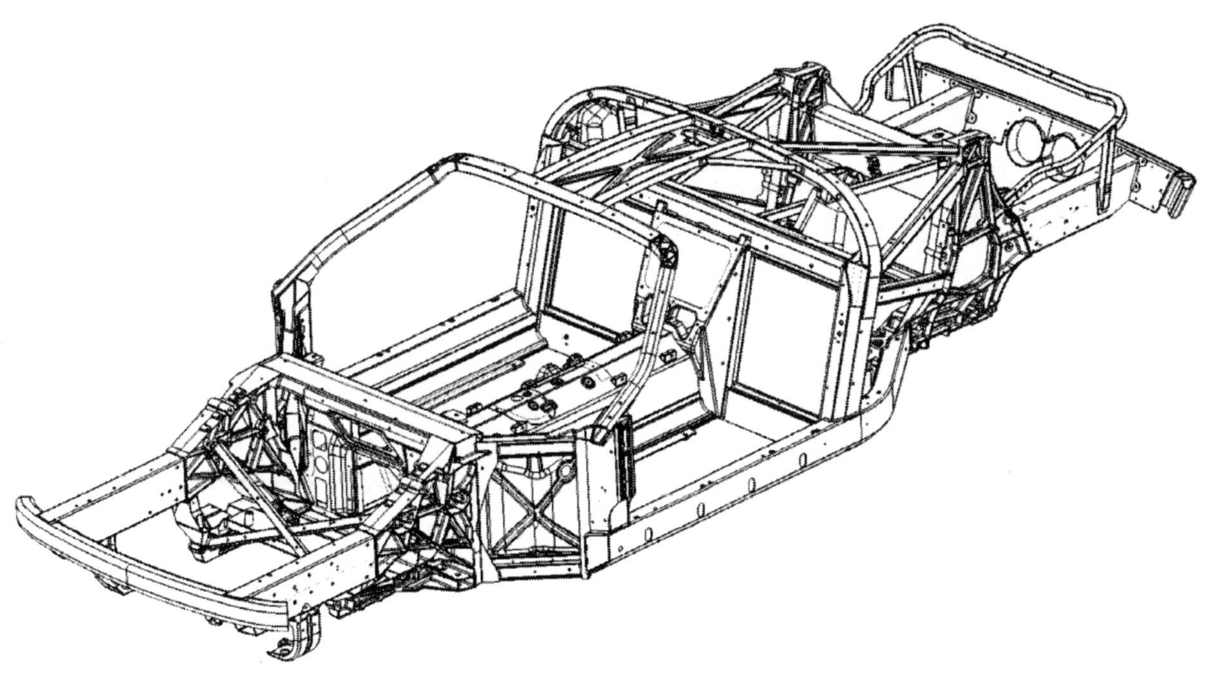

Figure 1. Ford GT Space Frame

To achieve these goals, an aluminum spaceframe was designed incorporating 35 different extrusion cross-sections, 5 complex castings, 4 smaller node castings and numerous aluminum stampings. A highly iterative design process which utilized continuous CAE analysis of the stiffness, crash and NVH performance and cross sectional properties of the various extrusions was used to balance the frame performance with manufacturing, cost and weight requirements. The resulting spaceframe design features a large central tunnel to house the centrally mounted fuel tank, highly optimized yet non-intrusive rocker sections, a series of stiffener bars removable for engine installation and service and an upper structure designed to transfer crash, bending and torsional loads around door openings which extend well into the roof area. The front and rear bumper structures including longitudinally mounted crush rails are bolted to the frame for ease of repair and a bolted-in magnesium cross member provides retention for the fuel tank. The remainder of the primary frame structure is mig welded while the numerous stamped panels are riveted and bonded to allow the thinnest possible material to be used.

INTRODUCTION

In order to achieve the extremely high performance goals of the Ford GT sports car while maintaining the styling and powertrain requirements established by the GT40 concept car, an aluminum spaceframe was chosen as the basic structure of the vehicle. Although this was not the first aluminum spaceframe developed by Ford Motor Co. the complexity and size of the vehicle required some unique design and manufacturing techniques in order to achieve the low weight, high stiffness and package requirements. The resulting design (see Figure 1) is a hybrid construction consisting of castings, extrusions and stamped, flat and rollbonded panels. Castings are used at the suspension attachment locations as well as at the interface of the rear bulkhead and central tunnel. Numerous extrusions with varying cross sectional properties connect the castings and form the remaining structure. All structural connections between the extrusions and the castings are welded except for the front and rear crush structures that are bolted to the respective castings to facilitate ease of repair. Numerous panels close off the front and rear bulkheads, form the floor system and close off the central tunnel.

This paper will describe the design and analysis process that was utilized during the design and development phase of this project and the resulting design elements that comprise the Ford GT spaceframe.

DESIGN CONSTRAINTS

As the primary structural element in the vehicle design, the frame is effectively a large, complex bracket onto which all other vehicle systems are attached. As such, the frame must accommodate these systems while at the same time providing sufficient strength and stiffness to allow each system to function properly and to allow the crash performance of the vehicle to meet the required guidelines.

In the Ford GT, as in most other vehicles, the systems that represented the most significant constraints on the frame design were the powertrain, occupant package and crash performance.

POWERTRAIN

In order to maintain a low vehicle center of gravity, the powertrain was mounted as low as possible in the vehicle. This prevented the packaging of a structural cross member below the engine, which resulted in cantilevered engine mounts. The overall size of the powertrain and mid-vehicle location presented numerous challenges to attaining the necessary overall torsional and bending stiffness as well as local attachment stiffnesses. In addition, the centrally located fuel tank required a large tunnel and prevented the use of larger side rails due to their effect on the occupant.

OCCUPANT PACKAGE

Although the vehicle provides seating for only 2 occupants with a minimum of luggage, the occupant package still presented frame design challenges in order to provide adequate leg and headroom, ingress/egress and outward visibility. Each of these necessitated the smallest possible frame sections and the use of narrow rockers and windshield pillars.

CRASH PERFORMANCE

The crash performance of the vehicle had to meet all the usual requirements for a road vehicle within the tight package constraints created by the suspension systems, center mounted radiators, door architecture, as well as the need for low overall vehicle weight.

In addition to these common constraints, the Ford GT included an additional constraint: styling. This was manifested primarily on the famous door design first seen on the original Ford GT racecars 40 years ago. These doors prevent the usage of a fundamental load path from the windshield header up through the cant rails and down the "C" pillar to the rear structure. An alternative shared load path was developed to attain the necessary structural stiffness and crash integrity.

DESIGN EVOLUTION

The frame design went through 2 main evolutionary steps. It was determined that the team needed a prototype build for development (work horses). The packaging of major systems (like the HVAC) were still in development; and stiffness targets still needed to be defined and met. In order to facilitate the build, a design level was frozen. The design was simplified to limit frame investment by using prototype processes and finding off the shelf extrusions to replace the more complicated design intent parts. A total of 13 extrusions were used to build the entire frame (production is 35). The manufacturing team created a single fixture to manually weld the frame. Additionally, preliminary designs of the suspension mounting castings were sand cast for the prototype vehicles. The prototypes were instrumental in understanding manufacturing issues and build issues. Data was used from this build to determine the manufacturing trends with the frame and try out design concepts for immediate feedback.

The second evolutionary step was the transition to a production level frame. Major improvements were in integration and optimization of the load paths. Hundreds of FEA runs were done to optimize the strength/stiffness vs. weight. Load paths were redefined; extrusion thicknesses were modified to put the mass in the most efficient locations, shear plates were optimized in construction/gage, and extrusion nodes were modified. Additionally, the vehicle package had progressed and the build level evolved. The tooling was developed for the production parts simultaneously with the tooling of parts for Job 1-3 cars built for the Ford Centennial. This helped give an early indication of the improved design and allowed for revisions earlier in the process. These improvements then went into the next build design.

LOAD PATHS

As a sports car with an emphasis on performance, the frame needed to be stiff so the suspension could perform correctly. Due to the constraints outlined above, this presented a unique problem. Without the ability to transfer load adequately to the large tunnel, and with small rockers and the door cut out, there was compromise in load paths. Combined with the lack of permanently attached rear body structure (due to the clamshell design), there was little room to strengthen the vehicle. The team accepted the challenge, then defined and utilized the existing paths. Certain load paths contribute significantly to the design optimization of the vehicle. The main areas of interest are:

1. Rocker
2. Tunnel
3. Roof
4. Floor
5. Removable beams
6. Castings

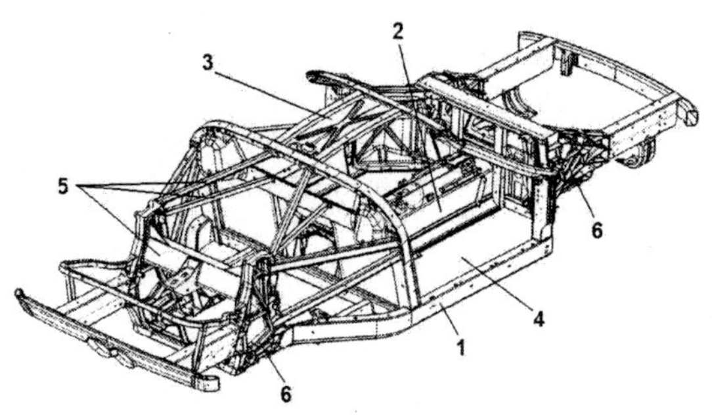

Figure 2. Primary Load Paths

Rocker

The rocker is the most significant lower load path. As the most efficient mass location and with potential to support bending, crash, and torsion loads, it was key to integrate the structure into the castings. The original design concept allowed the cooling hoses to run on the inboard side with a separate styling driven cover. By repackaging the lines under the sill cover, the extrusion was increased in size and inherent efficiency. This allowed the trim part to be integrated in the die design, eliminating parts, assembly and weight. Additionally it allowed for a better optimization of the cross-section and internal rib structure.

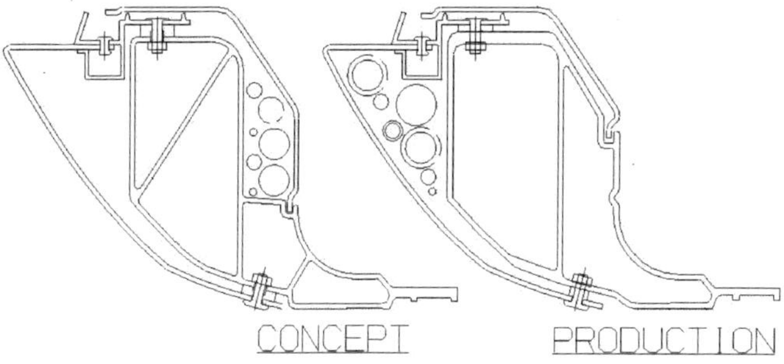

Figure 3. Concept and Production rocker cross-sections

To remove the inherent discontinuity in the rear bulkhead and to develop a larger support for the rear casting, the rocker was extended through the pillar to the rear engine mount. The 12" rocker section was bent using a custom tool to achieve the tolerances and contain the complicated shape. The complicated machining was offset by the large improvements in overall structure, stiffness, modal and crash response. The large and complicated shape also created a challenge for the die manufacturer. Although a multiple piece rocker was investigated, it turned out to have issues, the main design did not. The excessive welding caused dimensional problems on the frame, and weaker strength around weld heat affected zones. Through three main die trials, the supplier not only managed to overcome the difficulty in creating the shape but they also managed to control the tolerances better than expected.

Tunnel

The fuel tank package created the largest single section in the tunnel.

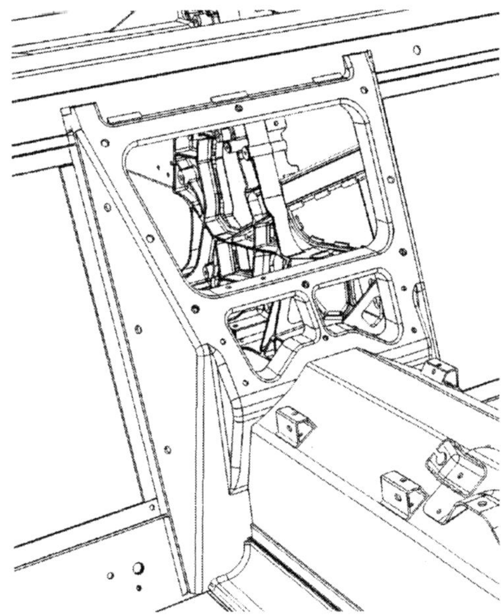

Figure 4. Center bulkhead casting

In order to utilize this central structure several steps were needed. To facilitate this as a load path, the front and rear tunnel to frame attachments were reinforced. To transfer load more effectively from the front suspension, the magnesium IP was attached to the dash cross member and tunnel below the radio, effectively bracing the dash against the tunnel.

In the rear, a central casting was designed to collar the tunnel section. Not only did this casting allow for a more continuous connection to the tunnel, it also increased the packaging freedom at the bulkhead. The casting allowed for service access to the water pump and hoses, pass through locations for the wiring and shifter cables and attachment locations for a variety of components (See figure 4). From a manufacturing side, the section was created out of a simple press-brake upper with two small-extruded sections and friction stir welded (See figure 5). This once again allowed the stylized trim to be integral to the structure but it also reduced distortion and eliminated the need to seal the joint.

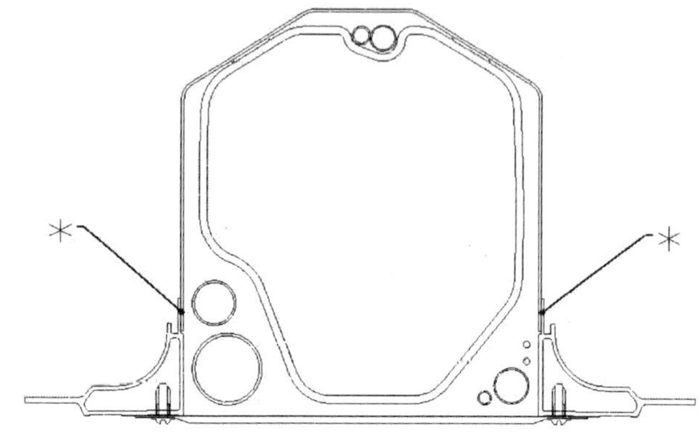

Figure 5. Tunnel cross-section

Roof

In order to meet the bending stiffness targets the roof needed to be a major component and load path. 50% of the roof structure was removed for the unique door opening, which eliminated the outer roof rail load path; the center was reinforced to handle the loads. The body panel inner structure and a shear plate were utilized to create a box-section around two bent fore/aft extrusions. At the same time, the largest A-pillar section possible was packaged with cast nodes at the top and bottom.

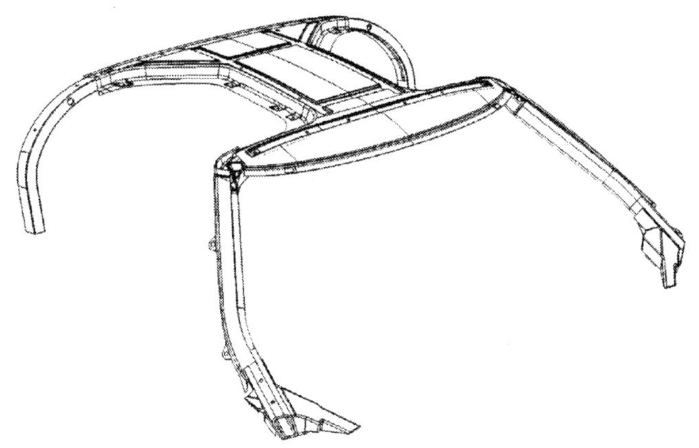

Figure 6. Greenhouse Structure

The cast design allowed a gusseted joint while simultaneously creating a fore/aft and up/down slip plane. As an additional step, the upper node was cast using a semi-solid process, which allowed for increased strength, elongation and tolerances. When the glass is combined with the roof structure, the load path is sufficiently reinforced to offer a 45% improvement in bending and torsion over a convertible.

Floor

The floor was separated into three sections, permanently attached left and right sections and a removable center section (shear panel). The center was designed to allow the fuel system and cooling lines to be assembled in the tunnel with service access. To lower the center of gravity and improve vehicle dynamics, the seats and fuel tank were put as close to the ground line as feasible. This only allowed 5 mm of space to package the floors. The main options investigated were:

1. 3 mm 6061-T6 AL (benchmark)
2. Sandwich sheets (Aluminum, plastic, and steel)
3. Carbon fiber
4. Aluminum honeycombs
5. Roll-bonded
6. Stampings

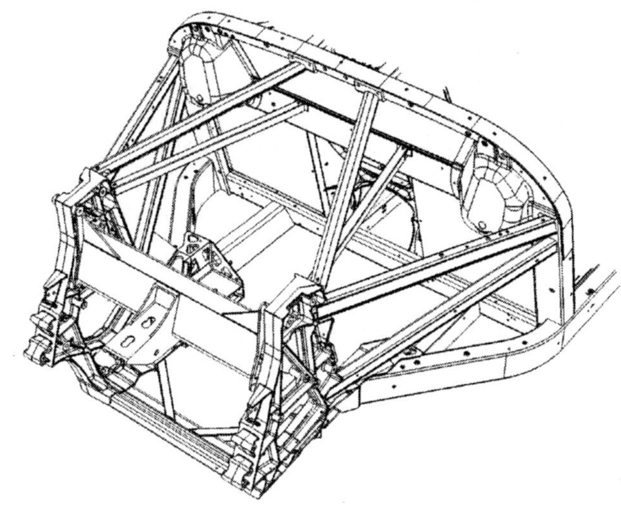

Figure 8. Engine compartment

Several materials could not be bolted without structural damage to the panels. The additional parts and operations to make those panels work was cost prohibitive. From a weight vs. stiffness/strength ratio, roll-bonding was the best choice. Through negotiations with the program management, the weight reduction offset the tooling investment

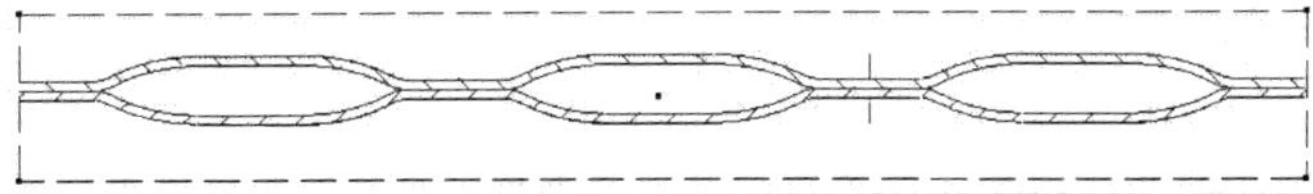

Figure 7. Roll-bonded floor section

Removable beams

The load from the rear castings needed additional support and better transfer. The castings tended to pivot around the engine axis without a strong central support. The back structure also had a tendency to rotate forward in car.

The workhorse was assembled with a single sub frame to facilitate service and expedite the build. In production, to achieve the stiffness and strength requirements, this part was separated into individual components. These parts allowed for a direct load transfer from the shocks to the roof structure and also to the tunnel. By putting these beams in a more direct path, they could be made lighter and more efficient.

The transmission support evolved through the same development. Locally, the rear castings were not stable with respect to each other. They parallel-a-gramed with suspension loads. With the limited amount of permanently attached structure, the cross member needed to do most of the work of holding the castings together. Although it is common practice, there was insufficient room to package a cross member above the air box (shock support). The low vehicle profile and engine package left no room for structure. Through a series of steps, a simple transmission support evolved with small gussets to support the suspension. By combining this with design revisions to the castings, a suitable support structure was created. The local suspension stiffness and engine mount stiffness were reinforced without a weight penalty.

Castings

The key to the structure was the design of the castings. They contain the majority of the weight, interfaces and function. The design required significantly fewer parts in manufacturing but presented several unique issues. The size and shape creates unique flow problems in the mold and they require an investment in machining time. After the initial design iterations to create the workhorse design, the team modified the concept into a truss like system. By creating a typical section with the vertical wall in the center there were several key improvements:

1. Load transfer characteristics
2. Increased material flow
3. Decreased tooling costs
4. Reduced weight

Load transfer was improved by allowing the mass to be placed near the ribs and the back wall, where it was needed. Specific load paths were created from the crush rails to the structure in front and rear. Additionally ribs were positioned to reinforce local suspension stiffness based on FEA direction. By putting the rib in the center, it created a truss like system to handle the load inputs. The amount, size and location of the truss beams could be changed as needed by the design team.

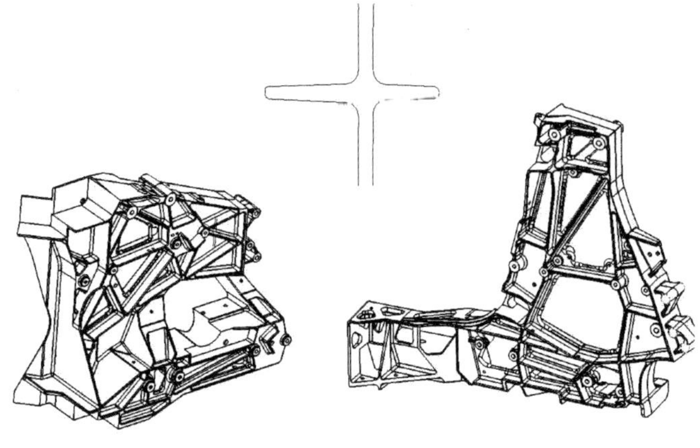

Figure 9 - Front and rear corner castings

As mentioned previously, the Job 1-3 vehicle castings had to be created at the same time tooling was being cut for the permanent molds to support the build schedule. Several key manufacturing limitations had been defined through the workhorse build and helped guide the design. In order to ensure program success under the timing plan, manufacturing worked closely with the design team through all stages of development. By designing the wall in the center, the cutting tools could be done without EDM burning of the ribs. This not only decreased the time to create the tool but also helped with mold release. Additionally, the central wall helped solidification and flow of material through an even distribution through the casting.

The end result of the increased design flexibility and manufacturing speed was a casting design, which incorporated many of the interfaces in one part while lowering the overall weight. The dimensional control was improved by limited welding and designed in slip planes. As an added benefit, the design met the styling request for a functional and technically compatible design with the overall theme of the Ford GT.

CAE

As discussed previously, CAE (Computer Aided Engineering) was used extensively as a design aid during the frame development. Quick turn-around of analyses allowed multiple design/analysis loops to be performed.

CAE PROCESS

The program team designed and implemented a process that maximized the contribution of CAE. Key to this was the relationship between the CAE team and the rest of the program team. Firstly, CAE engineers were co-located with the design engineers and designers. Secondly, communication between the CAE and design teams was encouraged at a routine working level, preventing the “walls” that sometimes separate the two. And thirdly, the program team contained both CAE engineers with a good understanding of engineering design as well as design engineers with a good understanding of CAE and an appreciation for its potential benefits and limitations.

Given the short duration of the design program, it was important to maximize the benefit of the various analyses performed. Analyses were therefore performed not just to determine the current performance of the design, but also to develop an understanding of the frame’s behavior so that the performance could be improved.

Software

The main commercial software programs used for the structural analysis of the frame were:

- HyperMesh and ANSA for mesh generation
- NASTRAN for stiffness, NVH and durability analysis
- LS-DYNA for impact analysis
- The Oasys LS-DYNA Peripheral software suite for pre-processing, model management and post-processing

In addition, several customized utility programs were used to maximize the efficiency of the CAE process. Particular attention was made to database management, massing and automatic post-processing of results. These will be described in the next sections.

Database Management

Although different finite element programs can have different preferences in terms of mesh density, connections etc., a decision was made to use a single mesh model for the frame. This minimized the resources required to update the frame mesh, and also allowed quicker feedback following design studies.

A single finite element mesh database was used for all the structural analysis disciplines (stiffness, NVH, impact and durability). This database was maintained centrally and used as a basis for iterations by the different disciplines. It was updated regularly so that the design improvements resulting from one analytical discipline (e.g. impact) could be evaluated for the other disciplines. One key area that went through several design iterations was the joint between the B-pillar and the rocker. The frame’s stiffness and impact performance were both very sensitive to this area of the frame, and so it was important that the effect of design changes on both stiffness and crash performance could be quickly evaluated in parallel.

During the evolution of the design, the complexity and size of the model increased. From an initial frame-only stiffness model at the start of the program, the model grew to include all additional components that provide stiffness, mass or loadpaths (e.g. body panels, closures, instrument panel, powertrain, seats, fuel tank, ancillary components, suspension, wheels, fascias and interior trim).

Not all components or sub-assemblies needed to be included in all models. Seats, for example, were included in side impact and seat strength analyses but not in roof crush analysis. An Oasys Primer database was used to manage the different impact loadcases. The Primer database contained all the vehicle sub-models in addition to all impact barriers, occupants and analysis specific data (e.g. initial velocities, restraints and termination times). For each analysis loadcase, a template was used to define the sub-models, barriers, occupants and analysis specific data required. The advantage of this was that only one central database needed to be maintained. Once the central database was updated, the analysis input files for the individual loadcases (front impacts, side impacts, rear impacts, roof crush, seat strength etc.) could be created in a matter of minutes.

For stiffness, NVH and durability analyses (where the number of model permutations was much smaller – limited to frame, frame-and-body and trimmed frame-and-body with closures), tailored scripts were written to create the various models from the main mesh database.

Massing

In order to maximize the benefit of CAE analysis, it is important that analyses are as consistent as possible. One important area is the vehicle or frame's mass. A-to-B comparison studies lose effectiveness if they use different masses. A customized utility program was used to ensure that all analyses were massed to a consistent level. Initially, masses were added to the frame to represent missing components such as seats, closures etc. As sub-models of these and others were added, so the massing program ensured that each sub-model was massed correctly and that the total vehicle mass remained constant.

In addition to ensuring that masses were applied consistently, this automated process removed the necessity to manually re-attach masses whenever part of the frame was re-meshed.

Bill of materials and material properties

The frame bill of materials (BOM) was maintained by the design engineering team. The information required by the CAE team (Part number, part name, material and gage) was automatically extracted from the BOM spreadsheet and incorporated into the CAE models on a regular basis. This minimized the risk that materials or gages could be modified manually (as part of a design study) and then not returned to the design-intent value.

A central LS-DYNA input file was maintained for the different materials used in the frame (and all other components). This file was automatically read into all analysis input files created, minimizing the potential for unauthorized or inadvertent modifications to be made. The material properties were derived from manufacturer's estimated properties and coupon tests.

Specific attention was paid to modeling of the weld heat-affect zones (HAZ) which, for aluminum, have reduced strength and ductility compared to the parent material. At each weld location the HAZ was modeled as a separate part with different properties from the parent material. For LS-DYNA, the *MAT_SIMPLIFIED_JOHNSON_-COOK_ORTHOTROPIC_DAMAGE material model was used. This material model allows definition of a tensile failure strain while still allowing higher strains in compression without failure. For durability analysis, where performance estimates were based on maximum stresses, care was taken to differentiate between high stresses in heat-affected zones and high stresses in parent metal.

Automatic post-processing, report generation and results tracking

Automatic post-processing and report generation allowed the CAE team to minimize the time spent post-processing the analyses and maximize the time spent *interpreting* the results and developing an understanding of the behavior of the frame and/or vehicle.

CAE ANALYSIS

Four main areas of CAE structural analysis will be discussed in this paper:

- Stiffness
- NVH (Local Stiffness)
- Durability
- Impact

Stiffness

Initial models, comprising the frame only, focused on static torsional and bending stiffness. The stiffness of the vehicle was evaluated with respect to the stiffness values and also the deformed shape.

Displacement analysis was used to assess the deformed shape. For torsional stiffness, the optimum structure is a tube, which will deform uniformly along its length. A twist plot (angle of rotation of the main structural members plotted against vehicle longitudinal co-ordinate) will in the ideal case be linear between the front and rear suspension mounting locations. For automotive bodies or frames, locally weak areas that limit overall stiffness manifest themselves as discontinuities in slope.

A similar plot can be produced for bending stiffness, with vertical deflection plotted against vehicle longitudinal co-ordinate. Again, discontinuities of slope indicate areas of weakness. Figure 10 shows a bending displacement plot from the Ford GT. The results of two analyses are shown. The first ("Initial Concept") is from an analysis performed early in the frame development. It shows significant discontinuities of slope just forward of the B-pillar (near the point of maximum displacement) and just forward of the rear suspension. The second analysis ("Optimized Design") shows the bending displacement after the frame design had been optimized. Not only has the displacement magnitude been reduced (indicating significantly increased bending stiffness), but there are fewer slope discontinuities, indicating a much more uniform and efficient structure.

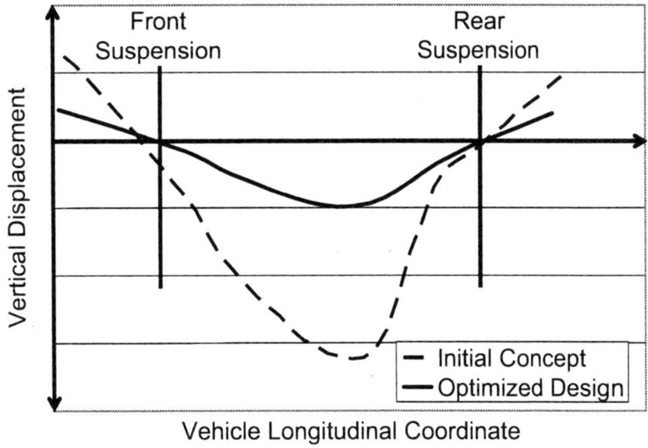

Figure 10. Bending Displacement Plot

For both loadcases, contour plots of strain energy were also used to identify areas of potential improvement. High values of strain energy indicate areas of the structure that need reinforcing. Low values of strain energy density indicate areas where less structure is required, suggesting the potential for mass reduction.

As the design matured and more information became available, so the focus on *static bending* stiffness was reduced. Increased analysis effort was given to the *dynamic* stiffness or modal performance of the trimmed frame and body. (*Static torsional* stiffness was retained as a primary target to ensure that the frame would be sufficiently stiff to allow suspension tuning for vehicle ride and handling).

The trimmed body model contained finite element representation of the frame, body panels, fixed glass, closures, instrument panel, steering column and seats. All other components rigidly attached to the frame/body were included as discrete mass elements.

The trimmed body model was used to predict the natural frequencies of the vehicle body. Using inspection of the mode of deformation and strain energy density plots, structural modifications were made to eliminate local panel modes and to increase the frequencies of the global body modes (vertical bending, horizontal bending and torsion) to achieve their target values.

NVH (Local Stiffness)

The NVH performance of the frame was assessed using point mobility (forced response) analysis. This analysis model was based on the trimmed body model from the dynamic stiffness analysis. The local dynamic stiffness was measured at the suspension and powertrain mounting locations. At each location in turn, a sinusoidal force was applied in each of the X, Y and Z directions over a range of frequencies. The displacement at the same point was measured and the local stiffness (at each frequency) and in each direction) was calculated as the ratio of the excitation force and the displacement.

For the Ford GT, the frame NVH targets were defined in terms of 2Hz point mobilities – i.e. the local stiffness of the mounting locations under a 2Hz sinusoidal force input. The target values were set to be significantly stiffer than the suspension bushings and powertrain mounts.

Results were analyzed using the same criteria as the dynamic stiffness results. Contour plots of strain energy density and mode shape visualization enabled the 'weak' areas of the frame to be highlighted. Numerous CAE studies were carried out to improve these areas until the targets were achieved.

As discussed earlier in the paper, the main areas of the frame that required considerable effort to achieve NVH targets were the powertrain mounts.

Durability

Because no proving ground road load data was available for the Ford GT, a simplified approach was used. The model used for this analysis was based on the trimmed body model but included the additional mass from the powertrain (applied through the powertrain mount locations). Loads calculated from a kinematics model were applied at the suspension mounting points. The models were restrained using inertia relief to represent actual loading conditions.

Maximum stress values were compared to target values derived for each material (including heat affected zones) from empirical test data.

Impact

Because of the importance of safety and the nature of the main energy absorbing structure (aluminum extrusions rather than the welded steel stampings more commonly used in automotive structures), it was decided that a correlation exercise should be performed at the earliest opportunity to verify the analytical approach used. A sled impact test was therefore performed on a front crash assembly (rails and bumper beam) at the workhorse level. Using accurate material data taken from tensile tests on extrusions from the same batch, *blind* predictions of the expected force/deflection were made. When the LS-DYNA predictions were compared to the test results, the agreement was considered to be significantly better than the level required to give confidence in the analyses. See Figure 11.

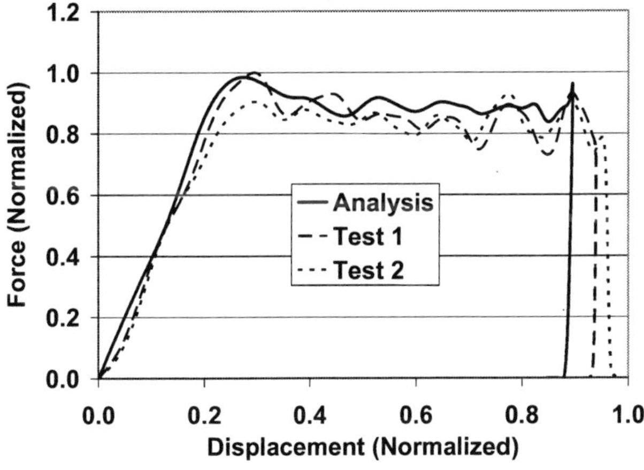

Figure 11. Front End Correlation Study

As with the stiffness analyses, the complexity of the impact models increased as the design matured and additional information became available. The finite element models eventually comprised more than 750,000 elements. These models consisted of the frame, closures, body panels, fixed glass, seats, powertrain, instrument panel, suspension, fascias, fuel tank, ancillaries (including wiper mechanism) and dummies/barriers where appropriate. This level of detail was required as the program timing did not allow numerous development tests. Figure 12 shows the full vehicle impact model.

One of the most important enablers for the impact analysis (particularly towards the end of the design phase, when last-minute design improvements needed to be assessed for many different loading conditions) was the Oasys Primer database management described in an earlier section of this paper. Multiple loadcase models could be quickly created from the master database.

Figure 12.

CONCLUSION

Through the use of the latest CAE, design and manufacturing technologies combined with lightweight materials, the Ford GT spaceframe has been successfully designed to provide the stiffness, crash, weight and package performance necessary to meet the vehicle dynamics requirements of a modern high performance sport car. The spaceframe meets all the US crash performance standards while helping to reduce repair costs through the use of bolt-on crush structures. The result is a lightweight structure that provides the necessary stiffness and package efficiencies as required by the other vehicle systems while also maintaining the vehicle styling established in the GT40 concept car.

2004-01-1256

2005 Ford GT - Interior Trim & Electrical

Alex Zaguskin and Charles Curmi
Lear Corporation

William Clarke
Ford Motor Company

ABSTRACT

Driven by a tight vehicle development schedule and unique performance and styling goals for the new Ford GT, a Ford-Lear team delivered a complete interior and electrical package in just 12 months. The team used new materials, processes and suppliers, and produced what may be the industry's first structural instrument panel.

INTRODUCTION

From the beginning of the design and development cycle, Ford prioritized key requirements:

1. Interiors and Electrical Systems must be developed within twelve months.

2. Low production volume meant tooling expense must be reduced as much as possible.

3. Because vehicle performance was the most critical target of the Ford GT, the weight of the car and its interior components was very important.

4. Although interior noise levels in this performance car did not need to meet typical luxury car standards, a certain quality of noise or "acoustic signature" was important.

Since the interior trim and the electrical system are so closely integrated, the Ford-Lear team was focused on making them work together.

The objective of the Ford GT electrical and electronics design was to provide a high performance system using as many carryover components as possible to minimize the total investment. Because the system had to be designed, manufactured, and delivered within 3 months of system freeze, the team was able to focus on only those tasks in the design cycle that were the most value-added.

MAIN SECTION

TIMING

Given critical vehicle timing, Lear as a supplier of the interiors and electrical systems, dedicated a team of engineers and designers who had prior extensive r performance vehicle experience. This team was located with the rest of the vehicle team to insure prompt communication and team spirit. Simultaneously with materials and process selections, the team was able to select local suppliers and quote the best possible tooling timing. The participation, dedication and innovative approaches by Lear's suppliers played a critical role in this development.

TRIM

In order to meet low-cost tooling targets, the team did extensive research on existing materials and processes.

The team first looked at Azdel SuperLite®, a low-density glass mat thermoplastic (LDGMT) that had been used extensively for headliner applications. LDGMT is composed of glass fibers bound together by resin particles and the supplier had been working to expand its uses to other interior applications. Lear engineers were convinced that LDGMT could be used structurally for instrument panels, doors, etc., although coverstock applications and attachment methods were not yet developed.

The Lear team decided to take a calculated risk to use this material for the Ford GT because it offered weight savings of up to 30%, and it could be compression molded, requiring only 50 percent of the capital investment of injection molding tooling.

In order to manage the risk, the team decided to tool a small but critical section of the IP and prove out the process using that part – before the entire design was complete. This trial tremendously expedited development work and built team confidence in the material. At the same time the team did parallel development work with a known alternative technology, Woodfiber, in which Lear is an expert. For a few months both approaches were considered equally until LDGMT technology was deemed ready for production.

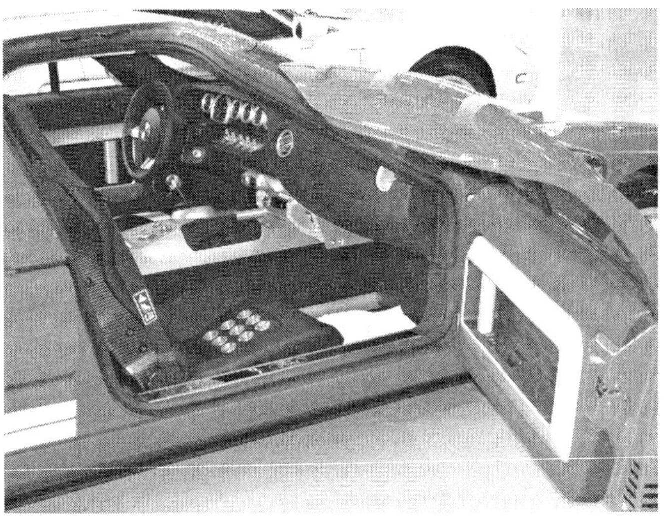

Figure 1. Ford GT interior

As the team gained confidence in LDGMT as a primary material for interior systems, they selected a local supplier to develop manufacturing processes. WK Industries, a prototype tooling supplier from Sterling Heights Michigan, won the contract.

Lear Corp., WKI, and Azdel together developed new processes for LDGMT applications using different coverstocks, including TPO and PVC skins and leather wrap. Considering low vehicle production volume and the fact that the compression molding process requires only 10 psi for compression, the team decided to use composite tooling (See Fig 2).

Fig 2. Composite tooling

This approach saved approximately 70% of the conventional injection molding tooling cost. The phenomenal success of LDGMT in Ford systems applications is a result of the work of the Lear GT interior, Azdel, and WKI team.

INSTRUMENT PANEL / CROSSCAR MAGNESIUM BEAM

The Ford GT's unique vehicle construction required the IP to be a structural member meeting stiffness, durability and crashworthiness targets. The IP structure had to be designed to carry an 11,000-lb. load. The design process integrated knowledge of analytical simulation, product engineering and design, safety, and manufacturing.

A multiple-piece design of the IP (figure 3) reduced cost while presenting several challenges that were successfully solved.

In order to meet tough design requirements, the team decided to use magnesium alloy AM60 for a cross car beam in the IP. This not only maintained cross car structure but also significantly reduced weight, a key vehicle design goal, and led to what may be the industry's first structural instrument panel.

Precise program management and supplier dedication led the team to need only 8 weeks for tooling timing for the magnesium casting verses typical 16-18 weeks for other materials. Northern Diecast did the magnesium casting.

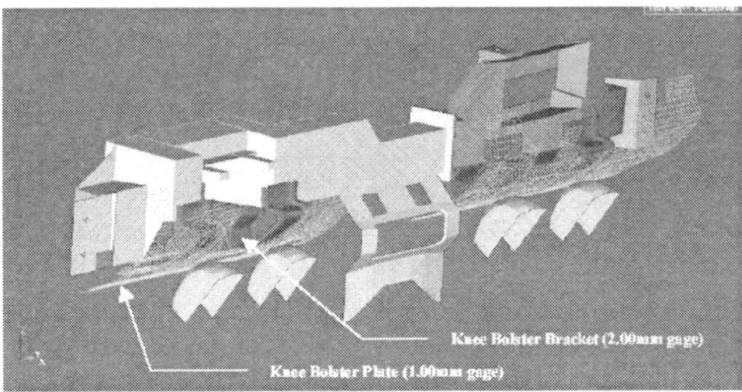

Fig 3. Instrument panel assembly

UNUSUAL TRIM APPLICATIONS

The vehicle theme mandated use of true materials. For example, all visual aluminum trim, such as headliner, door rings, switch bezels in the IP are made from the real metal, since no metal-like materials were allowed.

During selection of the materials and processes, the team found another local supplier – Quality Metalcraft Corporation (QMC). QMC had developed unique processes for producing class-A finish *exterior* parts with zinc tooling. Applying this technology to the *interior* application allowed the team to achieved unique and relatively inexpensive effects with excellent quality.

The zinc tooling process reduced the tooling cost by nearly half and tooling time by almost 75% compared to typical steel tooling.

ACOUSTICS

Acoustics and overall passenger compartment sound quality are critical for any vehicle. In a performance car, the distinctive "quality of noise" or acoustic signature is critical to its brand character, and so the design of the Ford GT's acoustical package presented significant challenges.

Sound insulators are the first choice, but are typically heavy. Since weight correlates so closely to vehicle performance, other creative methods had to be employed.

In order to select proper acoustical treatments, the car was divided into several subsystems. One of the most critical subsystems was the rear bulkhead, or firewall, and rear glass which separates the passenger and engine compartments. In order to control sound levels, the glass had to be upgraded from the traditional single pane. Multi-pane designs were evaluated for sound control and package space. A dual-pane design with defined air-gap was selected.

ELECTRICAL

The electrical and electronics system was built around eight major modules. Six of those modules were carryover and the other two: Lear's Smart Junction Box (SJB) and Stoneridge's Cluster Module (CM), are new and provide the backbone of non-powertrain system function.

The new Cluster Module was driven by the need to meet the Instrument Panel styling cues from the original Ford GT of the late 60's, which included a set of standalone analog style gauges. No Ford carryover module met this criterion.

The new Lear SJB provides direct integration of programmable electronics, microprocessors, and software - key to the functioning of the vehicle's feature set. The SJB also contains the components and connectors for the purpose of circuit overcurrent protection, splicing, and interconnections. The use of Lear's new SJB allowed the team to reduce weight, package space, and cost by combining the fuse box, harness interconnects, remote keyless entry receiver, beltminder module, generic electronic module and several miscellaneous relays.

The Ford GT SJB also controls the majority of the vehicle's user feature set. The most significant SJB challenge was the interface to several components. Those components include a new wiper system, push button start, bi-functional HID Headlamps, LED tail lamps with outage sensing, electronic door latching, and electronic trunk release.

The use of several carryover components & modules, coupled with the design flexibility of the Smart Junction Box, allowed the team to bridge all electrical and electronic system non-compatibilities.

CONCLUSION

Development of Vehicle interiors and electrical distribution systems typically takes three to five years. But within twelve months the Lear team was able to bring a complete system to production while simultaneously completing several complex research and development projects.

Selecting the right team, extraordinary creativity and, dedication, detailed program management, and calculated risk-taking were the key enablers in executing this program.

ACKNOWLEDGMENTS

WK Industries, Quality Metalcraft, Northern Diecast, Adzel, Inc., and Stoneridge

REFERENCES

1. Interiors Systems program timing
2. SAE Paper 2004-01-1258 – Kip Ewing, J. G. Hipwell, and E. J. Benninger
3. SAE Paper 2004-01-1261 – Scott Pineo, Naiyi Li, Xiaoming Chen, Tim Hubbert

CONTACTS

Alex Zaguskin, azaguskin@lear.com

DEFINITIONS, ACRONYMS, ABBREVIATIONS:

IP – Instrument Panel

EDS- Electrical distribution system

SJB – SMART JUNCTION Box

LDGMT - low-density glass mat thermoplastic

2004-01-1257

2005 Ford GT- Maintaining Your Cool at 200 MPH

Curtis M. Hill and Glenn D. Miller
Ford Motor Co.

Michael R. Evans and David M. Pollock
Roush Industries, Inc.

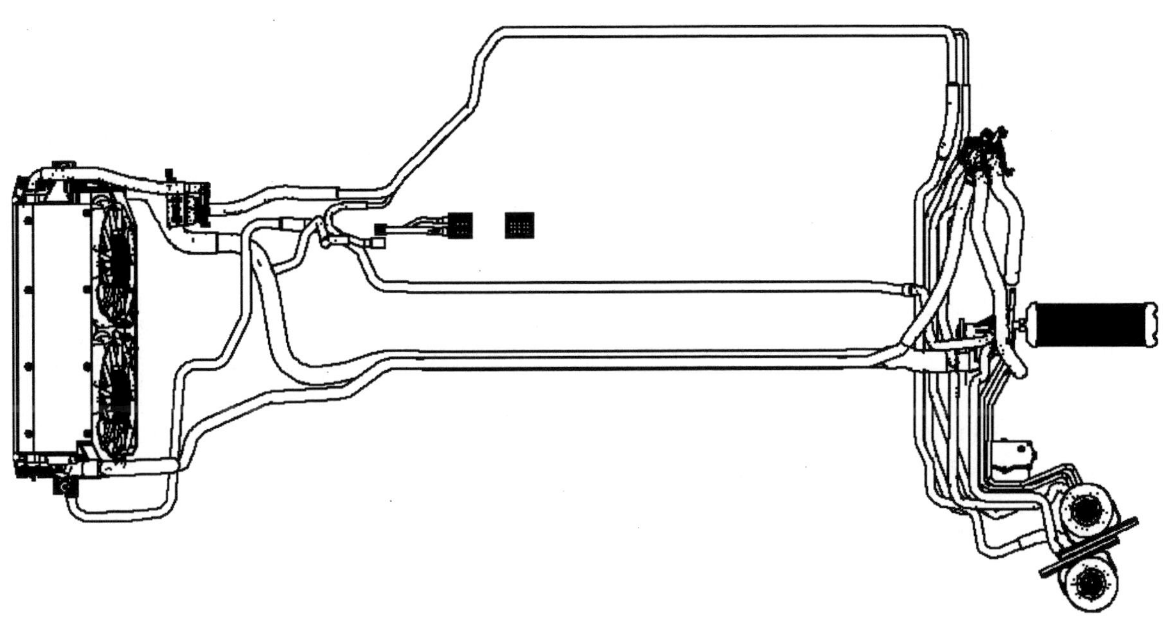

Figure 1. Plan view of the cooling system.

ABSTRACT

An integrated engineering approach using computer modeling, laboratory and vehicle testing enabled the Ford GT engineering team to achieve supercar thermal management performance within the aggressive program timing. Theoretical and empirical test data was used during the design and development of the engine cooling system. The information was used to verify design assumptions and validate engineering efforts. This design approach allowed the team to define a system solution quickly and minimized the need for extensive vehicle level testing.

The result of this approach was the development of an engine cooling system that adequately controls air, oil and coolant temperatures during all driving and environmental conditions.

INTRODUCTION

The Ford GT program was approved with several major challenges built in: the appearance and overall package size was defined by the show car, performance targets were consistent with world class supercars, and the program timeline was extremely short with aggressive milestones.

The GT architecture reflects the basic layout of the historic GT40: a mid-engine V8 with transaxle, central fuel tank, and front mounted cooling module. The supercharged, intercooled 5.4L aluminum engine target was 500+ net horsepower, while meeting emissions, driveability, and quality standards consistent with a modern road car and achieving race car performance levels. (See Paper #2004-01-1252 Ford GT Powertrain – Supercharged Supercar)

COOLING SYSTEM TARGETS

Cooling system performance targets were likewise defined consistent with world-class supercars. The Ford GT was to meet corporate requirements for high and low temperature operation under road car and racecar duty cycles. These include high ambient temperature operation at high engine speeds typical of spirited driving on twisty, mountain roads or road courses, continuous high vehicle speed at or near maximum velocity, and stop and go city traffic.

Performance targets are shown in this chart:

	Maximum Allowable Temp (°F)	Ambient Condition (°F)
Engine Coolant Outlet	245	110
Engine Oil Outlet	300	110
Combustion Air	200	110

COOLING SYSTEM ASSUMPTIONS

The Ford GT cooling system design includes the "normal" engine cooling components for water, oil and interior heating and cooling plus the intercooler circuit.

The front-mounted cooling module is made up of the AC condenser, intercooler radiator (low temp. radiator-LTR), and engine coolant radiator (high temp. radiator-HTR) Also mounted in the front is the electric LTR coolant pump.

The engine compartment components include the engine-driven engine coolant pump, the thermostat housing, LTR and HTR de-gas bottles, and the engine-mounted oil cooler. (The Ford GT engine uses oil to cool the pistons.)

Tubes and hoses connecting front and rear components are routed through the central tunnel and the right side rocker panel.

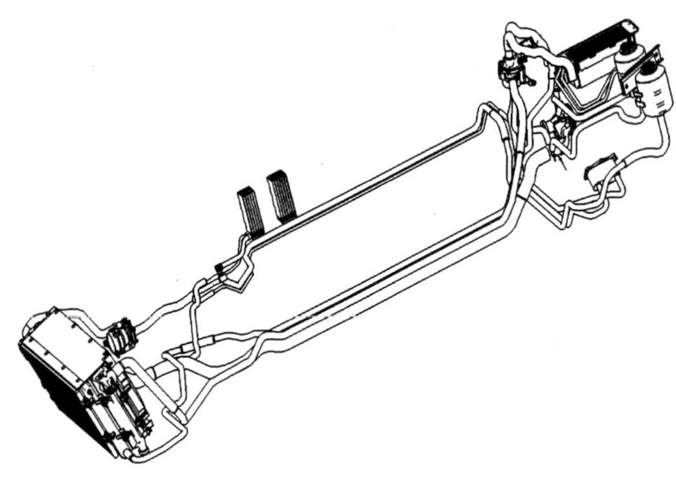

Figure 1a. Cooling system layout

SYSTEM DESIGN

The challenge of balancing the engine cooling requirements and engine power levels was separated into the following engineering tasks:

- Current design practices
- Vehicle performance requirements
- Vehicle package constraints
- Engineering Projections
    - Engine heat load estimation
    - Airflow estimates and practical flow optimization
- Cooling system performance predictions
    - Engine
    - Intercooler
    - Oil cooler

CURRENT DESIGN PRACTICES

The goal was to leverage existing base engine cooling system technologies and ensure the Ford GT would achieve program performance targets.

Competitive and corporate vehicles were evaluated against the program design targets. Using proposed engine power levels and current calibration strategies, existing thermal system performances were recalculated using the increased engine power levels and compared to program design targets. Calculated deficiencies in the existing base engine cooling system at these elevated power levels led to identification of the problem areas and allowed the scope of design and development work to be identified and prioritized.

VEHICLE PERFORMANCE REQUIREMENTS

Initially two drive cycles were identified for analysis. One cycle was classified as low-speed and high-power driving. The other was driving with high-speed, high power.

The term Vmin was used to describe the conditions of low average speed and repeated bursts of power, during zero to 70 mile per hour accelerations. This driving condition could occur on small to medium sized road courses or during aggressive over the road driving.

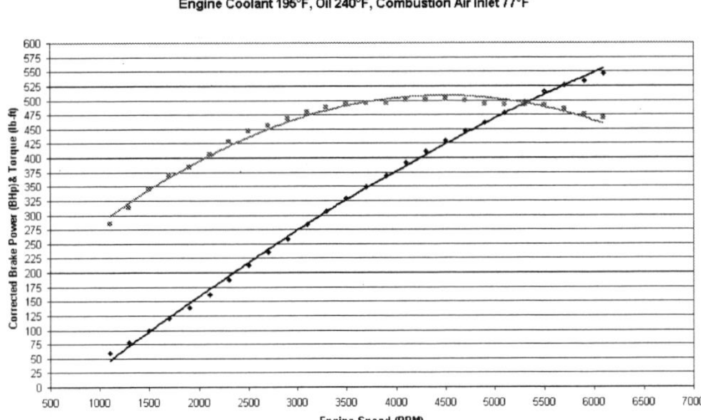

Figure 2: Projected Ford GT Engine Output

The other was Vmax, identified as a driving condition that would occur during high vehicle speed, high-power application. This type of driving might be typical of medium to large road courses with long straight sections and average vehicle speeds greater than 70 mile per hour.

	ENGINE SPEED (RPM) IN EACH TRANSMISSION GEAR					
VEHICLE SPEED	1st 2.6 RATIO	2nd 1.7 RATIO	3rd 1.23 RATIO	4th 0.95 RATIO	5th 0.76 RATIO	6th 0.625 RATIO
5						
6.23	650					
9.53	994	650				
10	1043	682				
13.17	1374	898	650			
15	1564	1023	740			
17.06	1779	1163	842	650		
20	2086	1364	987	762		
21.32	2224	1454	1052	812	650	
25	2607	1705	1234	953	762	
25.93	2704	1768	1279	988	791	650
30	3129	2046	1480	1143	915	752
35	3650	2387	1727	1334	1067	877
40	4172	2728	1974	1524	1219	1003
45	4693	3069	2220	1715	1372	1128
50	5215	3410	2467	1905	1524	1254
55	5736	3751	2714	2096	1677	1379
60	6258	4092	2960	2286	1829	1504
62.3	6498	4248	3074	2374	1899	1562
65		4433	3207	2477	1982	1630
70		4774	3454	2668	2134	1755
75		5114	3701	2858	2286	1880
80		5455	3947	3049	2439	2006
85		5796	4194	3239	2591	2131
90		6137	4441	3430	2744	2256
95		6478	4687	3620	2896	2382
95.6		6519	4717	3643	2914	2397
100			4934	3811	3049	2507
105			5181	4001	3201	2632
110			5427	4192	3354	2758
115			5674	4382	3506	2883
120			5921	4573	3658	3009
125			6168	4764	3811	3134
130			6414	4954	3963	3259
131.8			6503	5023	4018	3304
135				5145	4116	3385
140				5335	4268	3510
145				5526	4421	3635
150				5716	4573	3761
155				5907	4725	3886
160				6097	4878	4011
165				6288	5030	4137
170				6478	5183	4262
170.6				6501	5201	4277
175					5335	4387
180					5488	4513
185					5640	4638
190					5792	4764
195					5945	4889
200					6097	5014

Figure 3. Vehicle/Engine speed in each gear

It was apparent from inspection of the engine performance curves (Figure 2) and transmission gear ratios (Figure 3) that the proposed engine torque would be more than adequate to achieve most of the possible rpm-vehicle speed points.

This resulted in two additional concerns:

The engine power level required to overcome the parasitic losses in the engine could cause an overheat condition if the vehicle were driven at moderately low vehicle speeds with high engine speeds and minimal ram air effect.

The available coolant flow rates would be insufficient for the available airflow and the car could overheat if the engine speeds were low enough even if the ram air was adequate.

The resulting 4 drive cycles would be evaluated as soon as the Attribute Prototype (AP) cars were available.

VEHICLE PACKAGING

The mid-engine vehicle layout provided distinct design challenges. Because the Low Temperature Radiator (LTR) and High Temperature Radiator (HTR) cooling circuits have very long coolant piping runs, good plumbing practices were paramount in the design and packaging of the system. Pumping losses would play a significant role. The vehicle concept itself is a very compact package. The cooling system flow rates needed to be proportionally greater than any cooling system in current corporate passenger vehicles. Any deficiencies in plumbing and their resulting coolant flow restrictions would negatively impact engine performance and durability.

ENGINEERING PROJECTIONS: HEAT EXCHANGERS AND AIRFLOW

An initial analytical model of engine cooling heat loads was developed from the current 5.4L engine. The engine operating characteristics were extrapolated to provide an estimate for the Ford GT engine coolant and oil heat rejection levels. The estimated heat loads to engine oil, coolant, and intercooler were summed and balanced for various conditions.

Using component information supplied from prospective heat exchanger suppliers, the analytical model was created to balance the engine coolant requirements with the heat exchanger characteristics.

The following table summarizes the program assumptions of heat loads that needed to be managed through the cooling module to sustain Vmax at 200 MPH:

Component Heat Loads	Heat Load (Btu/min)
Condenser	500
Intercooler Radiator	2100
Engine Radiator	8000
Oil cooler	2800
Total Airside Load	13400

The minimum requirements for sustained airflow were calculated to be 7500 CFM at an ambient temperature of 110 degrees Fahrenheit. See the following chart:

Ambient Temp (F)	Total Heat to Air (Btu/min)	Airflow (lbm/min)	Airflow (scfm)	Air Density (lbm/ft^3)
80	13400	429.5	5276.5	0.0750
85	13400	446.7	5955.6	0.0750
90	13400	465.3	6203.7	0.0750
95	13400	485.5	6473.4	0.0750
100	13400	507.6	6767.7	0.0750
105	13400	531.7	7089.9	0.0750
110	13400	558.3	**7444.4**	0.0750
115	13400	587.7	7836.3	0.0750
120	13400	620.4	8271.6	0.0750

A study of air velocities achieved by current production vehicles concluded the effective airflow through the heat exchangers, expressed as a percentage of the available free stream velocity, was in the range of 12-18 percent.

Using an effective air speed of 18 percent, the following equivalent airflow were calculated based on a one square foot grille opening.

Vehicle Speed (MPH)	Velocity Ratio (Ve/Vf)	Effective Air Velocity (MPH)	Effective Air Flow rate (CFM)
30	0.18	5.4	475.2
60	0.18	10.8	950.4
90	0.18	16.2	1425.6
120	0.18	21.6	1900.8
150	0.18	27	2376
180	0.18	32.4	2851.7
200	0.18	36	**3168**

These results show that regardless of cooling module fin densities and grille open area, the effective air flow would not be sufficient to cool the engine. At a sustained vehicle speed of 200 MPH and a velocity ratio of 18 percent, the resulting airflow is 3168 cubic feet per minute (CFM) versus the required flow of 7500 CFM.

The next table depicts the unrealistic velocity ratio that would provide the requisite airflow.

Required Velocity Ratio for 7500 CFM

Vehicle Speed (MPH)	Velocity Ratio (Ve/Vf)	Effective Air Velocity (MPH)	Effective Air Flow rate (CFM)
30	0.43	12.9	1135.2
60	0.43	25.8	2270.4
90	0.43	38.7	3405.6
120	0.43	51.6	4540.8
150	0.43	64.5	5676
180	0.43	77.4	6811.2
200	0.43	86	7568

An investigation into optimizing airflow rates through the cooling module and bodywork of the GT was initiated. A full-size aerodynamic buck was created and various cooling module configurations and ductwork were evaluated in the wind tunnel. (Figure 4)

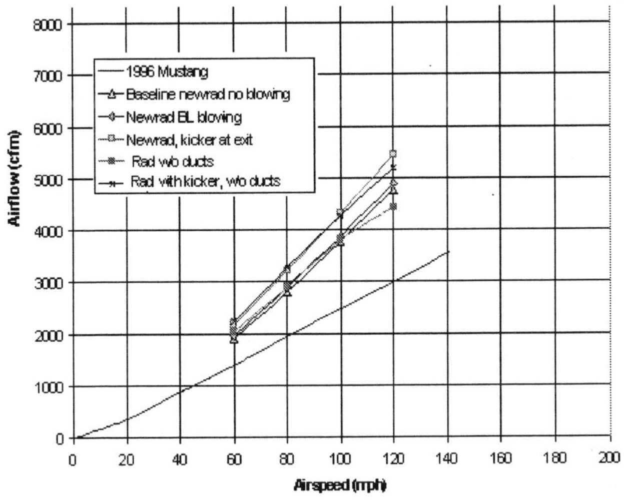

Figure 4: Measured Wind Tunnel Airflow

The airflow study results indicated that airflow could be a major issue. Further analysis would require in-vehicle testing when prototypes were available.

OPTIMIZING COOLANT MODULE PERFORMANCE

Engine cooling performance analyses were developed by studying the balance point of the engine coolant temperature and assumed airflow.

There is a direct relationship between the average air discharge of the cooling module and the discharge engine coolant from the radiator. This relationship can be used to describe thermodynamic balance problems between the coolant and air heat exchanger mechanisms. (Figure 5)

If the relationship between the average radiator discharge air and the coolant is known, then a maximum radiator air discharge temperature can be used to describe a borderline overheat condition. By constraining the radiator air outlet to maximum of 210° Fahrenheit, a solution matrix of minimum acceptable air mass flow rates could be solved. This solution matrix was generated for a range of cooling module loads and ambient operating conditions.

Sum of heat loads in cooling module (Condenser, HTR, LTR)

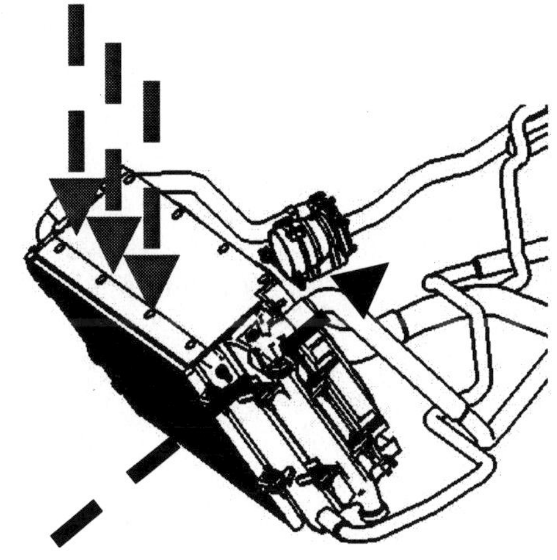

Airside balances all heat from cooling module

Figure 5: Cooling module heat load balance

COOLING PERFORMANCE PREDICTIONS

COOLING CAPACITIES

Computational fluid dynamics (CFD) computer modeling confirmed that the current engine cooling flow rates were inadequate. The system required a minimum flow rate of 100 gallons per minute in the engine, and a minimum flow rate of 85 gallons per minute to the radiator. (Figure 6)

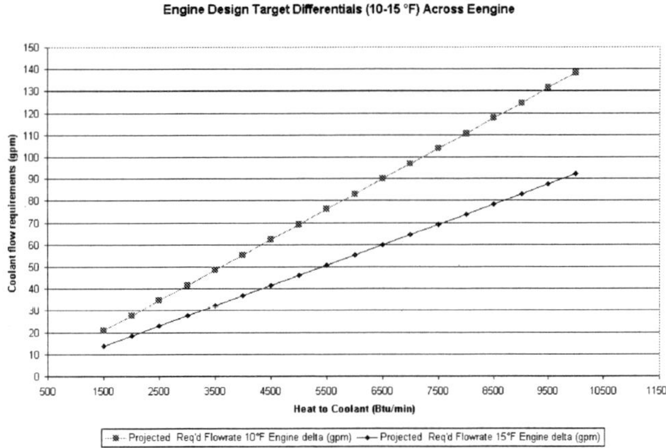

Figure 6. Required engine coolant flow

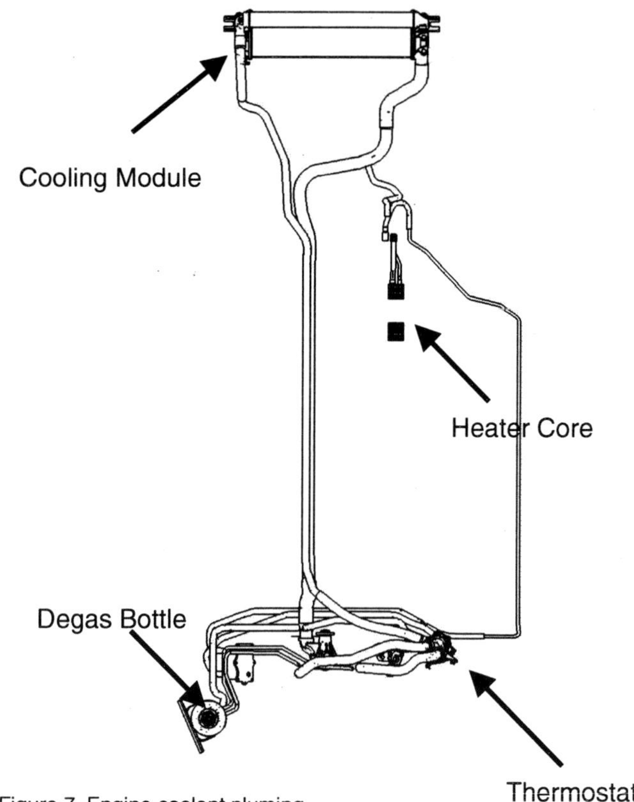

Figure 7. Engine coolant pluming

COOLANT FLOWS

All proposed cooling components were evaluated and matched to the Ford GT environment to optimize system design:

- Plumbing of circuits and flow balancing
- Engine coolant passages
- Thermostat and housing
- High temperature radiator
- Engine coolant degas circuit
- Heater core
- Water pump
- Intercooler circuit
- Oil cooler

Engine Coolant Plumbing	Figure 7
Cooling fill capacity	8.31 gal

Engine Block and Heads

Coolant passage flow paths were analyzed using CFD and modified to minimize coolant pressure drop and increase total coolant flow.

The first major action was to change from the production block's side-entry water passage to a front entry. This resulted in a flow improvement of 73.5% without increasing the pressure drop.

The next task was to revise the water jackets to adequately cool the engine without coolant flow between cylinders. The goal was to achieve equal flow to both banks and a 60/40 split to the exhaust side of the right bank and 50/50 on the left.

Extensive CFD modeling resulted in changes to the passages around the impeller, added flow diverters and revised cup plug locations. The new cylinder head was designed to increase flow through the exhaust valve bridge area compared to the production head.

Final CFD analysis predicted 51/49 right/left distribution. Engine dyno testing confirmed a 50/50 split throughout the pump RPM range.

Thermostat

Thermostat housing was redesigned to handle the increased engine coolant flows and to control the flow to the cooling circuits. See Figure 8

- Radiator Coolant Feed Line
- Heater Core and Oil Cooler Feed lines
- Engine Coolant Bypass Return Line

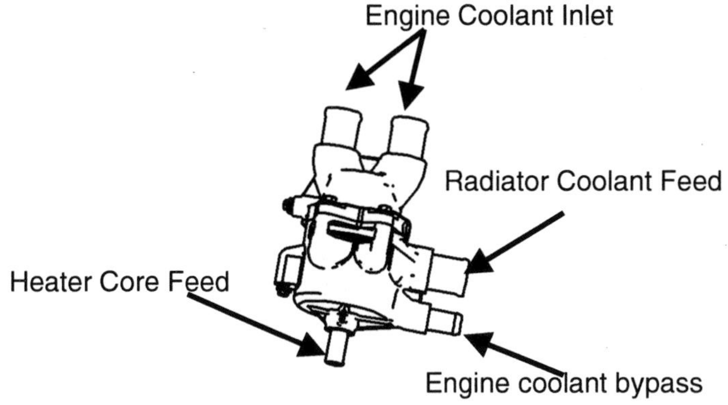

Figure 8. Thermostat housing

Thermostat	Figure 8
Thermostat	180°F Start to open
Material	Aluminum
Engine coolant inlet	2 inlets, 1.0 in
Bypass coolant outlet	0.75 in
Heater Core Outlet	0.5 in
Radiator Feed	1.25 in

High Temperature Radiator (HTR)

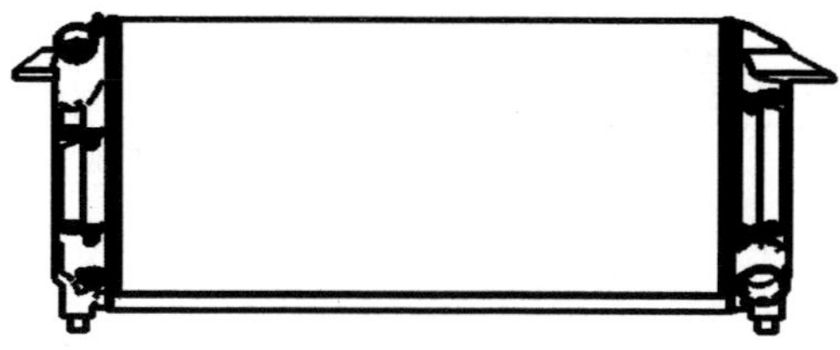

Figure 9. High Temperature Radiator (HTR)

Radiator (HTR)	Figure 9
TYPE	Crossflow 2 Row Core
Core Material	Aluminum
End Tank Material	Aluminum
Inlet Hose Barb	1.25 in (ID)
Outlet Hose Barb	2.00 (ID)
Core rows	34
Core Tube	0.084 in
Fin Density	16 FPI
Core Area	424 in^2 (2.94 ft^2)
Core Height	16 in
Core Width	26.5 in
Core Thickness	2.75 in

Engine Coolant Degas Bottle

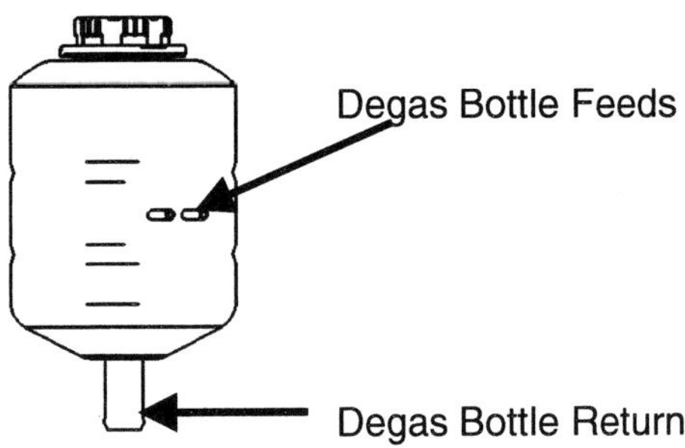

Figure 10. Engine coolant degas bottle

Engine Cooling Degas Bottle	Figure 10
Pressure Cap	16 psi
Engine coolant inlet Hose Barbs	2 Hose Barbs, 0.25 in
Engine Coolant Outlet Hose Barb	1.0 in
Coolant capacity to fill	0.396 gal

Heater Core

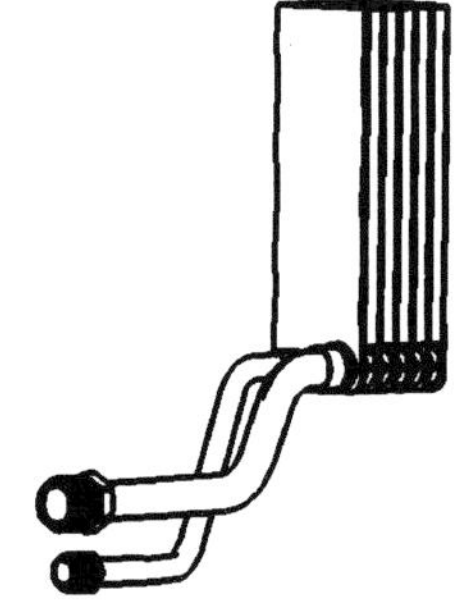

Figure 11. Heater Core

Heater Core	Figure 11
Fin Density	18 FPI
Number of tubes	21
Tube diameter	0.625 (ID)
Size	10.125 x6.85 x0.885 in
Heat Rejection	19400 Btu/hr @ 40 °F air inlet (200°F water temp) 160 cfm

Water pump

The water pump is a reverse rotation, front entrance design. The change from the corporate pump design was driven by the Vmax requirements of a minimum 100 GPM at peak power.

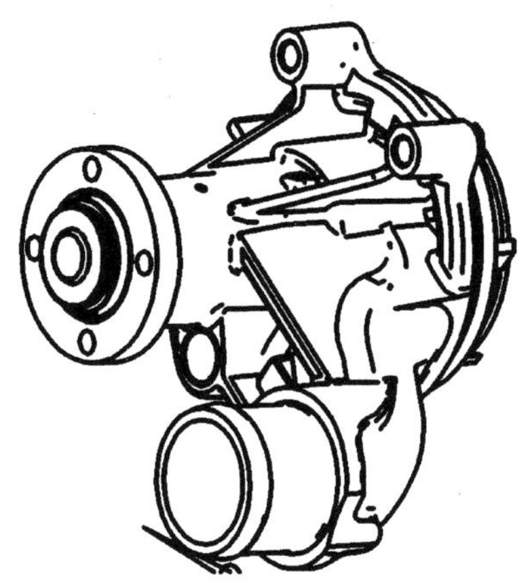

Figure 12. Water pump

Water Pump	Figure 12
Rotation	CCW
Inlet	2.0 in
Blades	7
Impeller Diameter	4.25 in

INTERCOOLER CIRCUIT

The Intercooler circuit was another design challenge. The net power target of approximately 100 horsepower per liter, required cooling airflow rates of 65 pound per minute to target temperature of 160 degrees Fahrenheit at 200 mph.. This requirement meant larger coolant flows and heat exchanger capacities than previously available in production.

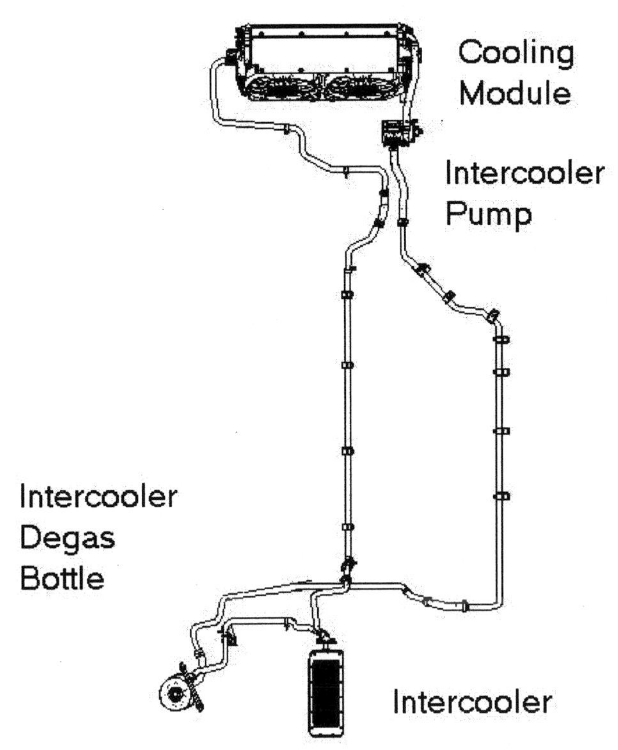

Figure 13. Intercooler circuit

Intercooler circuit Plumbing	Figure 13
Coolant Fill Capacity	1.27 gal
Coolant Plumbing Run	459.6 in

Intercooler

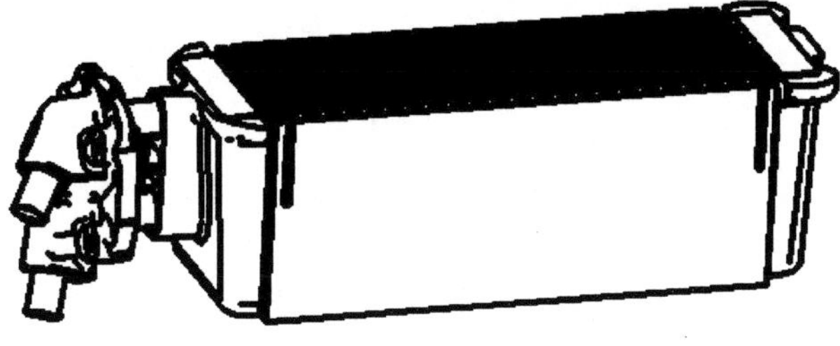

Figure 14. Intercooler

Intercooler	Figure 14
Size	11.625 x 4.5 x 0.5 in
Tubes	11 rows
Coolant Inlet Tube	0.75 in
Coolant Outlet Tube	0.75 in
Fin Density	22 FPI
End Tanks	5.0 x 4.25 x 0.75 in
Heat Rejection	2163 Btu/min
Coolant Flow rate	7 GPM
Combustion Airflow	61 lbm/min

Intercooler Electric pump

A higher capacity version of the corporate intercooler pump was needed. Program performance requirements set coolant flow targets at six GPM in to order cool the combustion air during Vmax, 200 MPH. (Figure 15)

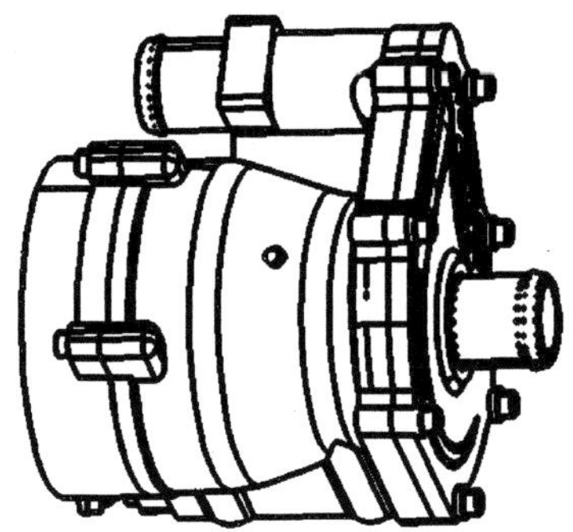

Figure 15. Intercooler electric pump

Intercooler Pump	Figure 15
Coolant flow rating	10 GPM @ 13.5 psi
Voltage input	9-18 vdc
Max current rating	20 amps

Low Temperature Radiator (LTR)

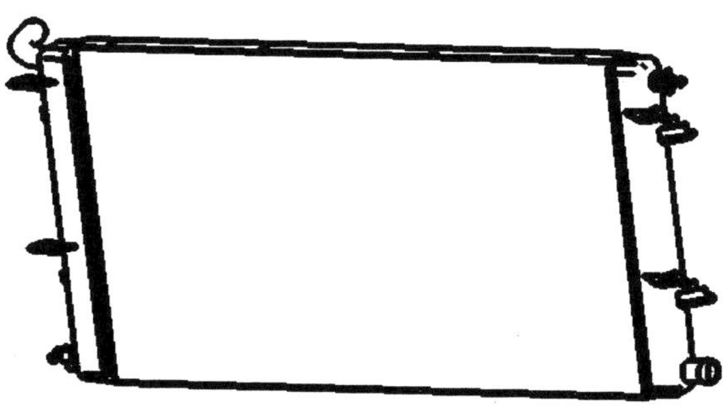

Figure 16. Low temperature radiator

LTR (Intercooler circuit)	Figure 16
Type	Single Core Crossflow
Core Material	Aluminum
End Tank Material	Aluminum
Inlet Hose Barb	1.00 in (ID)
Outlet Hose Barb	1.00 in (ID)
Core Rows	34
Core Tube	0.084 in
Fin Density	12 FPI
Core Area	424 in^2 (2.94 ft^2)
Core Height	16 in
Core Width	26.5 in
Core Thickness	1.25 in

Intercooler Degas Bottle

The coolant flow rates required to achieve the combustion air targets at Vmax required coolant flow rates through the Degas bottle nearly three times the normal. A significant design feature is the introduction of coolant to the bottle below the coolant fill level to eliminate coolant aeration at high coolant flow rates.

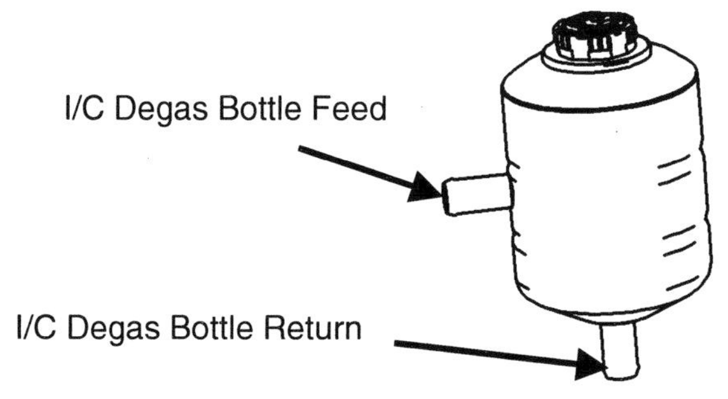

Figure 17. Intercooler degas bottle

Intercooler Degas Bottle	Figure 17
Pressure Cap	10 psi
Coolant Inlet Hose Barb	1.0 in
Coolant Outlet Hose Barb	1.0 in
Coolant fill capacity	0.396 gal

Oil Cooler

The engine employs an external dry sump oiling system with internal crankcase mounted oil squirters to provide piston cooling. Oil cooling is accomplished through liquid to liquid heat exchange using a plate cooler.

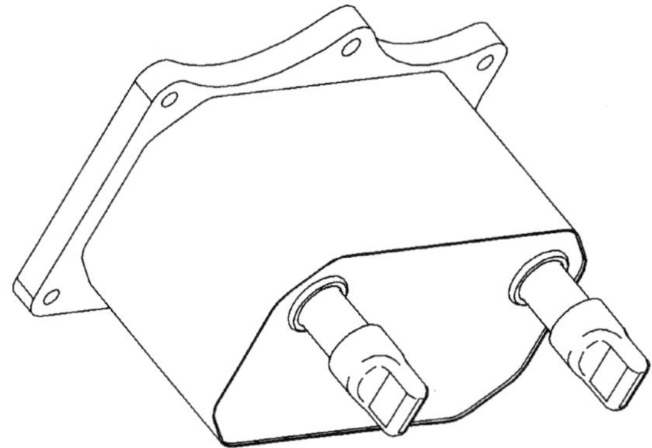

Figure 18. Oil cooler

Oil Cooler	Figure 18
Oil Cooling Plate	10- 10 x10 x0.015 in
Coolant Plates	10-10 x 10 x0.015 in
Inlet Tube (Oil circuit)	.0625 in
Outlet Tube (Oil circuit)	0.625 in
Inlet Tube (Coolant circuit)	0.6875 in
Outlet Tube (Coolant Circuit)	0.6875 in

SYSTEM VALIDATION

LABORATORY TESTING

THERMAL BENCH

The goal of these evaluations was to establish coolant flow rate and pressure drop performance maps for engine cooling designs. The methodology for these evaluations was to assemble a complete engine cooling system on a test stand, using in-vehicle positions, elevations and minimum modification to hardware. The engine coolant was heated with an electrical heater and pumped throughout these circuits using the water pump, driven externally by an electric motor. The cooling circuitry was tested through temperatures from 195 to 245° F. This methodology allowed quick evaluation of the following:

- Coolant flow rates in each coolant circuit
- Coolant pressure drop through each component.
- Water Pump sensitivity to boiling and cavitation during sustained high speed operation and rapid accelerations
- Individual component performance
- Total system flow rate and pressure drop

After each battery of tests, system performance was evaluated and changes to components were made.

Plumbing configurations, hose and tubing sizes, coupled with flow restrictors were part of the experiment. The goal of this approach was to achieve the optimum solution, which provided the best flow balance with minimal pumping losses.

Corporate Baseline Radiator Flow versus Ford GT

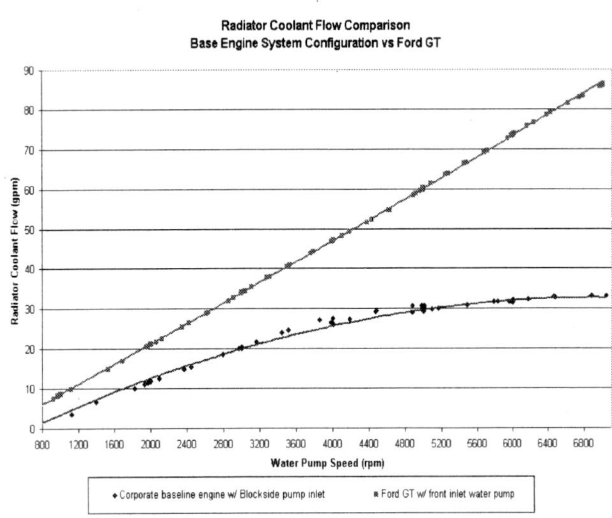

Figure 19. Radiator flow comparison

Radiator Pressure Drop and Flow Comparison

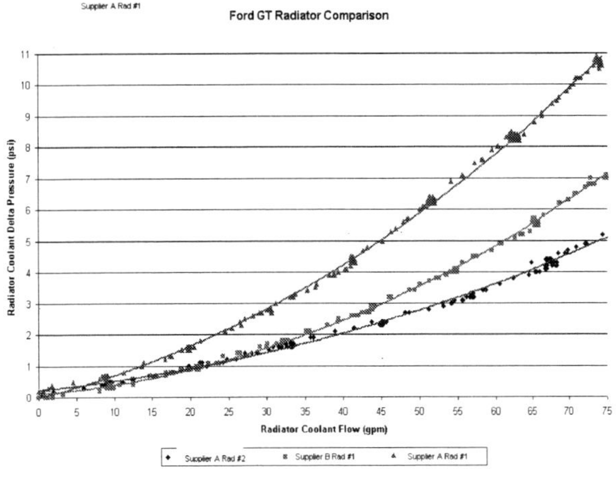

Figure 20. Radiator performance comparison

Cooling circuit balancing

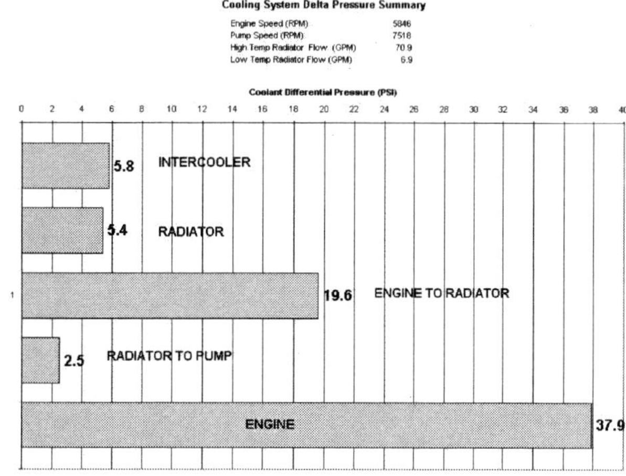

Figure 21. Cooling component pressure drops

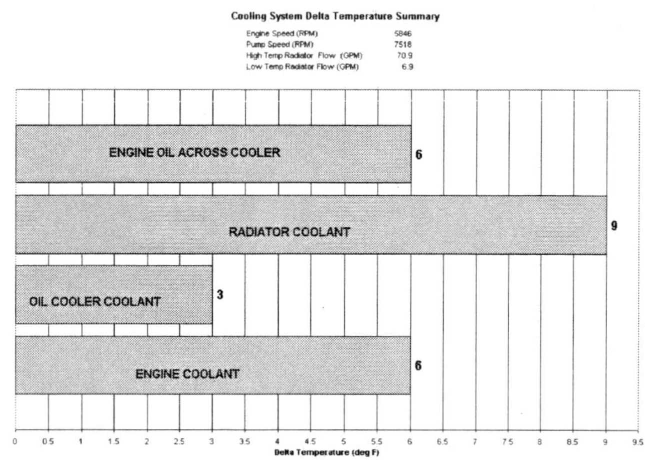

Figure 21a: Cooling component delta temperature

The thermal bench was a valuable tool to assist the vehicle packaging efforts, which often resulted in competition for the same package space in the vehicle. The information gathered from these tests helped define and rank the importance of each component to the overall system performance.

DYNAMOMETER

Engine dynamometer tests were conducted using early prototype engines to validate heat load assumptions made at the beginning of the program.

The results shown below validate the initial program design assumptions and the engineering models. The thermal bench and dynamometer testing provided confidence that the vehicle cooling system would perform to target.

WOT Heat rejection testing

Engine Control Targets for Testing.		
WOT 1100 to 6100 RPM in delta 200 RPM steps		
Engine Coolant Outlet Temp (°F)	Combustion Air Inlet Temp (°F)	Engine Oil Outlet Temp (°F)
195	77	240
210	77	250
220	77	260
230	77	270
240	77	280

WOT Heat Rejection

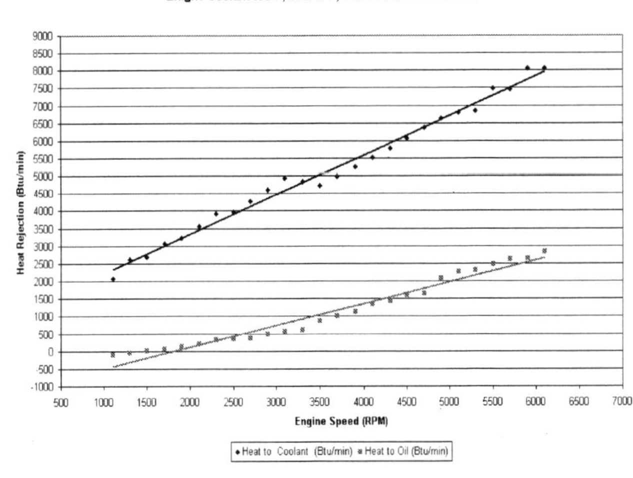

Figure 22. WOT heat rejection

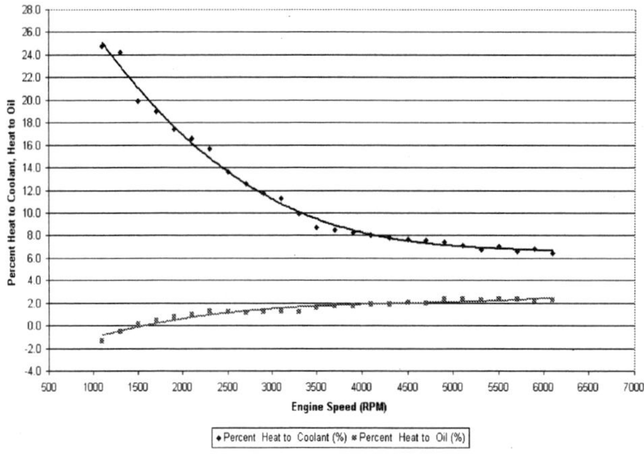

Figure 23: WOT heat rejection to coolant and oil

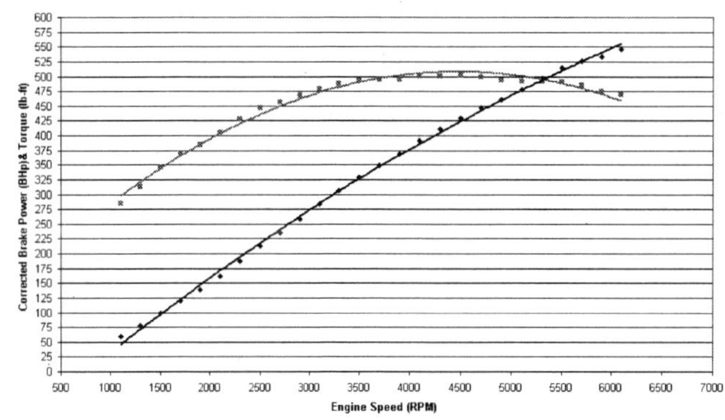

Figure 24: WOT Power and torque

Engine coolant flow was also compared to the initial design target. The actual coolant flow meets the required flow to maintain the target 10-15° F differential across the engine.

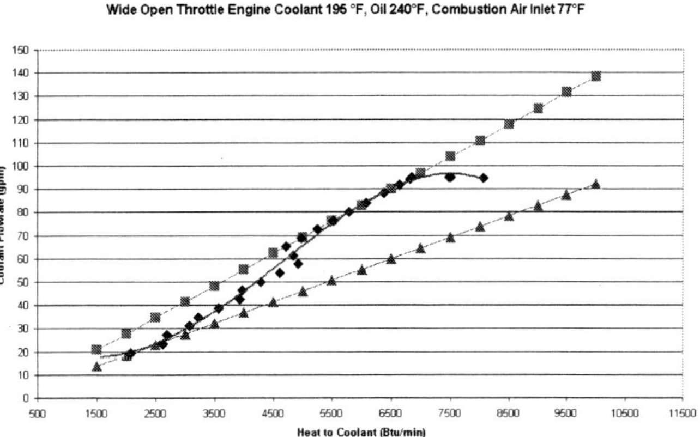

Figure 25. Actual vs. required coolant flow

CLIMATE CONTROL TUNNEL

A single Attribute Prototype (AP) vehicle was dedicated to evaluate the three thermal systems.

- HVAC and climate control systems
- Engine and powertrain cooling
- Vehicle thermal protection.

To reduce test time instrumentation was installed to allow simultaneous data collection for all systems. Total channel count exceeded 450, and packaging posed significant challenges due to the compact vehicle interior.

TEST PROCEDURES

As discussed earlier, customer drive cycles were the key to meeting performance targets. The standard validation test procedures needed to be modified to address the uniqueness of this vehicle and identify critical operating conditions that would cause failures in the thermal systems.

The testing strategy was to conduct the standard corporate tests and then evaluate vehicle performance under the 4 drive cycles identified previously..

Since the Attribute Prototype (AP) had all the required instrumentation to evaluate three separate thermal systems, the requirement to repeatedly conduct the same corporate tests was removed. With each return to the climate control tunnel, slight changes to the testing method allowed the newly updated system components to be evaluated to their design limits. This testing increased engineering confidence in the final vehicle design.

These tests followed the standard and modified steady state tests. One of the new and aggressive tests was a rapid acceleration after the prescribed vehicle soaks at temperature, similar to the Vmin cycle discussed earlier. Additional heat was built up into the thermal systems by dynamic and aggressive driving. The vehicle was accelerated on the chassis rolls as quickly as possible to a target speed. After reaching the target speed, different gear selections were made while maintaining that speed. This testing was intended to simulate high ambient temperature road course drive cycles.

ON-ROAD TESTING

On-road testing was employed to evaluate each of the three thermal systems during real world driving conditions.

The thermal systems were evaluated under extreme and varying driving conditions in the high ambient conditions in the southwest U.S.

This testing was important because the system designs might present problems under less than aggressive driving conditions than that evaluated in the tunnel. The question of whether the cooling system design was robust in different climates and altitudes needed to be addressed.

TRACK TESTING

The thermal Attribute Prototype (AP) car was also tested during road-race course operation. The road course is a two-mile hilly road course with 10 corners. It includes uphill and downhill sections and a 3200-foot straight.

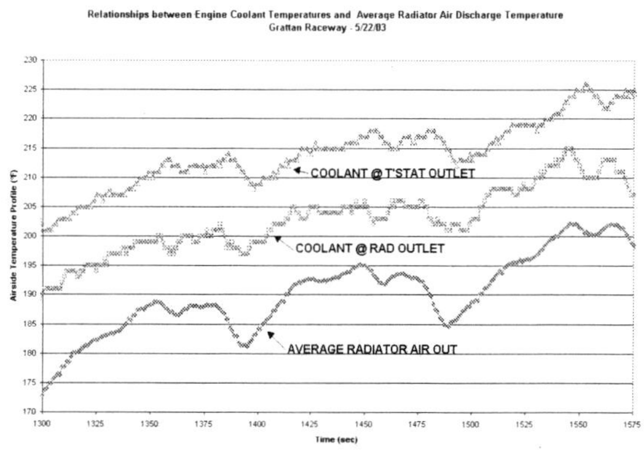

Figure 26. AP racetrack cooling performance

This portion of the test data illustrates the relationship between the engine coolant outlet, Radiator outlet and average air discharge temperatures. The design models that were used in the beginning of the program made use of this relationship to estimate the thermal performances of the cooling modules and ultimately the design of the heat exchanger. From this testing the Vmin, driving characterization is correlated to the thermodynamic balancing of the system. The engine coolant outlet temperature tracked the average air discharge temperature. With each successive lap of the track, it was observed that the engine coolant temperature increased and was heading toward a failure. The root cause for the impending failure was inadequate airflow to balance the coolant loads.

AP TESTING SUMMARY

All three testing modes, tunnel, road and track, yielded consistent and similar results. Repeated WOT accelerations at high engine speeds, the Vmin drive cycle, would result in a continued heat build that would lead to an overtemp condition unless the cycle was interrupted by a higher gear cruise, for example. The root cause was inadequate airflow.

AP level cooling package conditions and grille and hood openings were not representative of the later design levels incorporated into the confirmation prototypes (CP). Alternative frontal grille opening treatments, air handling designs, electric fan packages and fan operating strategies were identified. The cooling module supplier was contacted to investigate methods of reducing the airside pressure drops and to reevaluate the heat transfer impact on the cooling module performance with these changes.

CP TESTING

Confirmation prototypes (CP) incorporated the latest design level for cooling system packaging, system components and air handling.

The confirmation prototype (CP) level cars were subjected to road course testing at the Dearborn Proving Grounds. This testing was more aggressive than that for the AP vehicle. The testing ambient temperature was hotter; the usage of only first and second gear limited the vehicle speeds and maximized engine cooling heat loads.

Initial test results were essentially the same as those for the AP. Air handling refinements and LTR design revisions improved airflow sufficiently to meet the test target – continuous operation on the DPG handling track without overheating. The results are shown below (Figure 27).

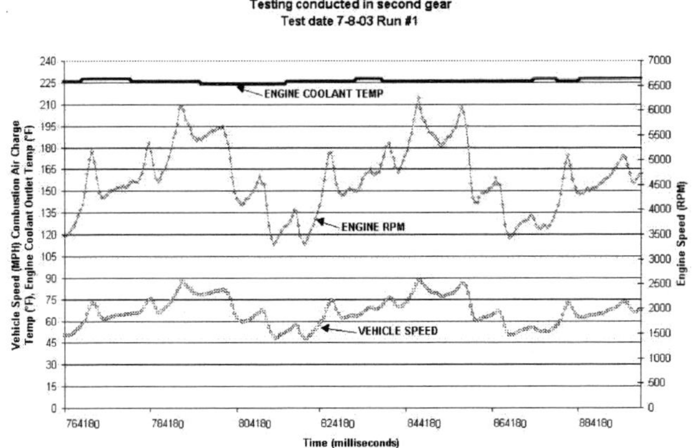

Figure 27: CP racetrack cooling performance

CONCLUSION

Through focused engineering efforts in CAE, CAD, laboratory and vehicle testing, the Ford GT thermal engineering team achieved supercar performance in record time.

Accurate definition of customer drive cycles made possible thermal system designs that provide predictable thermal control of the air, oil and water under all anticipated driving conditions

2004-01-1258

Light Weight Polypropylene Bonded Long Glass Fiber Technology – For Niche Vehicle Interior Trim

E.J. Benninger
Lear Corporation

J.G. Hipwell
Azdel, Inc.

Kip Ewing
Ford Motor Company

ABSTRACT

The Ford GT is a niche vehicle with a very limited production volume. A solution was needed that would minimize cost, particularly up-front investment, and produce a lightweight, high-class vehicle interior, while maintaining the standard quality, material and design requirements.

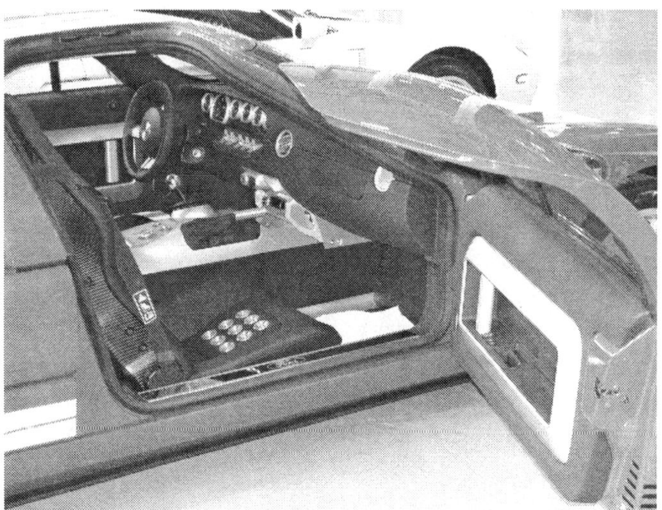

Figure 1 Ford GT interior

INTRODUCTION

In low volume, production-tooling costs can easily overwhelm variable cost. It is important to consider the total cost per the following equation.

Tooling Cost+Part Cost*Vehicle Volume+Requirements = Total Cost.

As well, It is also important to consider overall part weight and its final Class-A appearance. The use of "polypropylene bonded long glass fiber" material (produced under trade name Azdel SuperLite® - a low-density glass mat thermoplastic -LDGMT) proved to minimize tool and part cost, while producing a lightweight interior and meeting appearance and quality requirements.

Previously, LDGMT material was only validated to Ford as a headliner material. In order to use the LDGMT material as an interior trim material in the Ford GT, given its short development time, a certain degree of calculated risk had to be undertaken until it was validated for more than headliner usage.

MATERIALS & PRE-PROCESSING

The LDGMT material is composed of glass fibers bound together by resin particles. At production time, the hot material is nip rolled (pressure laminated) to a lower thickness to improve the bonding of the resin and the glass fibers. The LDGMT material cools down after being nip rolled and locks the glass fibers in place.

The unlofted LDGMT blank is under stress due to the cumulative bending of the individual glass fibers and bundles. These glass fibers are held in place by solidified resin particles.

Prior to molding, when the LDGMT blank is heated in the oven, the resin particles binding the glass fibers melt and the drop in resin viscosity releases the bending stress on the glass fibers. The glass fibers start to spring back, leading to overall lofting (growth in the z-direction of up to 170% increase in thickness) of the LDGMT.

The lofted LDGMT blank experiences redistribution of the resin and formation of new fiber interlocking held together by the redistributed resin, bridging the interlocking surface.

PHYSICAL PROPERTIES

Material (Tested At Room Temp)	Flexural Strength (MPa) SAE J949	Flexural Modulus (MPa) SAE J949	Specific Gravity ASTM D792	Multi-Axial Impact (J) ASTM D3763
700 GSM	7.8	863	0.338	4.7
800 GSM	9.0	1215	0.378	5.7
900 GSM	9.5	1450	0.418	6.8
1000 GSM	16.0	1825	0.458	7.0
1200 GSM	19.5	2030	0.504	7.6
1400 GSM	34.5	2670	0.603	9.3
1600 GSM	46.0	3470	0.711	11.9
2000 GSM	53.0	4800	0.848	12.9
Polypropylene	23-50	550-2410	0.9-1.24	1.2-29.6
Wood fiber	17-62	400-3000	0.8-1.1	4-30

GSM = Grams per Square Meter.

PROCESSING

A previous method of making a vehicle interior trim component or part, such as a door panel, involves molding an air-impermeable polypropylene substrate, and then drilling holes through the substrate so that air can be drawn through it. Alternatively, discrete holes may be molded into the substrate during molding. The method includes positioning of the substrate and a heated cover material over a vacuum mold so that the substrate is disposed between the cover material and the mold. An adhesive film may also be positioned between the substrate and cover material. Next, air is drawn through the holes to vacuum form the cover material to the substrate.

Because air is not able to pass through the substrate at locations where no holes have been formed, this method may result in air bubbles being trapped between the cover material and the substrate. Furthermore, subsequent to the vacuum forming, an additional procedure may be required to wrap the cover material around peripheral edges of the substrate, if desired. Such a procedure adds significant cost to the method.

The method of manufacturing the interior trim part with LDGMT material includes: Forming an air-permeable substrate into a desired shape; positioning the formed substrate and a cover material on a vacuum mold so that the substrate extends between the cover material and the mold, and vacuum forming the cover material over the substrate.

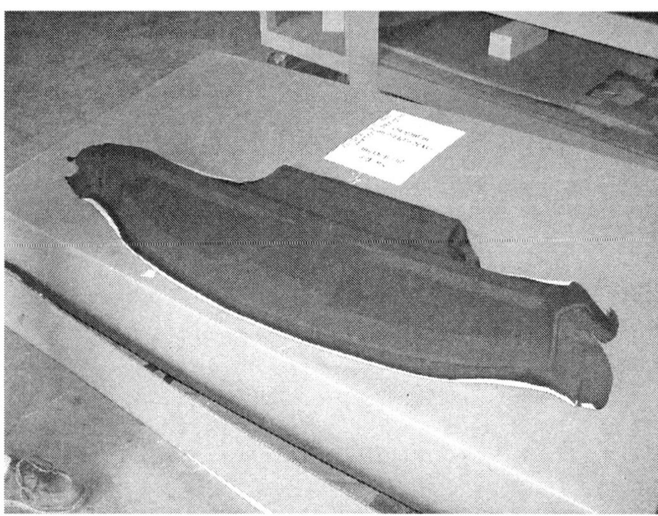

Figure 2 Semi-molded part (in-process)

BRIEF DESCRIPTION OF THE DRAWINGS

Figure 3 is a schematic view of a system or arrangement for manufacturing a vehicle interior trim part. The arrangement includes a vacuum mold that is shown prior to activation.

Figure 4 is a schematic view of the vacuum mold in an activated state.

Figure 5 is an enlarged portion of figure 4.

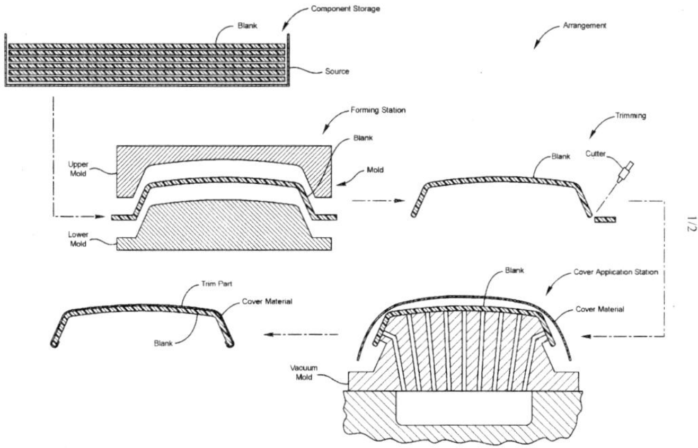

Figure 3 Schematic view of process

DETAILED DESCRIPTION

Figure 3 shows a system or arrangement for manufacturing a vehicle interior trim part, such as a door panel, package shelf, headliner, trunk panel, console panel, instrument panel, or any other suitable interior trim part. The arrangement includes a component storage area, a forming station, a trimming station and a cover application station. The arrangement may also have a conveyor system (not shown) that may include one or more conveyors, such as belt conveyors, for transporting components between two or more of the stations.

Following is a detailed description of manufacturing the interior trim part with LDGMT material using the production line arrangement outlined above. First, a main body or substrate, such as a blank, is selected from the component storage area, which includes a source of blanks. The blank is comprised of LDGMT.

Next, the blank is transferred to the forming station. In figure 3, the forming station includes a compression mold, (having first and second mold portions respectively). The mold portions may be mounted, for example, on platens of a press (not shown) that is operable to move these mold potions toward and away from each other. One or both of the mold portions may be heated in such a manner that the mold is capable of sufficiently heating the blank. Alternatively, or in addition to, the blank may be heated, such as in an oven (not shown) prior to being transferred to the forming station.

The mold is movable between an open position, shown in Figure 3, and a closed position (not shown). In Figure 3, each mold portion has a non-planar, contoured mold surface. For example, each mold surface may have at least one curved or bent section. Alternatively, one or both of the mold portions has a non-planar, contoured mold surface. When the mold is in the closed position, the mold portions have a non-planar, contoured mold surface. When the mold is in the closed position, the mold surfaces cooperate with each other t o mold or form the blank into a desired shape or contour. This forming step may be performed in such a way that the resultant shape of the formed blank corresponds to a desired final shape of the interior trim part. This forming step may also be performed such that the formed blank is provided with a non-planar configuration.

Next, the formed blank is transferred to the trimming station, where excess material may be trimmed from the formed blank. While the trimming station may include any suitable cutting apparatus for trimming the formed blank, in the embodiment shown in Figure 3, the trimming station includes one or more water jets. Alternative cutting apparatuses include, for example, knives and laser cutters.

The formed blank is then transferred to the cover application station, where the formed blank is positioned on a vacuum mold. Referring to Figure 4, the vacuum mold includes one or more vacuum passages that extend between the mold surface and a vacuum chamber. The vacuum chamber is in fluid communication with a vacuum source (not shown) that draws air from the mold surface through the vacuum passages.

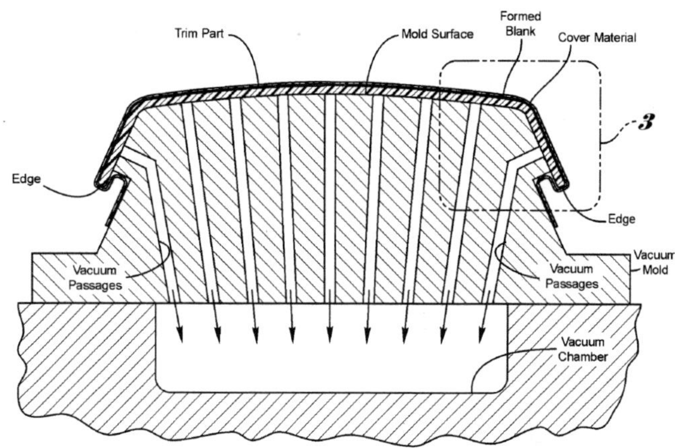

Figure 4 Vacuum mold

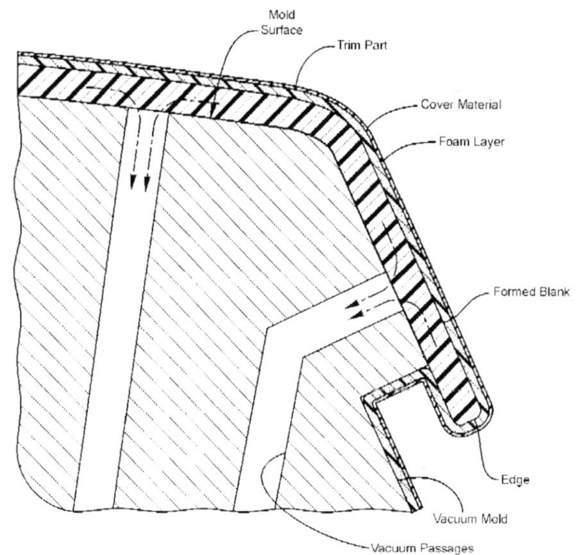

Figure 5 Vacuum mold detail

Referring to Figures 3 through 5, a cover material is also positioned on the vacuum mold and over the formed blank. The cover material may comprise any suitable material, such as a vinyl, thermoplastic olefin, or thermoplastic urethane skin. Furthermore, the cover material may be provided with or without foam backing. Alternatively, a separate foam layer may be positioned between the cover material and the formed blank. The cover material may also include an adhesive. Alternatively, or in addition to, one or more adhesive layers, such as adhesive films, may be positioned between the cover material and the formed blank. For example, one adhesive film may be positioned between the cover material and the foam layer and another adhesive film may be positioned between the foam layer and the formed blank.

The cover material, the foam layer and/or the adhesive films may be heated, such as in an oven (not shown), prior to positioning these components on the formed blank. Alternatively, or in addition to, the cover material, the foam layer and/or the adhesive films may be heated, such as with infrared heaters (not shown), after these components are positioned on the formed blank.

Next, referring to Figures 4 and 5, the vacuum source (not shown) is activated to vacuum form the cover material over the formed blank. More specifically, the vacuum source is activated to draw air through the formed blank so as to draw the cover material, the foam layer and the adhesive films against the formed blank and bond these components to the formed blank. The cover material, foam layer and/or adhesive films are preferably sufficiently air-impermeable so that these components are drawn tight against the formed blank. Because the formed blank is air-permeable, the presence of air bubbles between the cover material and the formed blank may be significantly reduced or eliminated. Furthermore, because of the air-permeable configuration, air may be drawn uniformly through the entire formed blank, including peripheral or end edges of the formed blank, using one or more vacuum passages. As a result, the cover material is able to wrap around end edges of the formed blank upon activation of the vacuum source. In the embodiment shown in Figures 4 and 5, the cover material is able to wrap around the front, side and back of each end edge. Consequently, subsequent operations are not needed to wrap the cover material around the end edges. Similarly, if the formed blank includes interior edges (not shown) then the vacuum forming step may be performed such that the cover material wraps around front, side and back portions of such interior edges. A secondary trimming operation may also be performed, if desired, to remove excess cover material from the finished interior part.

Because the cover material is attached to the blank after the blank has been formed into a desired shape, prominent features such as deep draws and/or sharp angles may be formed in the blank, using the mold or other apparatus, without risk of tearing the cover material. Moreover, because the shape of the formed blank may correspond to the final shape of the interior trim part, no additional shaping of the formed blank may occur before the cover application station.

TOOLS

The tools for the LDGMT process are considerably less expensive than that of typical manufacturing methods and their lead times can be less than one third as long. The tools for Ford GT were produced from the Huntsman Renshape material. The tools were operated in a Williams-White® 1000 Ton Horizontal Press with a fast close feature. This press was used for all of the Ford GT Parts.

Figure 6 Sample mold

FIELD DURABILITY

LDGMT, being a two-part material of Polypropylene and inert glass fibers, show the chemical properties of polypropylene and the physical properties of a glass composite.

The chemical composition of the material passes Ford Motor Company's, Odor, Fogging, Humidity, and Mildew Resistance requirements.

The physical properties of the Polypropylene and glass fibers passes Ford Motor Companies, Cold Impact, Shrinkage, Cold Flexibility, Dimensional Stability, Load Deflection, Humidity, Mildew Resistance, and Room Temperature Impact.

CONCLUSION

The LDGMT material parts are up to 30% lighter than an equivalent manufacturing method's parts.

The LDGMT material lends itself perfectly to low volume (below 20,000), high performance "niche" vehicles, where low vehicle weight, low tooling cost and short tool lead-time, are top considerations.

ACKNOWLEDGMENTS

1. Ford Motor Company, Dearborn, MI
2. General Electric Company, Southfield, MI.
3. Lear Corporation, Southfield, MI.
4. WK Industries, Troy, MI.

REFERENCES

1. Technology that Endures, Jesse Hipwell Azdel, Inc., Southfield, MI.,2002

2004-01-1259

2005 Ford GT Electrical & Electronics

Matt LaCourse
Lear Corporation

Dan Fisher
Ford Motor Company

ABSTRACT

The Ford GT Program Team was allocated just 22 months from concept to production to complete the Electrical and Electronics systems of the Ford GT. This reduced vehicle program timing – unlike any other in Ford's history -- demanded that the team streamline the standard development process, which is typically 54 months. This aggressive schedule allowed only 12 weeks to design the entire electrical and electronic system architecture, route the wire harnesses, package the components, and manufacture and/or procure all components necessary for the first three-vehicle prototype build.

INTRODUCTION

Team members from Lear Corporation and Ford Motor Company developed a strategy around a disciplined, well-defined program plan that was based on optimized tasks and included only those steps absolutely necessary to ensure a robust, quality product within the required timing at a minimum cost.

OPTIMIZED PROGRAM PLAN & EXECUTION

The key to developing the plan was a dedicated series of brainstorming meetings of experienced individuals representing the following disciplines: 2D & 3D CAD, electronic hardware, software, mechanical, electrical, systems, quality, manufacturing, key suppliers, and program management.

KEY CONSIDERATIONS TO THE PLAN

A plan was developed based around the following key objectives, enablers and processes:

- Top level objectives
- Design review process
- Co-location of key design team individuals
- Design flexibility
- Packaging feasibility
- Robust ground plan design
- High-performance wire harness protection
- Minimized costs using carry-over components
- Carry-over components vs. system design
- Team initiative, talent, & experience
- 3D CAD capabilities & data management
- Quality action plan
- Minimized complexity
- Local manufacturing for early builds

TOP LEVEL OBJECTIVES

The key to developing the program plan was a clear understanding of the objectives.

The Ford GT was to look great, handle exceptionally, and exceed the performance metrics with respect to its rival benchmarks. The vehicles design was based on many of the styling cues and appearance features of the original Ford GT of the mid to late 1960's.

The top level objectives helped the program team in its decision making with regard to choosing the appropriate requirements for each subsystem. In many cases, that meant exceeding the requirements typical of a mainstream OEM program.

DESIGN REVIEW PLAN

A three-hour, weekly 3D-based Digital Buck design review was an essential piece of the plan. All 3D design leaders and responsible engineers were required to be present for the meeting. The meeting followed a strict agenda based on open issues split by vehicle subsystems.

Additional weekly meetings were scheduled amongst the stakeholders to address all open issues identified in the weekly Digital Buck design reviews.

CO-LOCATION OF KEY TEAM MEMBERS

The majority of the Ford GT Team, consisting of employees of Ford and the Tier One suppliers, were co-located within a single facility in Dearborn, Michigan.

Co-locating these key design and engineering representatives was one of the most important factors in the team's success.

Whether issues were quality, design, safety, commercial, etc., decisions were made quickly and efficiently because the people responsible for making specific decisions were accessible in a moments notice. There was no need to write e-mails, place phone calls, and/or wait for weekly meetings to address urgent issues. Co-location was the single most significant enabler in keeping the program plan on track.

PACKAGING FEASIBILITY

The routing of the wire harnesses was challenging in that the development process is typically completed 8 to 16 weeks after the majority of the vehicle package is frozen. The pace of this program required that the wire harness routings and designs be frozen in parallel with the vehicle package.

As a two-seat, mid-engine sports car with limited interior and exterior package space, the vehicle presented the team with many challenges in the area of vehicle packaging.

The car uses a center-mounted fuel tank located between the seats. The fuel tank is surrounded by an aluminum structure that protects the fuel tank and provides vehicle stability from front to back. This structure provided the routing path and retention strategy for the large bundle of wire harness circuits that run from front to back.

The vehicle space frame is made primarily of welded aluminum extrusions. These extrusions have an internal web-type cross section to aid in stability and strength. This presented a unique challenge for the door harness routing. The wires could not be routed traditionally from the doors to the vehicle interior since there was no way to pass the door harness through the A-pillar. The team discovered a unique solution using two carry over grommets that actually route the wires around the a-pillars without need of any new components. Figure 1 illustrates this solution. The use of carry-over components was critical to this area, because a new trough or grommet would have been costly and would have caused a detrimental affect on timing.

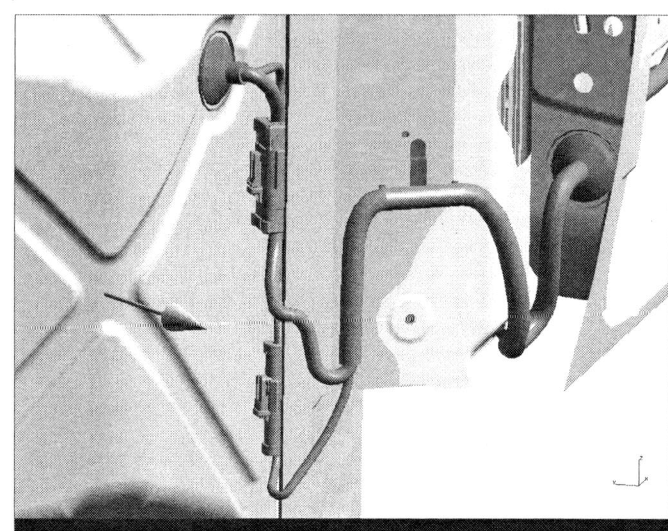

Figure 1 – Door Harness interface to Dash Panel

GROUND PLANE DESIGN

The vehicle is constructed of aluminum body panels over an aluminum frame. The frame interface to the ground eyelets required special consideration with respect to galvanic corrosion.

Typically, on frames made of steel, solder-dipped copper eyelets are attached directly to the frame by galvanized steel screws. However, to utilize the same method of the Ford GT would have ensured a breakdown of the ground interface due to the deltas in activation energy between the eyelet, screw, and aluminum frame.

To minimize the corrosive effects of the activation energy between dissimilar metals, the team utilized a series of M8 tin-zinc plated Riv-nuts, fastened to the aluminum frame. The Riv-nut was unique in that it was designed with a square outer diameter to prevent spin-out.

A standard M8 solder-dipped copper eyelet was attached directly to the Riv-nut by way of a S431 plated M8 bolt. The S431 finish has an oven-baked, dual-coating consisting of a chromate/phosphate compound that withstands 408 hours of salt spray and conducts effectively up to 645 ° C. The collar of the Riv-nut also prevents the eyelet from touching any part of the aluminum frame.

SDRC's I-DEAS CAD software was utilized to locate the Riv-nuts in manufacturing-friendly positions while preserving the grounding plane performance objectives (See Figure 2).

Fig. 2 – Ground Eyelet Interface

WIRE HARNESS PROTECTION

The mid-engine layout of the vehicle presented some challenges with respect to thermal protection of the Electrical Distribution System. To protect against the intense engine compartment heat, high-flex cross-link polyethylene UTMS 12519 specified wire rated to 150° C was used throughout the engine bay.

Thermflex[2] fiberglass wire covering rated to 700° C was utilized where harnesses routed within relatively close proximity of high-temp engine components. Quietsleeve[3] & Flexgaurd[4] wire coverings were used in areas where NVH, abrasion, and appearance caused issues of concern. All other areas of the vehicle were routed with ESB-M1L123-A thin-wall cross-link polyethylene wire.

MINIMIZE COSTS BY UTILIZING CARRY-OVER COMPONENTS

The major enabler in meeting the cost objectives for the Ford GT was the re-application of carry-over components. Ford Motor Company's large parts bins provided many of the electrical components used in the project.

CARRY-OVER COMPONENTS VS. SYSTEM COMPATIBILITY

Carry-over components were selected based on performance, packaging, and styling characteristics.

However, the use of the carry-over components opened up several challenges with respect to system compatibility.

One major area where new components were required was Lear Corporation's Smart Junction Box (SJB). The SJB was selected to reduce engineering effort, packaging space, manufacturing costs, and to bridge system incompatibilities. Lear chose to utilize one of its existing SJB's from a popular SUV. Although the majority of the hardware was preserved from the carry-over SJB, significant software and pin-out assignment modifications were required to drive the unique set of vehicle components and functions. Use of the carry-over hardware allowed the SJB to piggy-back on the design & process validation testing of the carry-over module from which it is based. However, a mini DV/PV validation plan was completed to prove out the new software and component interfaces.

The SJB contains components and connectors for the purpose of circuit over-current protection, splicing and interconnections. It also provides direct integration of programmable electronics, microprocessors, and software (See Figure 3).

The SJB controls and/or drives the following Ford GT features:

- Electronic door latching/unlatching
- Auto door locking
- Windshield wiper system
- Rear defrost
- Remote keyless entry
- Electronic trunk release
- Chimes
- Bi-functional high intensity discharge headlamps
- Fog lamps
- Front marker lamps
- Rear LED tail lamps with outage sensing
- Interior lighting
- Battery saver
- Delayed accessory power
- Perimeter anti-theft
- Courtesy lighting
- Driver configuration programming

The use of the SJB resulted in reduced weight, package space, and cost by combining the following:

- Fuse box
- Harness interconnects
- Remote keyless entry receiver
- Beltminder module
- Generic electronic module (GEM)
- Several miscellaneous relays

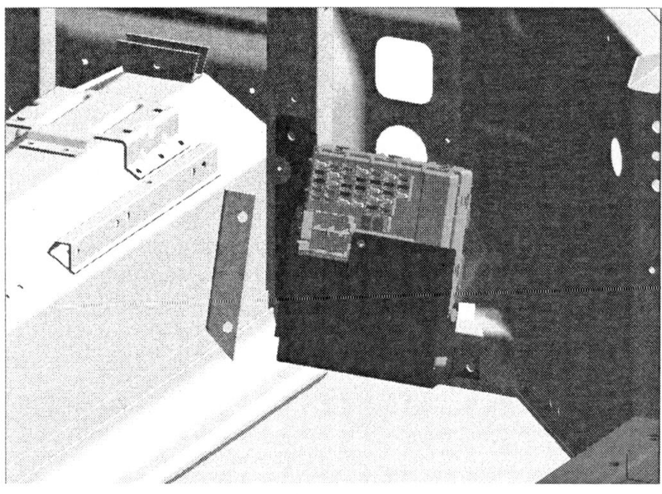

Figure 3 – SJB mounted in passenger footwell

Another area where new components were required was Stoneridge's Cluster Module. A driving force behind the Ford GT was its unique styling, and using a carry-over instrument cluster simply was not desirable. Stoneridge had been supplying parts to Ford for many years, but had never designed a full instrument cluster. The new Stoneridge cluster interprets signals from several sensors and devices throughout the vehicle to drive off-the-shelf, analog style gauges made by Autometer. This allowed for the unique, custom-looking instrument cluster.

TEAM INITIATIVE, TALENT, & EXPERIENCE

The Lear and Ford project management team was made up of experienced systems, wiring, software, and design engineers from Lear and other key suppliers to ensure robust, compatible designs that met the aggressive program requirements.

The team members were chosen not only on their experience, but also on their initiative, talent, passion to execute flawlessly, trustworthiness, proactiveness, and most of all, honor and integrity.

3D CAD CAPABILITIES AND DATA MANAGEMENT

The entire vehicle was designed, packaged, and tested through the virtual world of CAD before any prototype parts were ever produced. This decision saved tremendous time and money. Without the use of CAD, a program such as this one could never have been achieved within the allocated budget and schedule. Figure 4 shows a screen capture of the vehicle wiring and partial body.

Several of the suppliers chosen for this program utilized different types of CAD software packages. Although released data was managed in CATIA, software such as I-DEAS & Pro-Engineer played a major role.

The importance of data management cannot be overstated. Poor data management can have a major impact to the timing and budget on a program. The engineer and designer must be absolutely sure that they are designing to the latest level background data. Countless hours and money will be wasted if interfacing with old CAD data.

The program team brought in a small group of designers whose major responsibilities were to convert parts into CATIA data, archive the files into a main server, and continually report out part revision levels to all the supplier teams involved.

Suppliers potentially affected by updated data were required to complete a data upload form. The data was then uploaded to a secure web-based site where it could be retrieved.

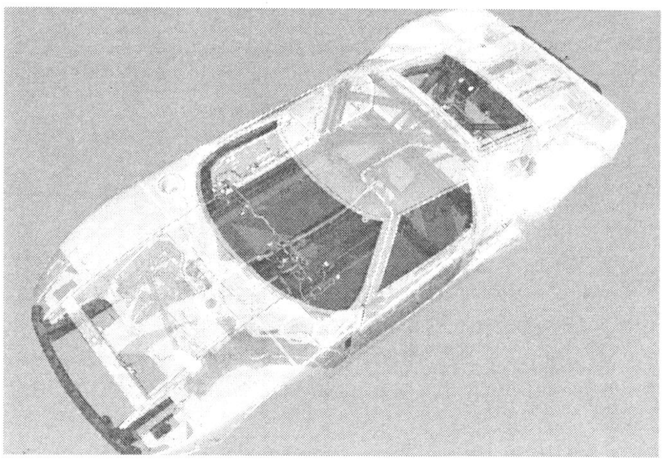

Figure 4 – Screen capture of wiring and body

QUALITY

Due to the agressive nature of the timing, attention to quality was paramount in the consideration of design alternatives. Actions taken to ensure quality and mitigate risk were as follows:

- Follow Ford Motor Company's SDS (system design specifications) where feasible
- Campaign prevent deep dive reviews (where public recalls were reviewed and plans made to keep them from happening for this vehicle)
- Warranty data analysis
- Ford approved DVP's (design validation plans)
- Completion of APQP deliverables including DFMEA, PFMEA, DVP, DFM, DFA, control plan, etc.
- Review Lear's "Lessons Learned" database
- System compatibility reviews
- Thorough weekly design reviews
- Utilize carry-over modules, connectors, terminals, etc.
- Vehicle process control meetings
- Breadboard and CP testing and analysis
- Fresh eyes reviews
- CAE / EDS fuse tool

Testing to verify the designs and to validate production reliability followed the normal testing process used for all Ford production vehicles. No unusual results were found due to either the unique nature of the vehicle or the projected customer usage.

MINIMIZE COMPLEXITY

The complexity in designing and managing multiple wire harness families with respect to optional engines, transmissions, features, and the like would have made managing this program much more difficult and increased the risk for mistakes.

For example, there are as many as twenty different instrument panel wire harnesses on any given mainstream vehicle. The program team reviewed and debated all potential features before reaching a single set. That single set of features was chosen on performance, weight, timing, and cost.

LOCAL MANUFACTURING FOR EARLY BUILDS

Lear Corporation made a strategic decision to outsource the workhorse and prototype stage of the wiring harness manufacturing to local specialty suppliers within a short distance of the vehicle assembly locations. This proximity proved to be very important as it allowed the wiring manufacturer to work very closely with Lear and the program team. It allowed real-time updates to vehicle wiring as changes were incorporated after system freeze.

CONCLUSION

To execute a program of this caliber within an unprecedented schedule requires:

1. Co-location of team members with initiative, talent, and experience
2. State-of-the-art 3D CAD capabilities
3. Well defined, well disciplined processes with a common team-oriented focus.

ACKNOWLEDGMENTS

Precision Electrical Services – Manufactured Workhorse level wire harnesses

Clements Manufacturing, L.L.C. – Manufactured CP (Confirmation Prototype) level wire harnesses

Lear Corporation – Manufactured production level wire harnesses

REFERENCES

1. Riv-nut – Textron Corporation
2. Thermflex – Bentley Harris
3. Flexgaurd – Bentley Harris
4. Thermflex – Bentley Harris

CONTACT

Matt LaCourse
Ford GT EDS Project Leader
Lear Electrical & Electronics Div.
Lear: 313.253.5085
E-mail: mlacourse@lear.com

Dan Fisher
Ford GT Design and Release Engineer
Electrical and Lighting
Ford Motor Company
Phone: (313) 59-42341
E-mail: dfishe19@ford.com

2004-01-1260

The Ford GT Transaxle – Tailor Made in 2 Years

Glenn D. Miller
Ford Motor Co.

Andrew Cropper, Bob Janczak and Steve Nesbitt
Ricardo UK Ltd

ABSTRACT

This paper describes the rapid development of the Ford GT transmission, from concept phase to production, where the technical challenges involved are implicit in the specifications provided. It presents the steps taken at a project management level to expedite development, as well as the tools used to design and rate components at the design stage. Examples of concurrent engineering are given as well as management techniques used to predict and address key risks. In addition, details of analysis and test procedures are given, underlining their contribution to the rapid introduction of the transmission to the market place.

OBJECTIVE

Ricardo Driveline and Transmission Systems, as a partner to the Ford Motor Company and Roush Industries, faced the challenge of designing, developing & manufacturing a six-speed manual transaxle for a prestigious high performance vehicle in an extremely short time scale. Critical to the program's success was the introduction of Purchasing, Quality and Manufacturing experience alongside Design Engineering from the earliest stages of the project, allowing key risks to be identified quickly and addressed accordingly. This approach greatly reduced critical paths to volume manufacture and allowed extremely short lead times for both prototypes and production-tooled components to be achieved, culminating in the on-time delivery of a world class transmission system.

Fast-track vehicle package management, combined with reliable duty-cycle generation enabled early definition of product design specifications before a team of design, analysis, development and manufacturing engineers were assembled to develop and integrate the driveline system into the Ford GT application. The two year timeline prohibited normal development procedures, while the limited number of vehicle builds and prototypes restricted the amount of testing and data acquisition available. Instead, specially developed proprietary software and advanced Computer Aided Engineering (CAE) techniques were used to predict mechanical and dynamic performance, and address assembly issues in advance of component manufacture. Vehicle usage during development was therefore limited to proving analytical calculations and assumptions, while demonstrating functionality.

In particular, CAE analysis allowed optimization of components for strength, weight and durability, while advanced dynamic methods were used to simulate and predict transmission characteristics such as shift quality and NVH. Gear-tooth contact analysis was instrumental in optimizing gear-tooth geometry to reduce contact error and mitigate against the effects of shaft and casing deflections under full load. In areas of prime risk, rig tests formed an important part of the design verification procedure and focused on validating certain analyses in representative mock-ups and assemblies. Finally, and in parallel, value analysis and design for manufacture activities were carried out to ensure cost targets could be met.

This paper investigates each of the engineering and managerial steps taken to enable the rapid introduction of a world class sports car, and highlights the technologies involved.

BACKGROUND

In March of 2002, Ford Motor Company (FMC) committed to a new GT production program to be introduced as part of Ford's centennial celebrations. Clearly, timing was of the utmost importance to coincide with the celebrations, and so Ford contracted Roush Industries for the development of the engine and integrated powertrain. In turn, Ricardo was approached as an "all-in-one" solution for the design, development and manufacture of the transmission and driveline system, including driveshafts and limited slip differential.

Key to Ricardo's selection was its ongoing in-house development of a low volume, high torque capacity, six-speed manual transmission. Prime performance targets were to be world class shift quality and refinement, and the ability for the transmission to be configured to suit a variety of vehicle and engine installation combinations. This background of work allowed a credible program plan to be defined at an early stage, from concept through to production and supply, while the project scope was still under development.

THE CHALLENGE

The project scope was to deliver a world class and tailored six-speed transaxle for the start of vehicle production in under two years. To achieve this, the main challenges were stretched performance goals, a limited number of prototypes for development and the commissioning of a new production facility, all in the same aggressive timescale as mentioned previously. As such, the key tasks that were managed for supply of the transmission and driveline are outlined in Table 1.

ENGINEERING • Specification of system and component performance targets • Management of package data • Concept and detail design • Design analysis and CAE • Mechanical development • Gear shift quality development
MANUFACTURING • Prototype manufacture • Tooling design • Facility design and commissioning • Production supply
PURCHASING & QUALITY • Cost and value analysis • Prototype sourcing • Production supplier selection • Advanced product quality planning

TABLE 1 *Project responsibilities for the Ford GT transaxle*

The short timeline meant that engineering processes had to be streamlined, in comparison with those of a typical volume program, while still delivering a world class product. Therefore time savings were sought in all phases of development, especially testing; here a number of advanced procedures were implemented as follows:

- Benchmarks and analytical duty-cycle derivations were largely taken from an existing database
- CAD data was used as an electronic "buck" to manage package data
- Extensive use of design analysis and CAE was made both as a replacement and enhancement to rig tests
- Simulation software was used for iterations and optimization to meet performance targets within design constraints
- Targeted rig tests were devised to reduce the need for in-vehicle tests and, where possible, to improve the fidelity of the associated simulation technique

In addition, Manufacturing, Purchasing and Quality departments were involved from the outset of the program to further accelerate the procurement of components, as follows:

- Early development of cost models removed the need for extensive cost-reduction activities so that engineering effort was focused on delivery into production
- Production sourcing was considered from the start of the project to enable fast supplier selection to reduce prototype and production tooling lead times
- Quality responsibilities were defined prior to project kick-off to identify key product risks for targeting in the engineering and manufacturing programs
- The flexibility of Ricardo's in-house manufacturing was leveraged extensively to achieve responsive lead times for transmission rotating components
- Planning of the new production facility began early in the project and was based on a fast set-up concept, using short lead time capital equipment
- Thus, by developing all strands of the transmission in parallel, the critical development path was greatly truncated and the lead-time reduced to a minimum.

DESIGN

The design of any transmission is clearly dependent on the performance requirements of the vehicle. In Ford's case the original specifications were as follows:

- Max. speed approximately 200mph
- Weight 1450Kg (estimated)
- 0-60 mph in <4 seconds (in 1st gear)
- 0-100mph in <8.5 seconds
- Supercharged 5.4 liter V8 developing
  - 353KW (500bhp) at 6500 rpm and
  - 712Nm (525ft.lb) at 4000rpm

Once Ford's tier 1 suppliers had been selected (and the co-operating partners on the project), initial efforts were turned to developing detailed Product Design Specifications (PDS). Additional targets such as weight and size were also defined based on benchmarking exercises performed both in advance of and during this project. The resulting PDS is shown below:

- 6 speeds (maximum speed in 5th)
- Ratios 2.6 (1st), 1.7 (2nd), 1.23 (3rd), 0.95 (4th), 0.76 (5th), 0.625 (6th), 3.364 (FD)
- Transaxle weight approximately 95Kg filled
- Shaft center distance 102.6mm
- 250mm diameter Crown Wheel, 14% hypoid offset
- 240mm diameter twin-plate clutch
- Clutch pedal load 178N, 140mm travel
- Triple synchronizer cones on 1st, 2nd, 3rd and 4th gear,
- Double synchronizer cones on 5th and 6th gear
- Oil cooler and integral pumped lubrication system
- Limited slip differential

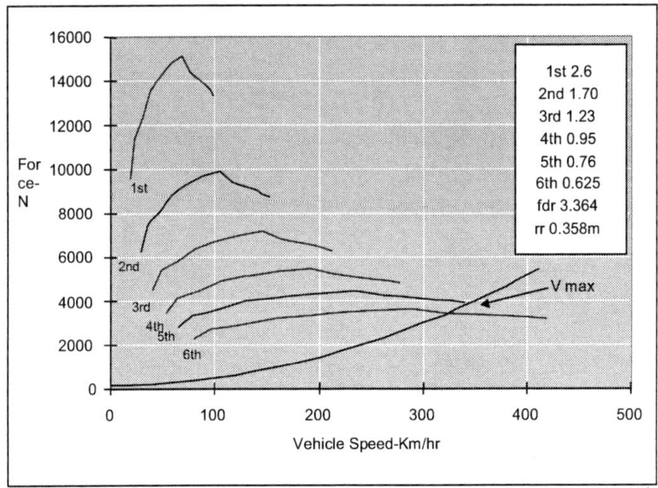

Figure 1 *Chart demonstrating tractive effort of vehicle as well as maximum speed in 5th gear.*

The clutch specification was significantly influenced by the requirement for the 0-60mph time to be achieved in 1st gear. One of the major implications of this was that the 1st gear ratio was comparatively high and thusrequired that the driver employ a considerable amount of clutch slip when pulling away from stationary. As a result, the high thermal loads that were predicted on the clutch plate meant that a large clutch diameter was required (240mm). However, the large clutch also required lower clamp loads, and thus helped towards the targeted pedal load of 178N, without the need to rely on spring assistance mechanisms, etc.

In designing the transmission, a large database of design data was employed to define duty cycles quickly and reliably. In addition, Ricardo was also able to source and modify an existing sports car transmission from stock to enable Ford to build a mule vehicle, which was to provide valuable road load data for all aspects of the vehicle.

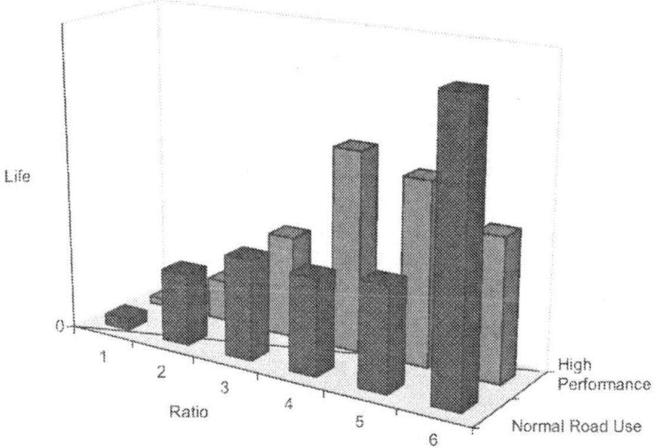

Figure 2 *Chart showing difference in duty cycle specifications for high performance and normal GT road usage*

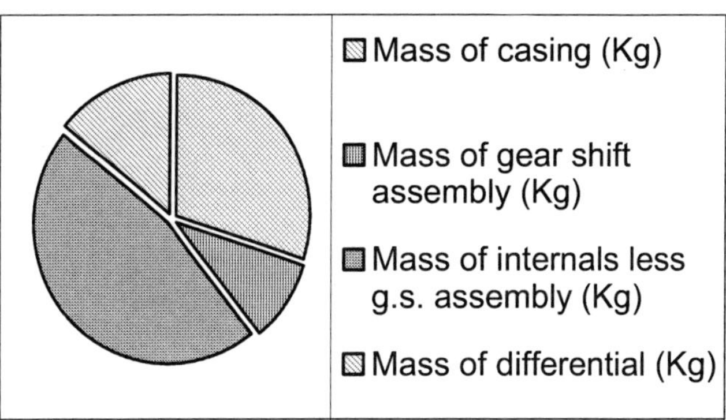

Figure 3 *Example of benchmark data showing the weight contribution of individual components in an equivalent transmission to that of the Ford GT.*

Overall project milestones were identified by Ford, with deliverables being specified for the support of workhorse; jobs 1, 2 and 3; CP (prototype); PV (pre-production) and production vehicle build phases. Around these, a project plan for the development of the transmission was created and a critical path defined.

For concept and final design, Ricardo co-located a dedicated design team, drawing on those engineers with the most experience in the design of GT transmissions. Where co-location was not possible, as between the engineers of co-operating companies, Ricardo installed empowered liaison engineers to smooth communications. These, in common with all steps in the development process, were in line with standard Advanced Product Quality Planning procedures (APQP) [1,2].

With regards to the departments that supported the design and development process, such as Manufacturing and CAE, these were already organized into dedicated function groups, which served a number of different transmission projects simultaneously. Therefore, to prevent communication breakdowns within the GT project, each department nominated a liaison engineer to interact with the design team, and ensure that recommendations were understood and implemented. Moreover, representatives from all departments attended regular design review meetings to ensure that the project remained on target and all departments were informed of the latest developments. A key example of this was the guidance provided by Manufacturing Engineers, in the cost management exercises, the outcome of which included recommendations with regards to ease of assembly, and cost savings as follows,

- Appropriate material selection
- Reduction of part numbers
- Minimizing the number of machining processes

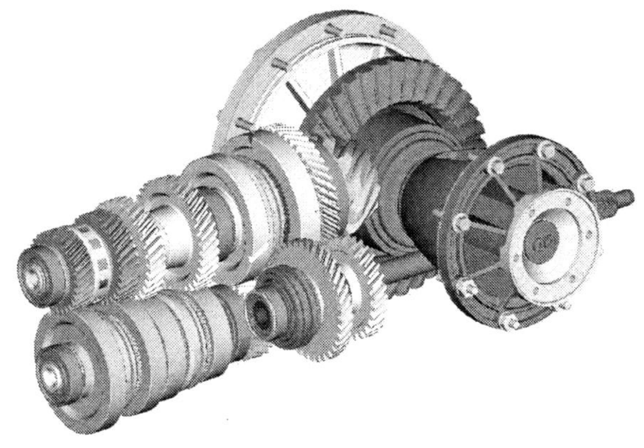

Figure 4 A CAD model of the internal components of the transmission

COMPUTER AIDED ENGINEERING

Once the product design specification was established, computer aided engineering expertise was then used to understand the performance of initial concepts, optimized both at a component and system level, and evaluate performance of the final design.

Initially an assessment was made as to a suitable duty cycle representing the predicted lives and torques to be seen in each gear while being subjected to normal road use. This was developed from database knowledge of current vehicle road data in conjunction with previous experience of high performance road vehicles. A secondary duty cycle representing high performance driving was also incorporated to establish the implications of such vehicle usage.

The first concepts were modeled as they were being generated to quickly understand the effect of shaft stiffness, bearing life, and gear geometry on the performance of the transmission. Durability and NVH concerns were also considered at an early stage in the concept development.

The bearing life was predicted in conjunction with the bearing supplier, where it was critical to ensure that the combination of high radial and axial loads could be accommodated without significant risk of bearing failure prior to durability requirements being met. High performance bearings were utilized at specific points of high load throughout the transmission.

The main aspects of modeling looked at assessing the shafts for the application with respect to deflections, resultant stresses from combined torsion and bending loads and whirl speeds. Shaft deflections were critical in ensuring gear misalignments were minimal and within specified guidelines for durability and NVH. In addition, analysis of splines was also performed to ensure root shear stresses, compressive stresses, tensile (bursting) stresses and pitch diameter shear stresses were below material limits.

Both parallel axis and hypoid gears were designed to ensure minimum noise while meeting bending and surface fatigue requirements, as defined in the duty-cycle specifications. Furthermore, the hypoid gears were designed to the minimum possible crownwheel diameter, for high performance and normal road use, to assist both package demands and weight requirements. With regards to the parallel axis gears, misalignments due to shaft and tooth bending deflections were accommodated by using proprietary software to define ideal levels of lead, crowing and tip relief along the gear tooth flanks. This allowed an assessment of the microgeometry modifications to be made for light load and full torque conditions, in order to maximize the size of the tooth contact patch and avoid any edge loading regardless of operating conditions. Further proprietary gear tooling software was also employed to check the manufacturability of gears.

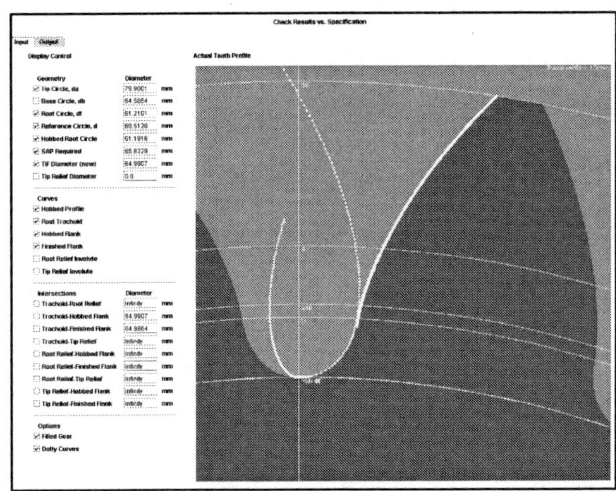

FIGURE 5 *Example of the Gear Tooling Software used to assess gear designs for manufacturability*

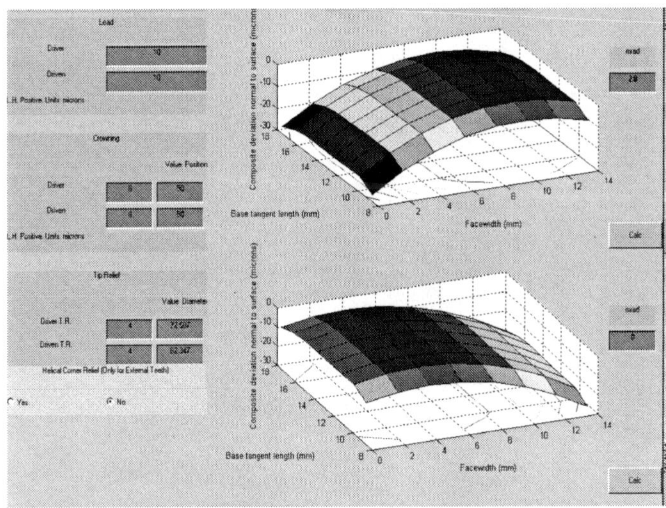

FIGURE 6 Example of gear tooth contact analysis software used to mitigate against effects of shaft deflections

Once the initial concepts had been generated and the vehicle package space around the transmission defined, this information was used in a topology optimization of the casings, based on the finite element method (figure 7) [3]. A model was constructed of the available package space for the casings at a stage when no casing design had commenced, other than to complete a package study: this greatly expedited the process. The topology optimization model was generated and attached to the vehicle mounting points with linear stiffness representation of the mounts, while the engine was modeled implicitly and attached to the topology finite element model. All operational loads were then applied to the model including all bearing loads from the concept phase and maneuver loads such as 3G bump in all axes. Frequency targets were applied to the model in order to ensure that the transmission major powertrain bending resonant frequencies were significantly above the maximum firing frequency of the engine, (6500rpm ≡ 433Hz).

The parameters of the optimization were carefully tuned to indicate the best reinforcing-rib patterns, while taking into account the ease of manufacture for the castings. The results of casings, covers and bulkheads were then checked to determine whether they were limited by their static structural performance or by their dynamic acoustic performance. Finally, a design engineer was given the recommendations of the topology optimization analysis, to design ribs that would significantly stiffen the castings without adding weight.

Once the design engineer had incorporated the recommended stiffening features, traditional finite element models were constructed of the entire powertrain, with the transmission attached to an explicit model of the engine (figure 8). The whole powertrain was then modeled as being mounted on linearised stiffness and the same loads used for the topology optimization exercise were finally applied to the assembly to check stresses and the deflections at the bearing bores. In addition, modal performance of the entire assembly was also checked to determine its first few natural frequencies, and where issues were found, mounting strategies were reviewed to minimize any transmission of vibration from the powertrain into the vehicle structure.

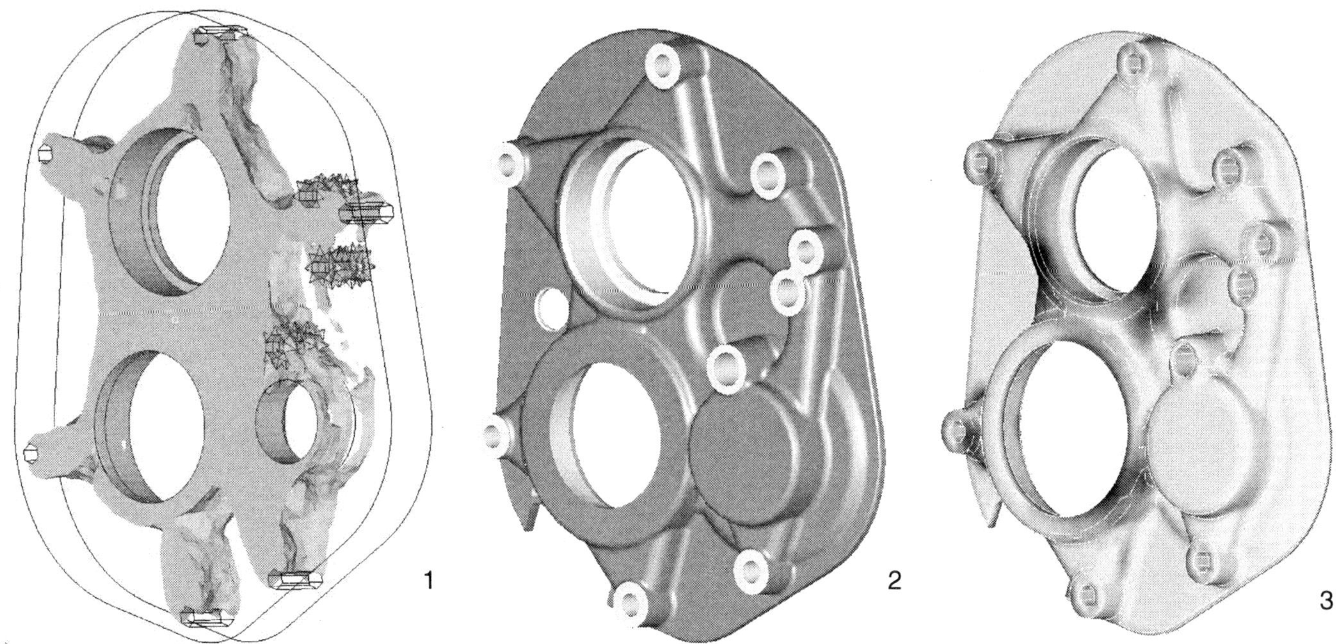

FIGURE 7 *Stages in the development of the transmission bulkhead from left to right: 1. Assessing the optimum locations for reinforcing material using topology optimization 2. An interpretation of these results by a CAD engineer. 3. The stress results from a conventional FEA analysis of the final component design (note the relatively even distribution of stresses within the structure)*

During development, noise and vibration rig-testing identified a potential high-frequency noise source from a vibration mode of a panel in the end cover of the transmission, which experienced low stresses. This was investigated using the finite element models and purely acoustic attenuating ribs were added to the internal wall of the casing. These had the effect of increasing the engine speed at which the vibration occurred and reducing the amplitude of noise generated. At the same time as casing analyses were being performed, gear-shift components such as the selector forks were also analysed with strict stiffness targets.

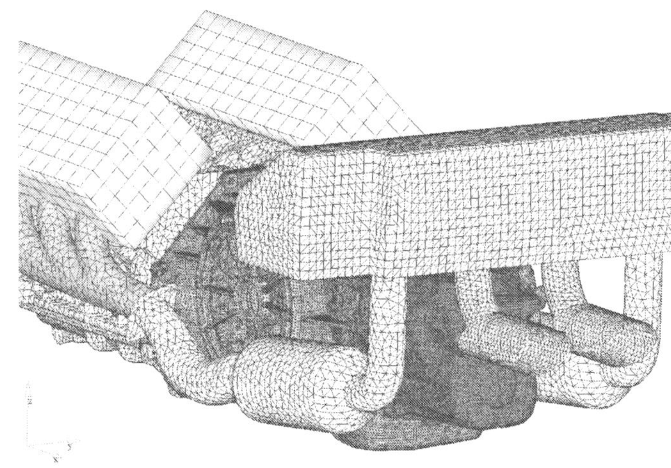

FIGURE 8 *The FEA model used to assess powertrain bend, utilizing models of the transmission, engine and exhaust system (engine model courtesy of Roush Industries)*

The entire driveline and gear selection mechanism was modeled, in which the complete effect of changing, for example, fork stiffness during a shift could be accurately predicted. From this analysis optimal values for fork stiffness were found and the shift forks tuned to give the desired shift quality using an iterative finite element process. This ensured that the forks were not over-stressed under extreme shifting loads, but had sufficient compliance to allow good shift quality.

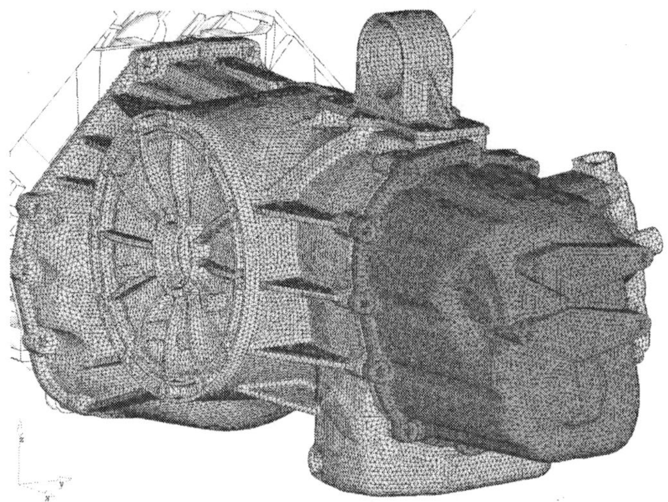

FIGURE 9 *The FEA model of the transmission alone*

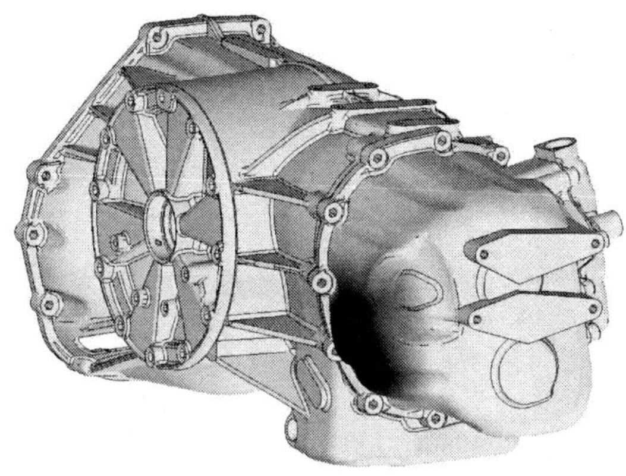

Figure 10 Results of an FEA modal analysis at high frequencies, following noise issues discovered in NVH testing.

SUB-SYSTEM TESTING

As previously stated, tight time constraints during development necessitated that numerous subsystems be bench-tested, before the first test cars became available. The best example of this was to be found in the development of the shifter mechanism, which received close attention from Ricardo's gearshift quality team [4,5].

In general terms, the selector is one of the most important points of contact between the driver and the vehicle and therefore makes a major contribution to the subjective feel of quality and brand values. In general it is found that the balance of transmission and lever ratios and cable routing has a big impact in terms of shift quality. The length of the cable and the routing greatly affect efficiency and drag, thus requiring optimization of the routing and correct cable supplier selection. The biggest problem faced was the required length (approximately 2.5m) and the location where the cable connected to the transmission. Nominal temperatures predictions along the length of the cable, however, were found to be unlikely to exceed 80°C while the maximum temperature allowed was 120°C.

To determine the best configuration for the cable routes and select the lowest friction components, an objective "GearShift Quality Assessment" System (GSQA) was used to enable measurement of cable efficiencies and drag, based on the various routings available. This was undertaken using mock-ups of the selector mechanisms, which took all dimensional parameters from the final CAD package model. Many different cable types were tested to determine the most efficient configuration. Upon completion of the exercise, cable efficiencies had been increased from initial values of approximately 46% up to 70% (equating to 5N drag), greatly enhancing feel and definition in the gear selection.

FIGURE 11 *GSQA system fitted to workhorse vehicle job 2*

FIGURE 12 *Selector mechanism test rig used to assess cable efficiency*

Steps have been taken to ensure that all production transmissions are tested for noise compliance and shift quality once production begins in January 2004. Automated end of line testing, based on the GSQA system used during the development phase, will ensure that all relevant parameters fall within specified tolerances. Typical parameters that will be tested are given below, where tests were made under predefined operating conditions, and examples of data taken from prototype transmissions are illustrated in figures (13 to 15):

- Synchro impulse
- Out of gear force
- Double bump ratio
- Cross-gate force
- Cross gate displacement
- In-gear void size
- H-gate overlap

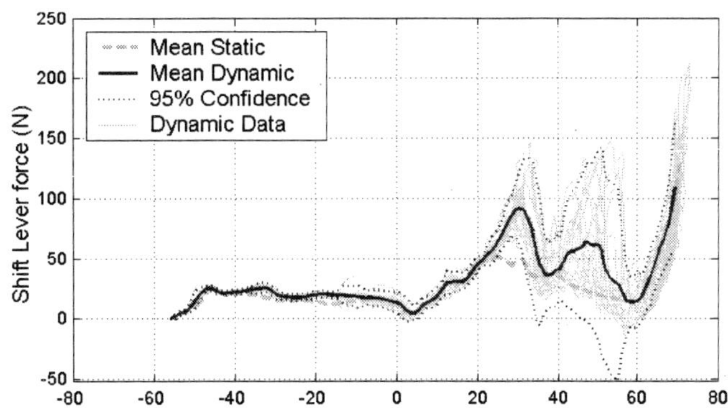

Figure 13 *Sample of measured handball force against displacement using GSQA system (CP1 4th gear selected)*

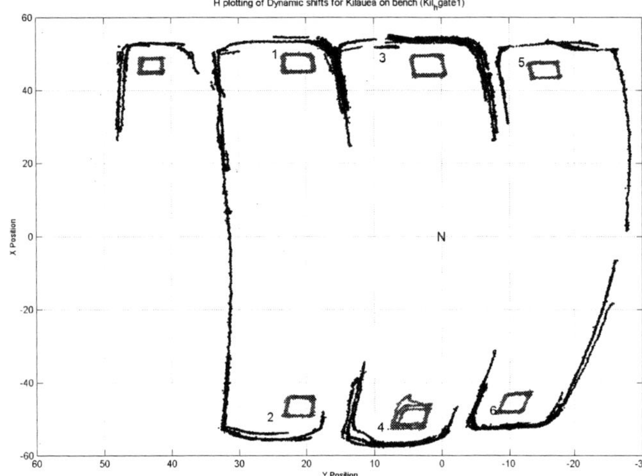

FIGURE 14 *Assessment of H-Gate overlap and In-gear void size using GSQA system (note that no H-gate overlap is present)*

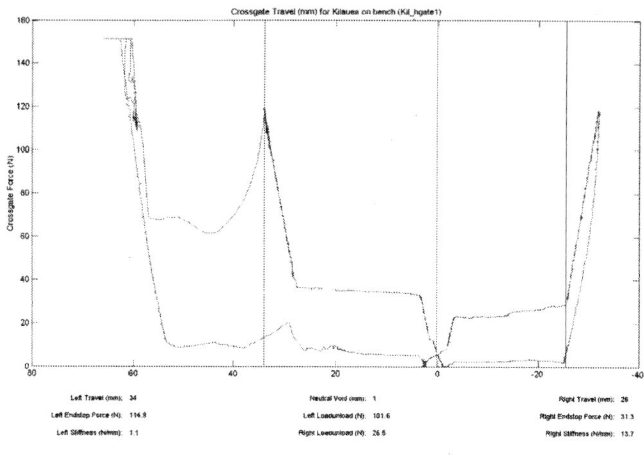

FIGURE 15 *Cross-gate displacement and force, measured using GSQA data acquisition software*

PROJECT DELIVERY AND QUALITY

The project timetable for delivery of the transmission, clutch and selector systems was extremely challenging, and included the following milestones:

- Project commencement - March 2002
- Completion of concept design – June 2002
- Completion of detailed design – September 2002
- Delivery of prototype parts – January 2003
- Prototype vehicle assembly – March 2003
- Engineering sign-off – November 2003
- Full transaxle production – December 2003
- Full vehicle production – March 2004

Strict adherence to the development schedule was essential, and so from the earliest stages of the project, weekly module meetings were convened between Ford and its 1st tier suppliers, during which all issues were discussed, and actions assigned. In addition to these, Ricardo also initiated a series of internal design reviews that, from the earliest stages, brought together all relevant disciplines to allocate resources and identify problems, thus enabling their swift resolution.

The biggest contribution to on-time/on-quality delivery was the high degree of concurrent engineering that took place between Ford, Ricardo and Ricardo's suppliers. The most obvious example of this was to be found in the use of an interactive package model, which was orchestrated by Ford and designed to co-ordinate the efforts of all engineering parties working simultaneously on the various vehicle sub-systems. The model automatically notified all parties of every evolution in the vehicle design and so all package issues, with respect to each sub-system, were identified, resolved and finalized in the vehicle layout by the 5th month of the project. This paid dividends when an early mock-up of the gear selector mechanism, based on the CAD package model, highlighted large inefficiencies due to the cable routing. Remedial measures were taken immediately, and by the time of the CP1 transaxle build phase (vehicle jobs 1, 2, and 3) the problem was resolved.

In addition to the use of advanced CAD modeling, Ricardo also implemented a second level of concurrent engineering between itself and its suppliers. This approach to component procurement was based on simultaneous concept development, supplier selection and cost-reduction exercises from the project outset. This had the benefit of bringing suppliers on board at an early stage, allowing them to allocate resources in good time, and in some cases make useful contributions to component design (as in the case of the design of the main casing for sand-casting techniques). In addition, this process also meant that suppliers produced parts for early prototypes, and were thus able to learn their lessons first hand from issues experienced throughout the project. Had a more conventional development plan been followed, whereby suppliers were selected at the engineering sign-off stage, it is certain the whole process would have been greatly elongated.

With regards to ensuring quality in the project, Ricardo had the responsibility for the selection and management of suppliers within its own development area, and so undertook to approach approximately 8 to 10 candidates for each of the likely component groups within the transmission. Each supplier was approached and asked to tender for the supply contract, based on a clear definition of what would be required. In addition, each supplier was asked to fill in and return a questionnaire, designed to establish the supplier's competence and suitability. Typical questions related to

- ISO 9000 accreditation
- "Standard and Poors" solvency rating,
- Experience in automotive component supply
- Quality processes
- Delivery performance
- Control of sub-suppliers, etc.

Thus a rating for each supplier was defined, and those found unsuitable on quality grounds were eliminated from the selection process. Where suppliers were selected, decisions were made based on cost and location, and in a number of cases, suppliers were asked to co-operate in sub-assembly procedures, before passing these components to Ricardo, for incorporation in the finished transmission.

Final assembly was undertaken at Ricardo, and the investments made reflected the short time-scale and low production runs available. In this respect, A highly skilled work force with minimal automation was opted for, as this eliminated the long lead times that were required to obtain automated equipment. In addition, the assembly process implemented a number of quality procedures such as "Takk-Time" and "Poke Yoke", to prevent errors in assembly. An example of this was the assembly-mount for the transmission casing, which featured a number of dowels designed to slot into each of the bolt-holes in the casing. If any of the bolt-holes were not drilled, this would become immediately apparent and would allow rectification before any expense was incurred in a wasted transmission build.

CONCLUSIONS

This paper presented the development-methodology and some of the analytical techniques used to bring the Ford GT transmission to fruition. Due to demanding time constraints, rigorous resource management was required, in line with best industry practice. A number of advanced procurement techniques, including concurrent engineering and supplier selection were also used. Additional time benefits were gained from the ability to accurately predict transmission behavior and durability by means of proprietary software tools and finite element analysis. This, in part, compensated for the limited access to prototype vehicles during development, while bench tests of simulated mechanical systems, such as selector mechanisms, ensured the validation of CAE analyses.

Cost and quality demands were met by means of rigorous cost-reduction exercises and attention to APQP procedures. Additional quality procedures were also implemented into the assembly routine, such as "Poke Yoke", to ensure continual improvements in process. Finally, end of line test procedures were also put in place so that every transmission assembled would be tested for compliance with noise, shift quality and functional specifications.

CONTACT

Steve Clarke, Director – Business Development, Ricardo Driveline and Transmission Systems, a Division of Ricardo UK Ltd, Southam Road, Radford Semele, Leamington Spa, Warwickshire, CV31 1FQ, UK

REFERENCES

1. Chrysler Corporation, Ford Motor Company and General Motors Corporation, "Advanced Product Quality Planning and Control Plan (APQP)", Corporate Reference Manual

2. Pugh S., "Total Design: Integrated Methods for Successful Product Engineering", Addison-Wesley Publishing Company, 1991

3. Jesse Crosse, "Engineering Accomplished", Racecar Engineering Magazine, February 2001

4. Kelly, D and Kent C, "Gear Shift Quality Improvement In Manual Transmissions Using Dynamic Modelling", FISITA 2000 World Automotive Conference Proceedings, 2000, F2000A126

5. Kelly D., Fracchia M., Harman P., "Manual Transmission Gear Shift Quality Development Using Integrated Dynamic Simulation and In-Vehicle Development Process", Global Powertrain Conference 2003 (In Press).

ACKNOWLEDGEMENTS

The authors wish to acknowledge the following parties for their assistance and contribution to this project:

- Ford Motor Company
- Roush Industries
- NTN bearings (suppliers of 4T® and ECO TOP® bearings)
- Altair (suppliers of Optistruct FEA topology software)
- Mathworks (suppliers of Matlab Simulink software)

2004-01-1261

2005 Ford GT Magnesium I/P Structure

Tim Hubbert
Lear Corporation

Xiaoming Chen, Naiyi Li and Scott Pineo
Ford Motor Company

ABSTRACT

This paper describes a new concept for a Ford GT instrument panel (IP) based on structural magnesium components, which resulted in what may be the industry's first structural IP (primary load path). Two US-patent applications are ongoing.

Design criteria included cost, corrosion protection, crashworthiness assessments, noise vibration harshness (NVH) performance, and durability. Die casting requirements included feasibility for production, coating strategy and assembly constraints.

The magnesium die-cast crosscar beam, radio box and console top help meet the vehicle weight target.The casting components use an AM60 alloy that has the necessary elongation properties required for crashworthiness.

The resulting IP design has many unique features and the flexibility present in die-casting that would not be possible using conventional steel stampings and assembly techniques. Section properties of the magnesium diecasting were easily modified to develop a structure with the necessary structural characteristics that balanced NVH, durability, stiffness, and crash performance. These targets were achieved in a timely fashion, based on an extremely short time budget.

INTRODUCTION

Magnesium, as a lightweight material, has grown quickly in automotive industry applications during the past decade. Many of the key Ford GT program directives revolved around system level specifications for vehicle dynamics and how this car would "feel" when compared to the best supercars in the world, for example, the Ferrari 360 Modena. The use of magnesium for the instrument panel structure was identified early in the concept stage of the program as a viable way to decrease vehicle weight and improve dynamics.

Several performance characteristics were identified prior to the conversion of the Ford GT crosscar beam to magnesium. First, additional design efforts were needed in the FEA model to assure the magnesium I/P beam would contribute significantly to the torsional stiffness of the vehicle. Also, the magnesium beam was considered critical to the support structure vibration characteristics of the steering column and would provide a specified natural frequency.

DESIGN PROCESS

In order to successfully execute the instrument panel (and faithfully represent the concept IP), the interior team of the Ford GT was assembled quickly and comprised a number of experienced design engineers, safety experts, and CAE analysts.

The design of the magnesium IP beam was driven by the functionality, package constraints, NVH, durability and safety requirements. The team completed the preliminary design through brainstorming, considering all requirements. The proposal, a closed section beam with rib reinforcements to meet NVH and crash safety targets, is shown in Fig. 1.

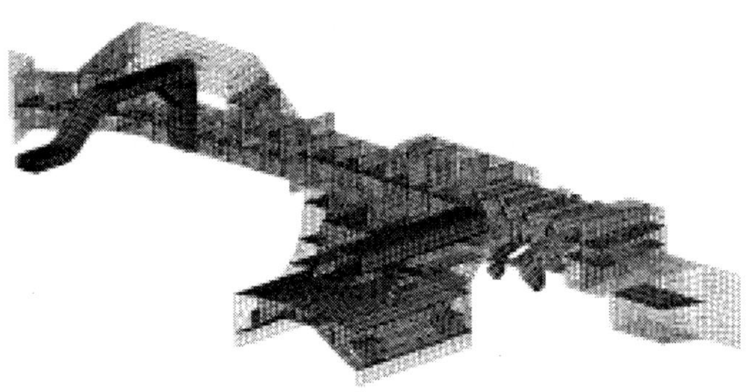

Figure 1. Design concept of the Magnesium IP

The team then applied finite element analysis as a major tool for design iterations. FEA gave directions and suggestions to modify the IP beam design to meet requirements.

Given that the class A surface of the IP was essentially established by the GT40 Concept car, the structure necessary to achieve the NVH and safety requirements needed to be contained under the pre-defined skin. In order to maximize the safety and NVH performance of the IP structure, targets for both systems were established and fed into CAE models for optimization.

The magnesium structural beam and its connections to the aluminum extrusion cowl provide load paths that are significant during crash. The I/P beam and its connections to the hinge pillar and cowl provide significant support to the crush forces generated during impact.

Safety targets included the amount of load from a particular frontal crash pulse that would need to be transferred from the front structure, into the cowl, outward from the IP structure, up through the A-pillars and toward the rear of the vehicle. It was essential that the occupant zone not be compromised during a crash event.

By running continuous CAE iterations of various IP structures and attachment methods the team would be able to predict what type of loads the IP system would be capable of handling.

The NVH targets established for the IP structure by the program were very important to ensure the driver would receive true dynamic feedback from the steering column and pedal package without unwanted system noise. By using a steering column with good initial NVH characteristics, the team wouldn't need to make up for a lack of stiffness in the steering column with the IP structure.

As the design progressed, the team identified the need for an integral load path from the center of the IP down into the tunnel. The resultant structure was designed to be an effective load path and a visible, stylish design element. Working closely with design studio personnel, the final center magnesium casting (radio box) provided an improvement in safety and NVH as well as a quality-integrated look. This center magnesium piece houses the radio, start button, passenger airbag deactivation switch and 12V power supply.

According to the vehicle chassis design, the distance between the cowl cross member and the IP beam was about 400 mm (16 inches). Two aluminum struts therefore were used to connect the two components, which provide additional load paths for frontal impact crashes and support for better NVH performance. The concept of the design is shown in Fig. 2.

Figure 2. Design concept for struts

NVH REQUIREMENTS

The team built the NVH models using HyperMesh [1]. They ran normal mode simulations using OPTISTRUCT [2] to check the steering column and the IP beam natural frequencies as well as stiffness. The simulations evaluated the design concepts, beam geometry and attachment. It was revealed that both the beam stiffness and the beam attachment to the tunnel were critical for NVH performance. We made design changes based on the FEA suggestions.

STRUCTURE INTEGRITY REQUIREMENTS

The struts connecting the cowl top cross member and the IP beam were designed as load paths to enhance occupant protection upon frontal impact. This made the IP beam a load-carrying component. It will take the load from the struts and then transfer it to the tunnel of the vehicle. The load target setting for each strut was 25 KN (5,600 lbs.) based on the energy management of the front structure. It means that the IP beam needs to take about 50 KN (11,000 lbs.) of load and maintain its integrity during a frontal impact.

The Ford GT magnesium IP beam design was a challenge because of two factors: a conventional IP beam is not typically designed as a load-carrying component, and magnesium is less ductile compared with steel.

The failure strain of the magnesium alloy AM60 varies in a large die casting component, while standard tensile coupons show about 6% elongation [4]. The IP beam had to be designed considering the material property variations.

Fig. 3 is the sub-system FEA model for the IP beam design. It has about 100,000 shell elements. The target load was applied to the two struts, while the hinge pillar, roof and the back of the tunnel were fixed. LS-DYNA [5] was used to run the analyses. Plastic strain was monitored.

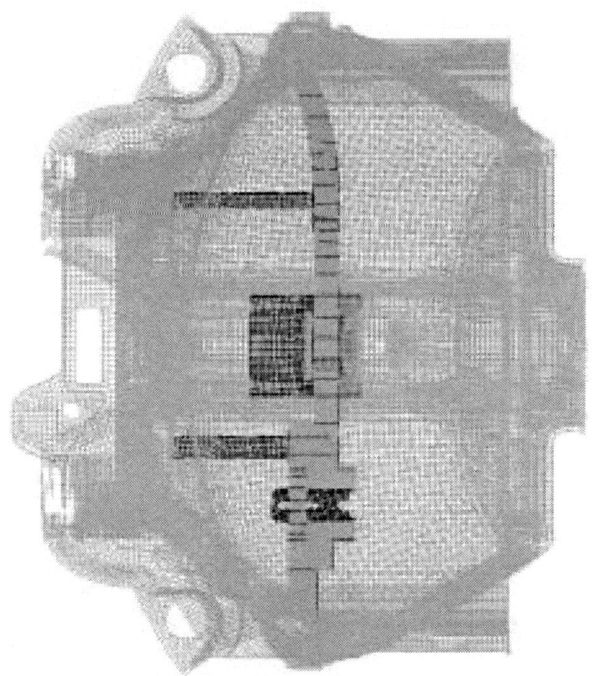

Figure 3. Sub-system FEA model

We observed the beam behavior through animations. Areas with plastic strain equal to 0.06 or above were considered failures. We established design concepts based on the findings, and beam section geometries, rib types, wall thickness and beam to body connections were defined throughout the process.

The final result showed that the magnesium IP beam was able to carry the desired load while maintaining its integrity. The simulation also provided a map based on the plastic strain contour of the beam, which later could be used as valuable information for casting process development.

Full vehicle simulations validated the magnesium IP beam designed by sub-system FEA. It was found that the beam could take the target load without losing its integrity, Fig. 4.

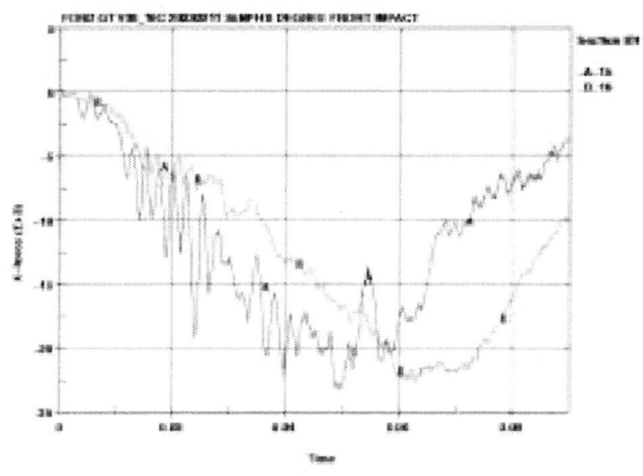

Figure 4. Loads going through the struts

EA BRACKET PACKAGE REQUIREMENT

The available travel distance for the energy absorption (knee bolster) brackets is critical to meet the femur load requirement. A die-cast part allows us to provide the EA brackets with two inches of crush distance by locally chamfering the rear bottom of the IP beam. This design maintains the beam cross-section size for stiffness and load carrying capacity and allows the EA brackets to absorb the energy that is required. The details are shown in Fig. 5.

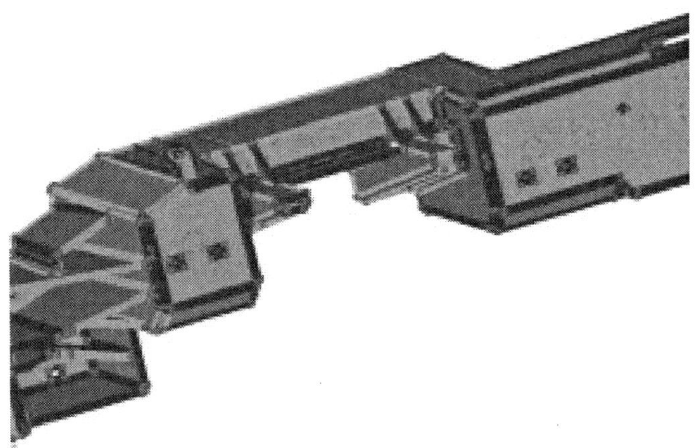

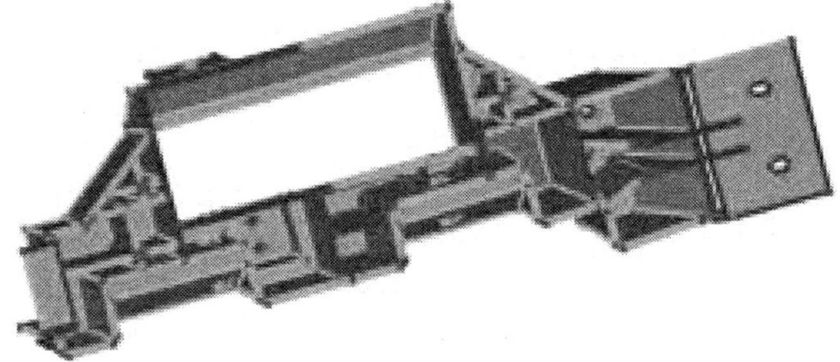

Figure 5. IP beam design for knee bolster distance

BEAM – COVER CLEARANCE REQUIREMENT

Another design consideration was to leave sufficient clearance between the IP beam and the panel cover to meet head impact requirements. The impact zone is shown in Fig. 6. A series of 10 impacts were performed in LS-DYNA. The simulation results indicate that the panel cover and the IP beam will not contact each other because of the space between them.

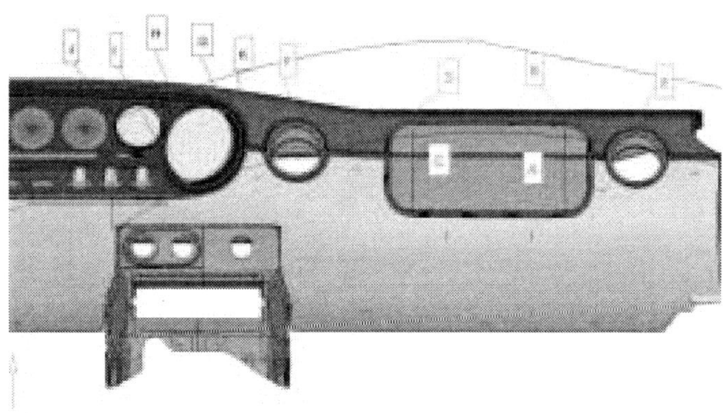

Figure 6. Impact zone definition

MANUFACTURING PROCESS

The Ford GT contains seven magnesium castings all of AM60B alloy. There are driver beam, passenger beam, radio box, console top and three cover plates as shown in Fig 7. Each casting was produced with the high pressure die casting cold-chamber process with die casting size up to 1200 ton.

Figure 7. Interior view

The design and manufacturing sequence are as follows. The design of these castings was completed using SDRC-IDEAS and FEA analysis utilizing LS-DYNA software later validated the casting design to meet its functional requirements. The die tools were prepared with gating and runners based on the tooling supplier's own experience. Due to the program's tight timeline, we could not fully use the casting simulation to support the tool fabrication.

Some results of fill and solidification analysis were shown in Figs. 8 and 9 where potential porosity may appear in the castings. Going through the casting simulation analysis, we may presume this will happen prior to production of the car and help to minimize casting defects caused by premature design.

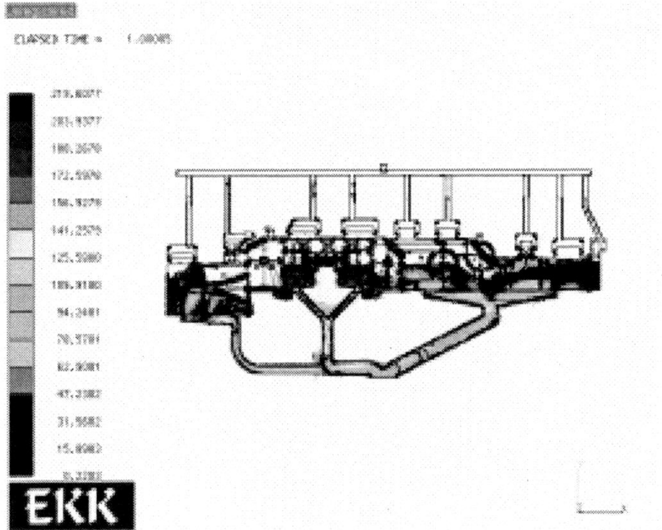

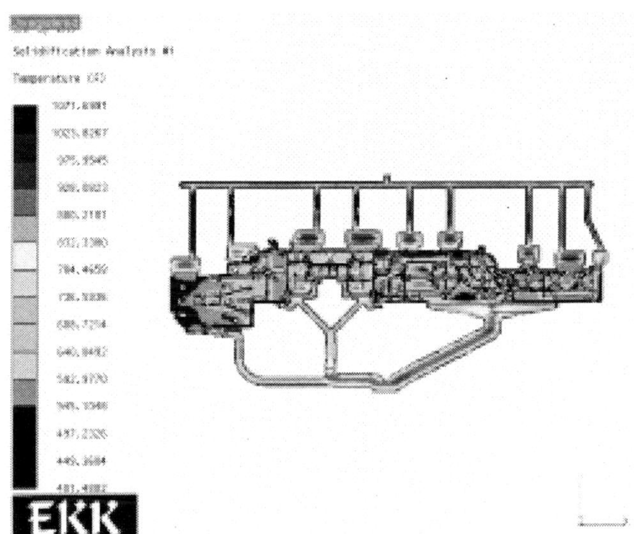

Figures 8 and 9. Solidification analysis

We also verified the mechanical properties of some smaller tensile specimens extracted from the castings, particularly the driver and passenger beams based on ASTM E575 specification. Among the properties, a critical one is the elongation where the beam has a design target of 8% for crash-sensitive zone.

As for casting quality, we used the 9 MV x-ray system at Ford to evaluate the casting per ASTM E505 specifications. Figure 10.These findings reflect the need to optimize the casting process window and the design to minimize casting defects.

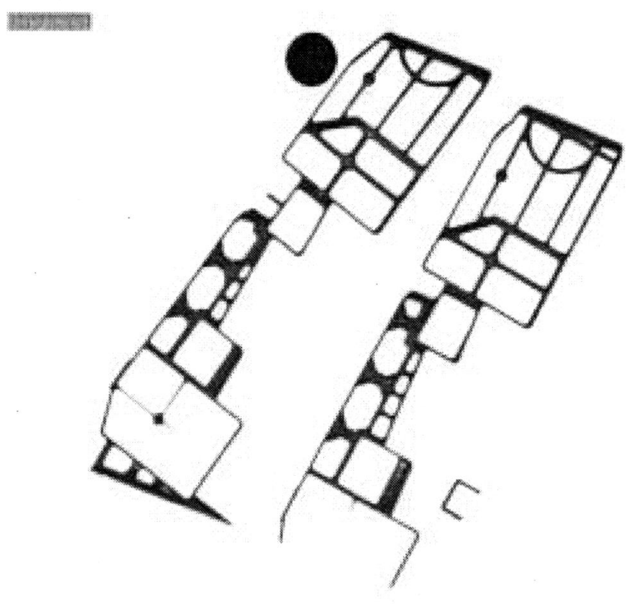

Figure 10. X-ray evaluation

DIMENSIONAL CONTROL / BUILD QUALITY

During the vehicle's development cycle, DVA's (dimensional variation analysis) showed that the dimensional capabilities of the magnesium die casting process, and of the GT IP in general, would improve IP fit-and-finish, and decrease gaps. We achieved this in several ways, including careful selection of datum structures for assembly methods, fixtures and design considerations.

ASSEMBLY

We used a two-component epoxy adhesive system for joining the cover plates and two beams as well as mechanical fasteners to aid during the curing process.

CORROSION PROTECTION

There is no general corrosion with the IP castings because they are not likely to be exposed to salt. To minimize galvanic corrosion we applied "e-coat" to the "Z" bracket to isolate the magnesium beam and the steel bracket near the aluminum A-pillar.

WEIGHT

The total weight of the seven castings in the IP assembly is 13lb. Since we don't have a carry-over item as a comparator, we can't tell exact weight saving for this design. However, in general, 30% weight saving from an aluminum version is common.

CONCLUSION

One year ago Ford Motor Company and Lear Corporation embarked upon a project to convert a portion of the Ford GT structure to magnesium utilizing the instrument panel's crosscar beam. Magnesium substitution in the automotive industry has increased steadily over the past decade. Parts such as instrument panel crosscar beams have been replaced by magnesium at a rate of 6 – 12% per year.

Ford's commitment to the use of lightweight materials throughout its entire range of vehicles has continued to increase. While the Ford GT will utilize magnesium components to lower weight to improve vehicle dynamics, Ford trucks and SUVs soon will benefit from the increased use of magnesium for body and powertrain components to improve fuel efficiency.

The use of magnesium in the IP structure of the Ford GT is a perfect way for the company to move forward with the integration of lightweight materials. By taking weight out of the structure of the car, the engineering team is better able to improve handling dynamics – so that it will accelerate faster, stop quicker, and corner with more precision and agility.

The Ford GT (figure 11) will be a collector car from the moment it is released and its successful use of magnesium in the instrument panel will be seen for years as a smart application of this lightweight metal in the name of performance and safety.

Figure 11. 2005 Ford GT

ACKNOWLEDGMENTS

Altair for the FEA component simulations. Arup for the full vehicle analysis. Jeff Webb for IP design suggestions. Para Weerappuli for Madymo information. GT magnesium IP team, personnel from Ford, Lear, Altair, ARUP, Mayflower. Ford GT program and R&AE management.

REFERENCES

1. Altair Computing, HyperMesh Version 4.1.
2. Altair Computing, Optistruct Version 4.0, 3003
3. MVSS 208 Occupant Crash Protection, Federal Motor Vehicle Safety Standard, 2000
4. X. Chen, T. A. May, J. P. Webb and D. A. Wagner, Magnesium Automotive Structure Component Design, Proceedings of 59th Annual World Magnesium Conference, PP101-106, 2002
5. Livermore Software Technology Corp. LS-DYNA Version 960 2002

HISTORICAL SAE PAPERS ON THE FORD GT-40

670065

THE FORD GT SPORTS CAR

by
ROY LUNN
FORD MOTOR COMPANY

Painting by Peter Helck, AUTOMOBILE QUARTERLY, Spring, 1964

INTRODUCTION

The origin of automobile racing cannot be established easily, but it can be assumed that it all started with the build of the second car. In studying the history of racing, it becomes apparent that the emotions that generate the desire to invent and create are synonymous with the desire to demonstrate the results of creative efforts in open competition. Conversely, the desire to win also has generated technical advances which have contributed to, and accelerated the evolution of the present-day passenger vehicle.

Although modern industrial methods have devised other systems for developing products, racing still provides a unique opportunity for the exploration of new ideas and materials. To keep abreast of competition and meet a racing schedule necessitates quick model turnover, and, because this type of work is executed by small groups, the individuals involved develop a wide versatility of skills and experience relating to total vehicle concept and development.

The major advantages of the racing program can, therefore, be summarized as follows:

- The main line product benefits from the development and usage of new ideas and materials.
- The engineers are afforded a wide range of experience in a concentrated time period.
- An excellent promotional medium is provided by demonstrating products in open competition.

It was with this foregoing basic philosophy that Ford Motor Company, in 1962, re-entered racing in the fields of stock, drag, and rallies and made plans for participation with Ford power at Indianapolis. Early in 1963, the decision was made to extend Ford's participation in competition to the highly sophisticated form of road racing known as GT sports. This arena of racing had been dominated by European manufacturers whose vehicles, over decades, had reached an extremely high state of technical development. Such famous marques as Bentley, Talbot, Lagonda, Aston Martin, Mercedes and Ferrari had figured prominently in the sport, while American efforts were left largely to individual enthusiasts.

The decision to enter this highly developed form of racing was influenced not only by the technical challenge involved but also by the fact that the cars had to be road-legal and raced mostly on commercial highways, such as the Le Mans circuit in France. Consequently, GT sports vehicles are closely allied to normal passenger vehicles and encounter all the problems of highway driving including handling, driver environment, braking, stability and safety. The main difference, however, is that under racing conditions these problems are accentuated, thus providing an excellent development ground for new techniques and innovations.

In the early stages of the program, consideration was given to acquiring an established builder of GT cars, but it was finally decided to create a unique Ford vehicle to challenge the established European supremacy in this form of racing. This paper, therefore, has been prepared to provide the automotive profession with an account of the conception of the Ford GT and its evolution leading to the victory at Le Mans in 1966.

THE TECHNICAL CHALLENGE

Like most product programs, the design and performance objectives for the Ford GT project were largely established by the status of the leading competition. It was evident, from an analysis of competitors in 1963, that top speeds in excess of 200 mph, average laps of more than 130 mph, and durability to sustain an average of more than 120 mph for 24 hours would be necessary to compete successfully at Le Mans in the ensuing years.

The racing objectives were also established. They required the cars to be potential winners in the long-distance races such as Daytona, Sebring, Spa, Nurburg Ring, Targa Florio, as well as Le Mans, and to be capable of winning the FIA World Championship for this type of vehicle. Added to these targets was the timing objective that required the cars to be racing within one year of starting the program.

Attempting to meet these objectives was a scintillating technical challenge, particularly starting from scratch; whereas, competition had reached its sophisticated product level after nearly 40 years of evolutionary development. It was therefore considered necessary to pursue a highly analytical approach to the design in its concept stage rather than rely on evolutionary development.

The magnitude of the engineering problems involved may better be appreciated by a look at the conditions that exist on a race track such as Le Mans. Figure 1 shows this famous circuit, which is made up of conventional roads that are closed to commercial traffic only for the race in June and a short practice session in April. The cars travel clockwise on this 8.3-mile track and encounter road conditions which test every aspect of a car's capabilities. In the 1966 event, speeds ranged from 215 mph on the main straight to 35 mph on the slowest corner, incurring severe braking, acceleration

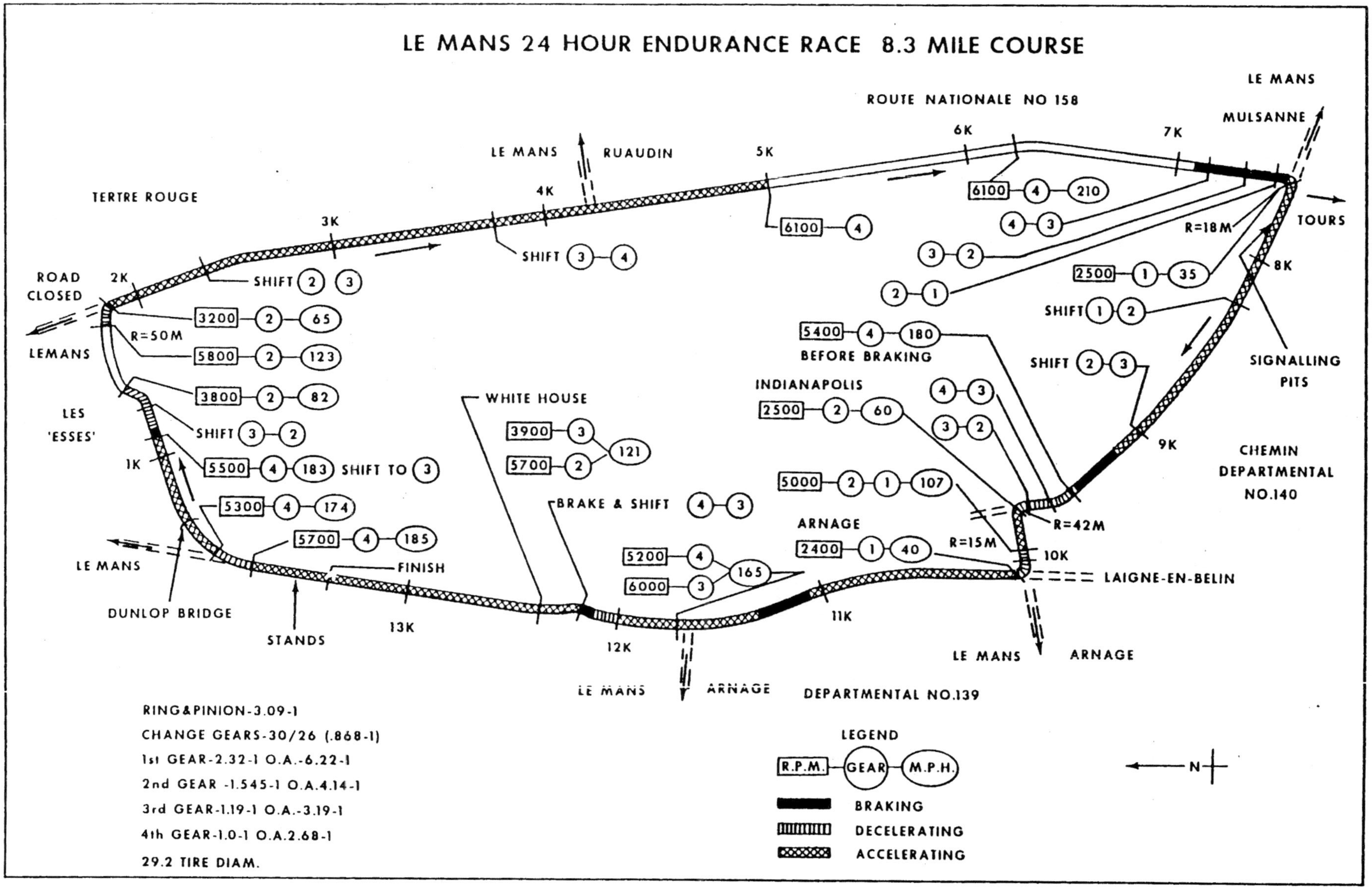

LE MANS 24 HOUR ENDURANCE RACE 8.3 MILE COURSE

LE MANS
ROUTE NATIONALE NO 158
MULSANNE
6K
7K
LE MANS
RUAUDIN
5K
TERTRE ROUGE
4K
6100 — 4 — 210
4 — 3
TOURS
3K
6100 — 4
R=18M
SHIFT 3 — 4
3 — 2
ROAD CLOSED
2K
SHIFT 2 3
2500 — 1 — 35
8K
2 — 1
3200 — 2 — 65
SHIFT 1 — 2
R=50M
5400 — 4 — 180
LEMANS
5800 — 2 — 123
BEFORE BRAKING
SHIFT 2 — 3
SIGNALLING PITS
INDIANAPOLIS
WHITE HOUSE
4 — 3
LES 'ESSES'
3800 — 2 — 82
2500 — 2 — 60
SHIFT 3 — 2
3900 — 3
3 — 2
9K
CHEMIN DEPARTMENTAL NO.140
1K
5500 — 4 — 183 SHIFT TO 3
121
5700 — 2
5000 — 2 — 1 — 107
5300 — 4 — 174
BRAKE & SHIFT 4 — 3
R=42M
ARNAGE
5700 — 4 — 185
2400 — 1 — 40
R=15M
LE MANS
10K
5200 — 4
FINISH
165
LAIGNE-EN-BELIN
6000 — 3
DUNLOP BRIDGE
11K
STANDS
13K
12K
LE MANS
ARNAGE
LE MANS
ARNAGE
DEPARTMENTAL NO.139

RING&PINION-3.09-1
CHANGE GEARS-30/26 (.868-1)
1st GEAR-2.32-1 O.A.-6.22-1
2nd GEAR -1.545-1 O.A.4.14-1
3rd GEAR-1.19-1 O.A.-3.19-1
4th GEAR-1.0-1 O.A.2.68-1
29.2 TIRE DIAM.

LEGEND
R.P.M. — GEAR — M.P.H.
BRAKING
DECELERATING
ACCELERATING

N

Figure 1

and constant shifting up and down through the gears. Other corners had to be negotiated at speeds up to 175 mph, and there was full power application for the major part of the circuit. The 24-hour duration of the race, with its night-day aspect and varied weather conditions, necessitates a fully-equipped road vehicle in every sense. For reference against the original objectives, the 1966 event was won at an average speed of 126 mph, despite considerable rain, and a new lap record was set at 142 mph.

VEHICLE CONFIGURATION AND PACKAGE

In 1962, a group from Ford Product Research and Styling areas had designed, constructed, and developed the Mustang I sports car (SAE Paper No. 611F). These same personnel were then assigned to the GT program, and the information which evolved from the Mustang I study served as a starting datum for concept work on the GT sports car.

The initial problem was to select a vehicle configuration which was likely to meet the performance objectives and could be packaged within the FIA rule limitations. The Mustang I exercise had clearly shown the advantages of using a midship engine configuration to attain a low, sleek vehicle silhouette. This arrangement also offered excellent weight distribution characteristics and had been well-proven in other spheres of racing, such as Formula I. It was therefore decided to pursue this same configuration for the GT car.

Initial package studies showed that the essential components could be installed in a vehicle silhouette of 156 inches long, 40 inches high (hence the name GT 40), and 95-inch wheelbase, and still meet the FIA requirements. The

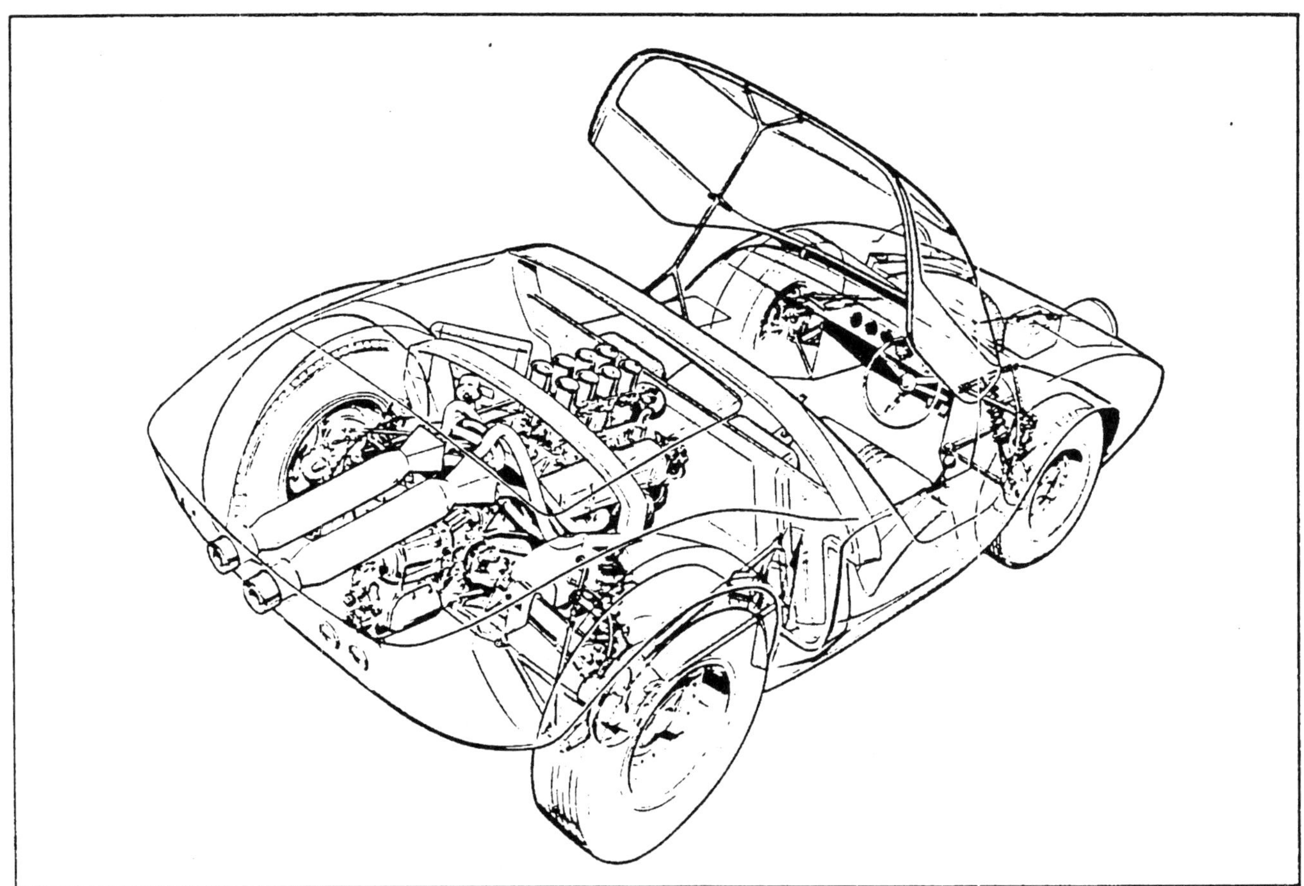

Figure 2 – Original Concept Sketch of GT 40

over - all arrangement included a forward - hinged canopy top; twin radiators located behind the seats with side-ducting; the 256 CID V-8 engine developed for Indianapolis; cross-over tuned exhausts; forward-located spare wheel, oil tank and battery; fixed seats and movable controls; side-sill gas tanks; and, because no suitable transaxle existed within the Company, a proprietary vendor - developed unit was selected. The original sketch showing this combination of ingredients is shown in Figure 2.

Concurrently with package development, a full - size clay was constructed for over - all shape appraisal. The essential requirement was to encompass the basic mechanical ingredients and meet the FIA rule limitations. With these exceptions, however, the choice of shape was largely determined by what seemed right at that time as there was no previous knowledge of road car forms developed for speeds in excess of 200 mph. The result of this original shape study is shown in Figure 3.

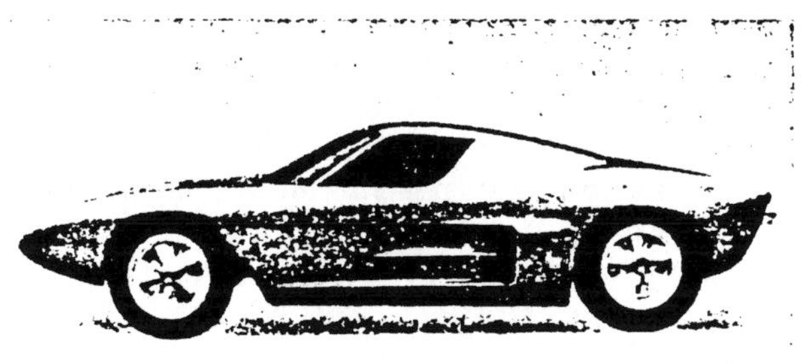

Figure 3 – First Concept Clay Model of GT 40

Subsequent analysis of side radiators showed heat dissipation to be marginal, and a forward-located unit was therefore adopted. The hinged canopy was also dropped in favor of two separate doors in order to clearly meet the FIA rule requirements.

Having arrived at a basic configuration and initial shape, an analysis program was then planned for the following areas:

- Aerodynamics
- Engine
- Transaxle — Driveshafts
- Body
- Suspension — Steering — Brakes — Wheels
- Interior — Driver Environment
- Fuel System

AERODYNAMICS

It was evident from the outset of the project that aerodynamics would play a major part in the program. With the exception of land-speed record cars, no vehicle had been developed to travel at speeds in excess of 200 mph on normal highways. The speeds involved were greater than the take-off speed of most aircraft, but, conversely, the main problem was to keep the vehicle on the ground.

Following initial package and shape studies, a 3/8 aerodynamic model was constructed, and a series of tests were carried out at the University of Maryland wind tunnel (Figure 4). Early tests showed that, although the drag factor was satisfactory, the lift at 200 mph was over half the weight of the vehicle. Subsequent tests with variations of nose height showed the low nose to have some advantage, but lifts were still totally unacceptable. The major improvement came with the addition of "spoilers" under the front end which not only reduced the lift to an acceptable standard, but, quite surprisingly, also reduced drag. A selective summary of these early test figures is shown in Figure 5. It should be remembered that these tests were conducted with a 3/8 model with equivalent speed readings of 125 mph. Results, therefore, had to be extrapolated to 200 mph, and ground effects could not be recorded.

Figure 4 – 3/8 Model in Maryland Wind Tunnel

TABLE OF LIFTS AND DRAGS IN POUNDS

Configuration	Front End Low-Positioned Spoiler	Drag	Lift Front Axle	Lift Rear Axle	15° Yaw Drag	15° Yaw Lift Front Axle	15° Yaw Lift Rear Axle
Basic Car	None	503	528	168	590	768	384
High Nose Shape	None	519	540	108	614	844	362
Low Nose Shape	None	507	445	199	596	704	422
Low Nose Shape	3.50 deep Flat Faired	488	236	272	591	309	343

Figure 5

Shown in Figure 6 are the total drag and rolling resistance curves plotted against available horsepower. This shows that the car should reach approximately 210 mph. In actual fact, the original GT 40's could only reach 197 mph in still air, although they did exceed 200 mph when passing other cars. The reason for this discrepancy was established when the actual prototype was tested in a full-size wind tunnel. It was found that 76 of the 350 horsepower available were being absorbed in internal ducting such as radiators, brake ducts, engine air, and interior ventilation; whereas, only 30 hp had been allowed for these items in the original calculations. Another item which did not show up in these early wind tunnel tests was the aerostability problem, which will be discussed later in the paper.

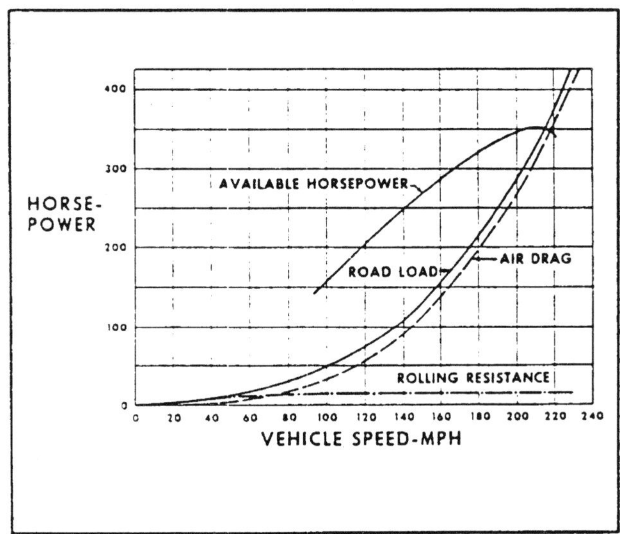

Figure 6

ENGINE

As previously mentioned, the engine selected for the GT 40 was the 4.2 liter (256 CID) unit that had been developed by Ford Motor Company for the 1963 Indianapolis Race. It was derived from the 289 Fairlane engine but included the use of aluminum block and heads, and a dry sump oil system, but, unlike the present Ford double overhead cam Indianapolis engine, still retained push rods. To adapt these units for road racing required detuning to run on commercial pump fuel; addition of full-sized alternator and starter systems; changes to the scavenge system for greater variations of speed and cornering; providing an induction system with greater flexibility for road use in adverse climatic conditions; and general detail changes to suit the package installation. These engines gave approximately 350 hp in their detuned state for long-distance races.

TRANSAXLE

The vendor-developed transaxle was packaged into the concept despite its disadvantage of having only four speeds and non-synchromesh engagement. This unit had been used previously on lightweight vehicles in sprint events, but analysis showed that it should be capable of handling the GT 40 power requirements. In addition, it was the only commercially-available unit that would meet the timing objectives.

Figure 7 shows the engine and transaxle combination, and Figure 8 shows the unit being installed in the vehicle with the cross-over-tuned exhaust system.

Figure 7 – 256 CID Engine and vendor-developed Transaxle – the Power Train Used in the Original GT 40's

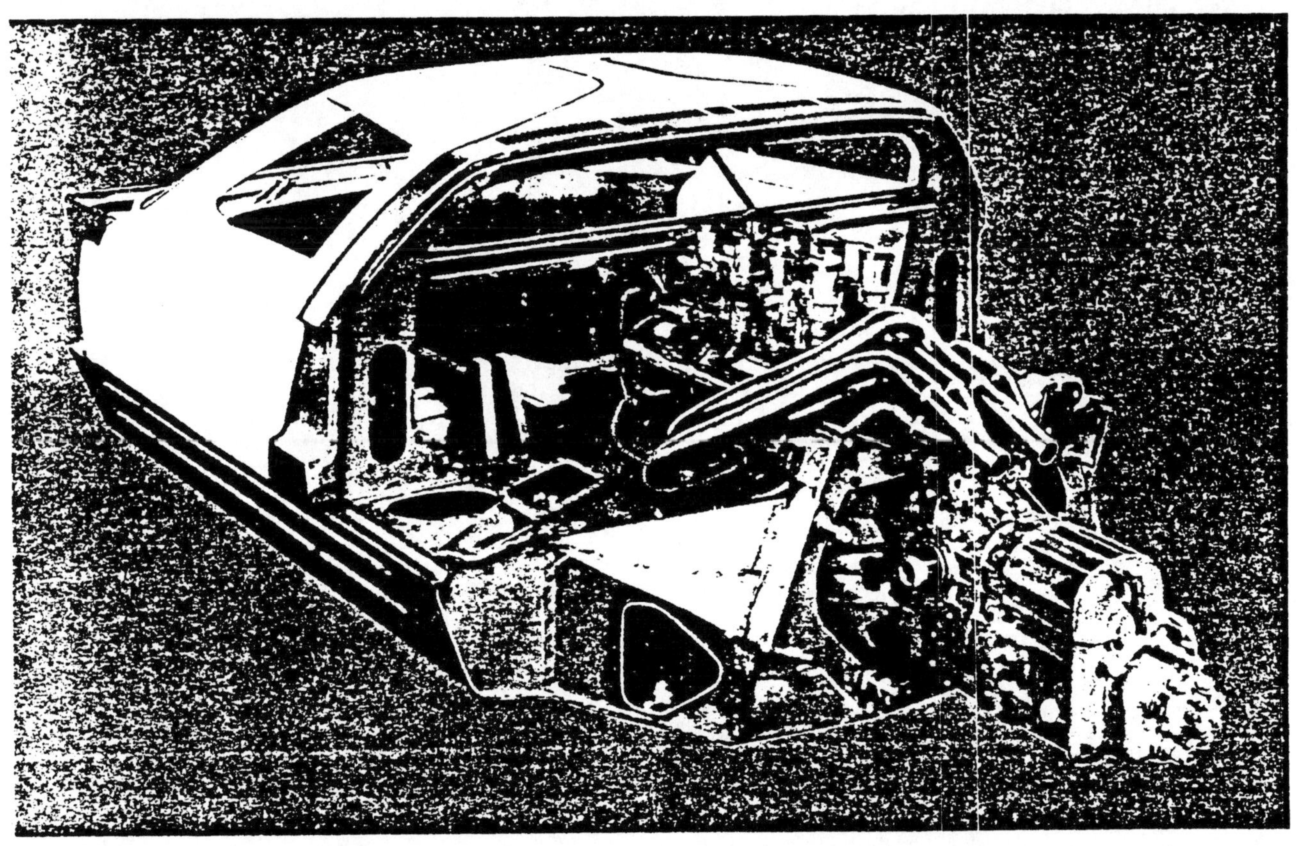

Figure 8 – Installation of 256 Engine Showing Tuned Exhaust System

The driveshafts were originally planned with single Cardan universal joints at the outboard end and pot joints inboard. Rubber couplings were later selected for the inboard end, mainly in an attempt to dampen out harshness and improve general driveline durability.

BODY

It was elected to use a thin sheet steel (.024"-.028") construction to avoid lengthy development of exotic lightweight materials. The strength-carrying structure consisted of a unitized underbody with torque box side sills to house the fuel cells, two main bulkheads, a roof section, and end structures to pick up suspension mountings. Front and rear substructures were attached to provide for body support, spare wheel, radiator, and battery mounting, and to give supports for the quick-lift jacks. The doors were cut extensively into the roof to provide reasonable entry and exit and, together with end sections and rocker panels, were made of hand-laminated fiberglass materials.

Great care was taken to design all fittings flush with the body panels, including the glass sections which were installed by adhesive techniques.

Figure 9

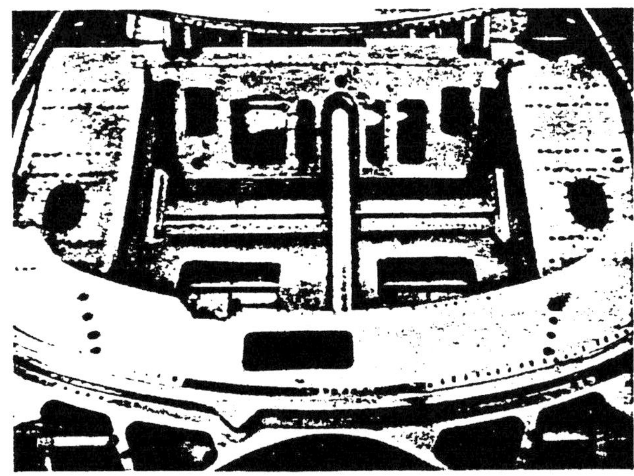

Figure 10

Figure 11

The use of steel sheet for the structure allowed normal methods of welding and brazing in the fabrication. Projection welding was used extensively because of the many blind sections in the structural members. The resulting structure provided an extremely strong unit, giving over 10,000 ft./lb. per degree in torsional rigidity. Figures 9 through 14 show the body structure in the process of fabrication and assembly.

Figure 12

SUSPENSION, STEERING, BRAKES, AND WHEELS

A number of factors governed design of the suspension units. The package size imposed space limitations; the lightweight structure required spreading the attachment points to minimize point loadings; the high-speed aerodynamic tests indicated the desired use of "anti" features; the units required adjustability to suit the varying circuits; and the resulting balance of compromise still had to provide for excellent road-handling characteristics.

The front suspension was designed as a double "A" frame, with the cast magnesium upright supporting the live wheel spindle and

Figure 13

Figure 14

the Girling aluminum brake caliper. The foot well and the position of the spare wheel necessitated an unusually short top arm. The support axes of the "A" frames were arranged to provide an anti-dive feature of approximately 30 per cent.

The rear suspension used double-trailing links from the main bulkhead and transverse links comprising a top strut and an inverted lower "A" frame. The angling of the "A" frames to the magnesium upright casting, combined with the arrangement of linkage geometry, provided anti-lift and anti-squat features of approximately 30 per cent.

These multi-link suspensions presented a problem in establishing wheel geometry. Extensive use of the computer was required with so many links moving in different planes and on canted axes. Once the basic configuration of suspension linkage had been established, a computer program was formulated that took into account all the factors involved. Curves could then be plotted in a matter of a few hours to meet a given condition — a process which speeded up the design period and aided the balance of compromise involved.

A rack and pinion was selected for the steering system, mainly because it was particularly suitable for the package conditions involved. The rack had a ratio of 16:1 which

Figure 15 — Front Suspension and Steering Assembly

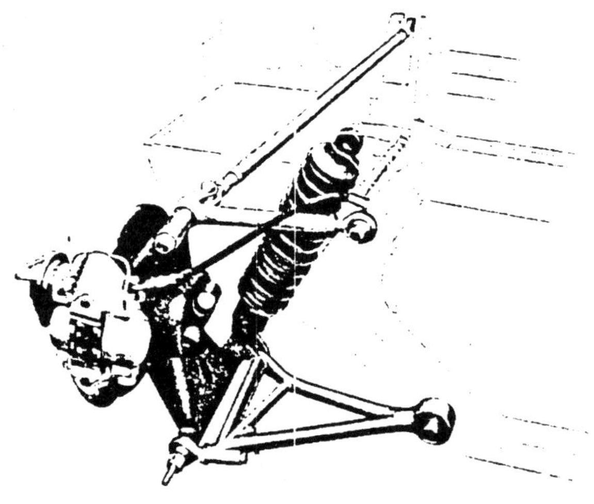

Figure 16 — Rear Suspension Assembly

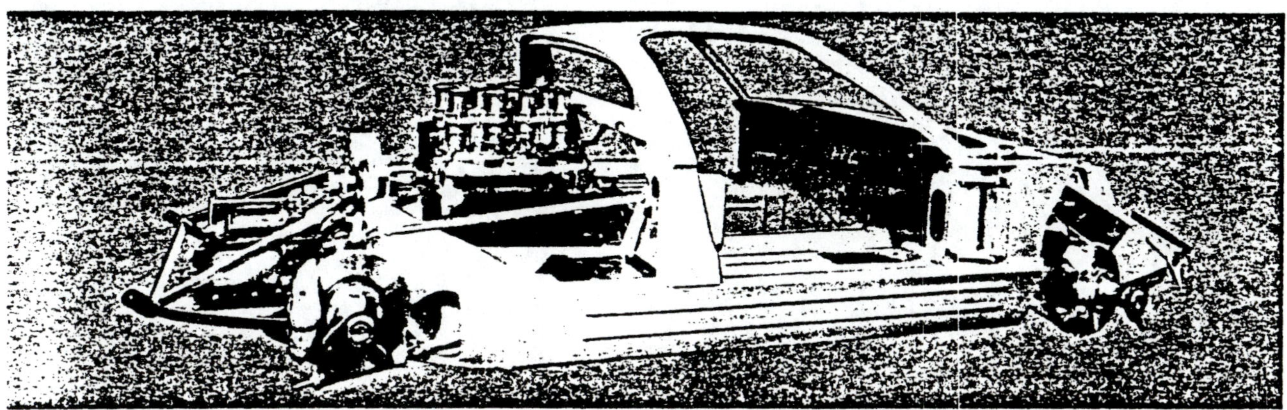

Figure 17 — Basic Structure with Suspensions and Engine Installed

in turn gave an over-all ratio of 2-1/4 turns of the steering wheel from lock-to-lock.

Girling CR and BR racing calipers were used front and rear, respectively, with solid cast iron discs, which were 11-1/2 x 1/2-inch thick. A dual master cylinder was employed for separate front and rear systems which incorporated a balance mechanism for adjustment of braking distribution.

Cast magnesium wheels were originally specified, but development problems precluded their use on the first cars. Prototypes were therefore fitted with wire wheels with alloy rims of 15-inch diameter with a 6-1/2-inch wide front rim and an 8-inch wide rear rim.

Figures 15 and 16 show the front and rear suspension units, and Figure 17 shows these units in an over-all context with the chassis.

INTERIOR — DRIVER ENVIRONMENT

Driver environment was a major consideration as long-distance races require maximum driver concentration for periods of up to four hours. An interior buck was constructed as a physical aid in developing seating conditions and to determine optimum positioning of instrumentation and controls.

The fixed-seat, movable-pedal concept was carried over from the Mustang I project. This arrangement offered structural advantages and provided snug support around the driver to help prevent fatigue from high-speed cornering effects. A nylon netting was used for the basic support medium and was covered with a pad containing ventilation holes to help evaporate driver perspiration. The pedals were mounted on a cast alloy member which could be adjusted for variation in driver size (Figure 18).

Figure 18 — Adjustable Pedal Mechanism

Instruments were positioned so that their faces pointed directly at the driver in order to minimize distortions and reflections. All switches and controls were located and formed so that they could be reached easily and recognized visually or by touch. Flow-through ventilation was provided, together with full protection from adverse weather conditions.

The general arrangement of the interior is shown in Figure 19.

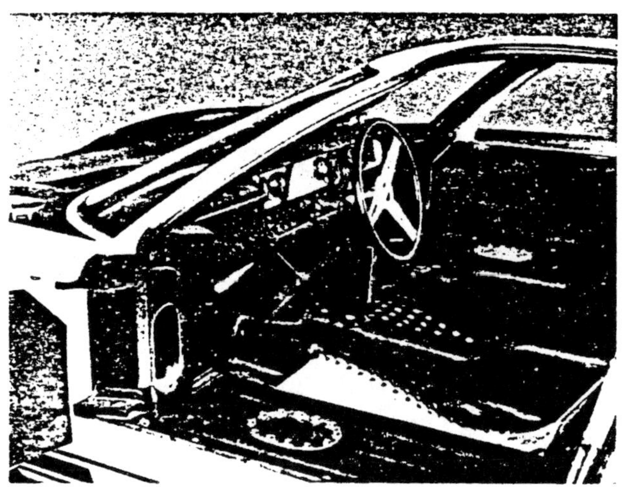

Figure 19 — Interior Showing Fixed Seat Arrangement. Passenger Seat Trim Removed to Show Nylon Net Support.

Figure 20 — Structure Being Prepared for Fuel Cell Installation

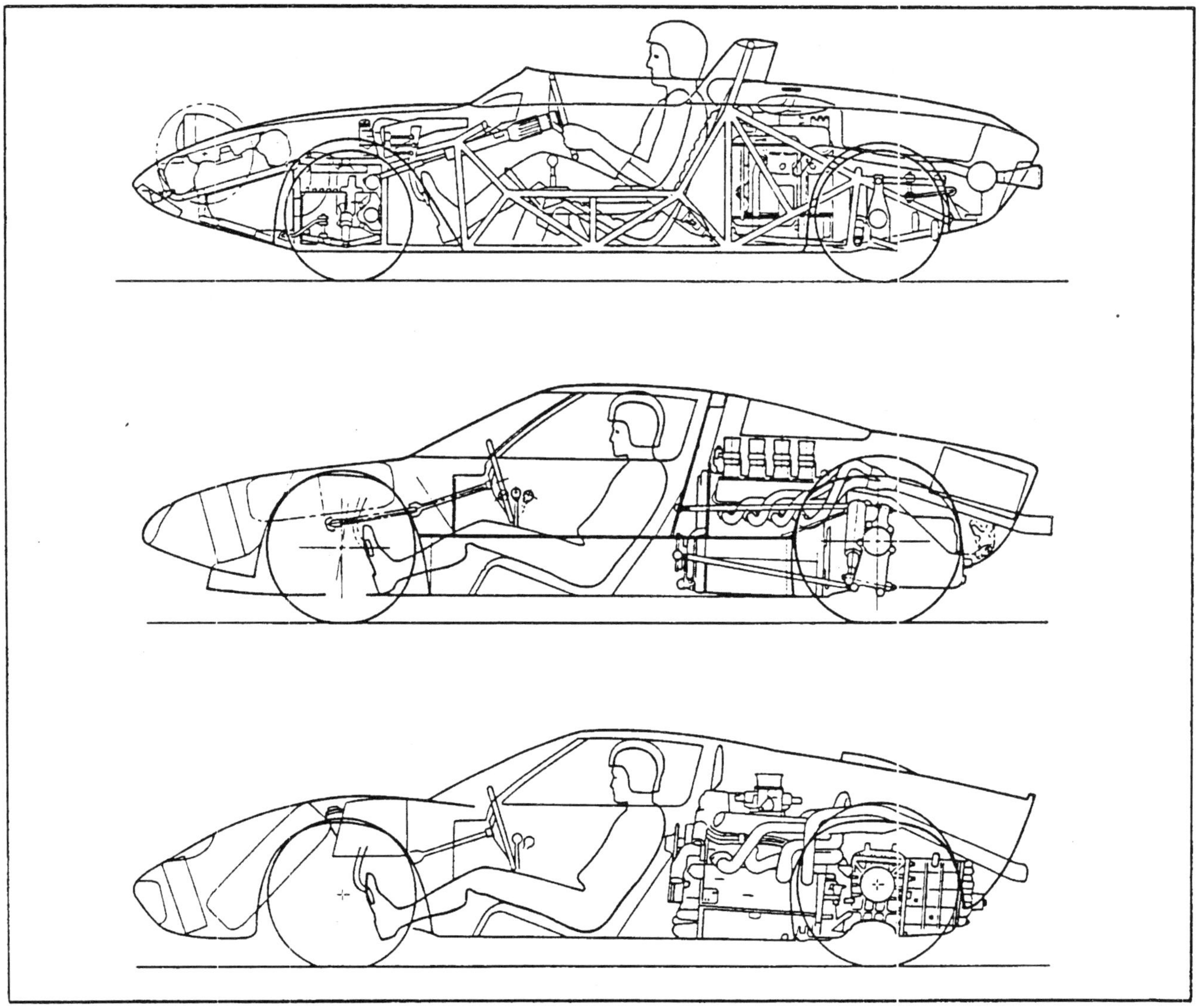

Figure 21 — Outline Packages Showing Evolution of GT 40 from the Mustang I. The Later Development of the MK II Package is Shown for Reference

FUEL SYSTEM

To contain the allowable 42 gallons of fuel in this small package, provide for rapid filling, devise a means of picking up the fuel, and provide adequate driver safety was a study within itself. The arrangement selected was two separate tank systems in the side sills, each with its own filler cap and fuel pickup box. These separate systems were designed with individual electric pumps feeding a common supply pipe to the carburetors. Provision was also made in one tank for a reserve pickup unit. The steel shell of the tanks was, of course, part of the main structure. In these were fitted neoprene bags to aid in crash safety. Baffling was attained by means of a plate supported from the top inspection cover. A fuel cell is shown prior to assembly, in Figure 20.

The findings and effects of each of the specialized studies were continuously reflected in the package layout and full-size clay model. The evolution of the package can be seen in Figure 21, which shows the Mustang I as a datum, the GT 40 package, and the later MK II layout is shown for reference.

PROTOTYPE BUILD

The design and analytical studies were completed during the summer of 1963, together with a clay model reflecting the package changes (Figure 22). The problem was then how and where to execute the final design build and development.

It was finally decided to execute this phase of the program in Europe, since many of the proprietary components were readily available in this area, as were experienced craftsmen in this field of racing. An arrangement was made with the Lola Company to use their resources and facility for one year, as they already had some experience in GT sports cars with a midship engine configuration. In forming this alliance, we were also able to use one of the Lola prototypes for the installation and development of the Ford suspension and driveline components.

In September of 1963, the center of activity was therefore moved from Dearborn to England, together with a nucleus of Ford engineers, car layouts, power pack components, and full-size models.

Component testing was completed by the end of November, 1963, and the remainder of that winter was spent in detailing and procuring items for the first prototype builds. The first GT 40 car was completed on April 1, 1964, some eleven months after putting pencil to paper in Dearborn. This car is shown in Figures 23 and 24. A second vehicle was completed ten days later, and hectic preparations were made to get both vehicles to the Le Mans practice on April 16. Bad weather conditions in England prevented any serious testing and the cars had an aggregate of only four hours running time with no high-speed experience before being shipped to France. The first day of practice also proved to be rain drenched and after very few laps, the first car was totally wrecked on the Mulsanne Straight when it left the road at over 150 mph. The second vehicle also experienced trouble and suffered a minor collision. Luckily, both drivers were unharmed,

Figure 22 – Developed Model of Original GT 40

Figure 23 — Original GT 40 Prototype

Figure 24 — Original GT 40 Engine Compartment

but obviously some stability phenomenon existed that had not been apparent during the design analytical phase. The problem and the solution were found within one week after returning to England, where further testing was carried out at the MIRA proving ground. The fault was found to have been an aerostability condition which caused a rotary motion of the rear end of the vehicle comparable to that of an arrow without feathers. The motion had increased with speed and, accentuated by the wet track, eventually resulted in rear end breakaway. Subsequently, it was found that the adaptation of a rear end "spoiler" not only had the effect of putting feathers on an arrow, but also slightly reduced drag. Apparently, the "spoiler" creates an airtail which artificially increases the vehicle's aspect ratio and moves the center of pressure rearwards. It also increases the adhesion of the rear wheels and, surprisingly, the effect of this small addition could be felt down to 70 mph.

Figure 25 — Prototype at Nurburg Ring 1964 with "Duck-Tail Spoiler" Added to Rear End

The second car from the Le Mans practice was modified by the addition of the "spoiler" (Figure 25) and was rebuilt in readiness for the GT 40's first race outing at Nurburg Ring on May 31, 1964. The car performed most favorably in practice and qualified second only to the fastest Ferrari. It also ran second in this 1000 Kilometer race in the early hours but retired after 2-1/2 hours. The reason for the retirement was a suspension bracket failure because of an incorrect welding process, but when the vehicle was examined, there were several other areas showing distress and near failure. The outing was, therefore, most successful as a development exercise, and the lessons learned were quickly incorporated in the three vehicles being built for the Le Mans race in mid-June, 1964. These vehicles were completed and weighed in at Le Mans scrutineering at 1960 pounds, less driver and fuel.

In practice, the cars qualified second, fourth, and ninth. During the race, one car held the lead for the early hours before retiring with a transmission failure. The second car retired after five hours with a broken fuel line, and the third car retired after 13-1/2 hours with transaxle problems but not before establishing an all-time lap record.

Every attempt was made to correct the transaxle problems within the limited time available before the next race at Rheims, France, on July 5, 1964. Again, the cars led the race in the early hours, set new lap records, but all retired with transaxle failures. In addition, the nature of this circuit showed insufficient cooling of the brake discs which remained red hot during the entire time the cars were running.

The GT 40's first season of racing in 1964, therefore, showed seven starts in major events with no finishes. The cars had demonstrated that they met the performance objectives but failed badly on durability aspects. The winter of 1964 was devoted to detail preparation of the cars for the 1965 season and at this stage, the responsibility for racing the vehicles was given to the Shelby-American racing team. Twenty-one modifications were executed on the transaxles, the rubber driveshafts were replaced with Dana couplings, and the decision was made to install standard 289 C.I.D. cast iron engines, using wet sump lubrication. The original cast wheels were also installed and increased to 8-inch front and 9-1/2-inch rear rims. Two of these cars made their first appearance in the 1965 season at the Daytona 2000 Kilometer Race on February 28, 1965. They finished first and third in this event, setting an average speed record of 99.9 mph for the distance in 12 hours and 20 minutes (Figure 26). Two vehicles also were entered in the Sebring Race in March, 1965, and finished second over-all and first in class, once more demonstrating that a fair degree of durability had been attained. These cars were raced by the Company once more in 1965 at Le Mans, but without success.

The decision was then made to manufacture 50 of these cars in order to qualify them for the production sports car category. These cars were completed in the 1965 period after detailed changes and the adoption of the 5-speed ZF transaxle. These GT 40's were sold to the public and, in the hands of private race teams and individuals, won the World Championship for production sports cars in 1966.

Figure 26 — GT 40's First Win at Daytona 1965

MARK II PROGRAM

In the fall of 1964, the engineering team relocated in Dearborn and started operations at Kar-Kraft, a Ford contracted facility. This team continued engineering on the GT 40 and also started a new experimental vehicle project.

The 1964 season had shown the prototype GT 40's were currently competitive on performance factors but lacked durability. Although work was progressing on correcting durability problems, it was obvious that the GT 40 performance, in the fast-moving racing field, would soon be outmoded. The problem was, therefore, how to get an improved power-to-weight factor and at the same time achieve a high durability level. The alternatives were to generate more power from the 289 C.I.D. series engine or adapt the 427 C.I.D. engine which had been developed for stock car racing. This latter approach would also involve the development of a unique transaxle to handle the higher power. The other indeterminates were whether the additional weight (some 250 pounds) for the larger engine and heavier transaxle and driveline would unduly deteriorate handling and accentuate braking problems. It was decided, however, to explore this approach by constructing a test vehicle and physically evaluating its performance. The program was initiated in the winter of 1964 and was designated the MK II project. At the outset, it should be emphasized that the exercise was intended to generate information for a future model, and there was no intention of racing the car.

Package studies showed the 427 C.I.D. engine could be accommodated in the GT 40 basic structure by modifying the seating position and rear bulkhead members (Figures 27 and 28). The basic suspension units were unchanged, but provision was made for 8-inch wide cast magnesium front wheels and 9-1/2-inch rear wheels. Housing of the wider spare wheel necessitated revising its position, and the new front end arrangement made provision for a remote engine oil tank on the bulkhead and a larger radiator (Figure 29).

A major problem was to generate a transaxle unit which would handle the 427 C.I.D. power and the extra weight of the vehicle. For expediency the gear cluster from the conventional 427 C.I.D. driveline was used but with completely new housings and axle unit. This approach resulted in a heavier and less efficient arrangement than a direct transfer box, but had the advantage of using developed components. The housings were designed in magnesium, and a pair of quick-change gears

Figure 27 – Installation of 427 Engine in MK II Chassis

Figure 28 — Original MK II Engine Compartment

transmitted the power to the pinion shaft. The resulting over-all package from these changes required new front and rear structures and body shells.

The first experimental MK II vehicle was completed during April, 1965, and was evaluated on the 5-mile oval at Ford's Michigan Proving Ground (Figure 30). After only a few hours of tailoring, the car lapped this circuit at an average speed of 201-1/2 mph and exceeded 210 mph on the straight-away. Subsequent testing on road circuits showed that handling had deteriorated only slightly. From the results of these tests, it was calculated that this vehicle should be capable of lapping the Le Mans circuit in 3 min. 30 sec. to 3 min. 35 sec. without exceeding 6200 rpm. If these

Figure 29 — Original MK II Front End Arrangement

Figure 30 – Original MK II Prototype with Long Nose Configuration

lap times could be realized at this relatively low engine rpm, the car would obviously have high potential to win at Le Mans. The decision was made, therefore, to attempt to run two of these experimental cars in the 1965 Le Mans event as an exploratory exercise. This decision was made at the end of April, and the cars would therefore be going to the event without the benefit of the practice week-end.

In the ensuing five weeks, the first car underwent initial testing and rebuild, and a second car was hurriedly constructed. The second car actually arrived at Le Mans without even having turned a wheel. Having missed the April practice, the first evening of pre-race practice was spent in tailoring the cars to the circuit. On the second evening, the car that had never turned a wheel before arriving at the

Figure 31 – Two Original MK II Cars at Le Mans 1965

track set an all-time record lap of 3 min. 33 sec. — an average of 141 mph.

One car qualified first, and when the race started on Saturday, both cars went out ahead of the field and comfortably lapped at 3 min. 40 sec. without exceeding 6000 rpm (Figure 31). Unfortunately, hurried preparation resulted in the cars being retired after two and seven hours, respectively, with non-fundamental driveline problems. One car had a speck of sand in the clutch slave cylinder which caused the piston to stick and generate heat at the throw-out bearing. The heat, in turn, softened an oil retaining ring in the axle ultimately resulting in loss of oil. The second car broke a gear which had been incorrectly drilled. The cars, however, achieved their purpose of establishing the capability of the engine-driveline combination. The potential indicated in this initial experimental outing resulted in the 1966 program being based on the MK II vehicle.

The following chart shows the MK II power-to-weight factor compared to the original GT 40 and the production version. Vehicle weights are quoted, less fuel and driver.

	Vehicle Weight	HP	HP/Lb.
MK II	2400 Lbs.	485	.202
Production GT 40	2150 Lbs.	385	.179
Original GT 40 Prototype	1960 Lbs.	350	.179

In preparation for 1966, a concentrated vehicle development program was planned using the Daytona, Sebring, and Riverside tracks. In addition, specialized component developments were initiated on items such as engine (SAE Papers No. 670066, 670067) ignition and electrical system (SAE Paper No. 670068), transaxle (SAE Paper No. 670069), driveshafts and brakes (SAE Paper No. 670070). Although some fundamental changes emerged from this development program, the main emphasis was on refinement to establish durability rather than improve performance. The final engine-driveline arrangement is shown in Figure 32.

A major contribution to speeding up development was originated by the Ford engine and transmission engineers. They evolved a dynamometer which could run the engine and

Figure 32 — Developed Engine and Driveline Assembly

driveline units under simulated road conditions that had been recorded on tape in an instrumented vehicle (SAE Paper No. 670071). This device allowed component testing to proceed independently of vehicle availability and climatic conditions.

Major changes that resulted from testing and development included:

- New shorter nose configuration to save weight and improve aerodynamics.
- Addition of external rear brake scoops.
- Higher efficiency radiators.
- Strengthened chassis brackets for durability.
- Live rear hubs for improved durability.
- Internal scavenge pump to minimize vulnerability and save weight.
- Generally improved ducting to radiators, carburetors and brakes.
- Crossover fuel system with a single filler neck.
- Ventilated brake discs to improve durability.
- Quick-change brake disc design to facilitate changes during pit stops.

All of these changes were incorporated in the vehicles that made their first appearance at the Daytona 24-Hour Race on February 5-6,

Figure 33 — Final Layout of Components for the MK II

Figure 34 — Winning MK II at Daytona 1966

1966 (Figure 33). The MK II cars virtually led the race all the way, finishing first, second, and third for their first victory (Figure 34). It is interesting to note that in 1965 when this race was approximately at the half-way point, the winning GT 40 averaged 99.9 mph. In 1966, the event was of 24 hours duration and the winning average was 109 mph. This indicates the fast-moving nature of this field of competition.

The second race in 1966 was the Sebring event, where MK II's finished first and second, setting new distance and lap records, and a GT 40 finished third over-all. The car that finished first was an open version of the MK II with an aluminum underbody that was designated the XI (Figure 35).

One car was entered in the Spa 1000 Kilometer Race and finished second.

After attending the practice session in April, eight cars were prepared for the Le Mans event that took place on June 18-19, 1966. MK II's qualified in the first four places and set a new lap record of 3 min. 31 sec. or 142 mph. The race took place in cloudy weather with intermittent showers during the 24 hours. MK II's finished first, second, and third and, despite the weather conditions, established a new record for the 24 hours of 126 mph (previous best was 122 mph on a dry track), (Figure 36).

As a result of winning Daytona, Sebring, Le Mans, and finishing second at Spa, the MK II also won the World Championship for prototype cars in 1966, thereby meeting the original objectives set forth in 1963.

Figure 35 – XI Car Winning Sebring 1966

Figure 36 – The Three MK II's Taking Checkered Flag at Le Mans 1966

SUMMARY

It required three years, new technology, facilities and financial backing to take the Ford GT from the drawing board to the checkered flag at Le Mans. But above all, it required personal effort, ingenuity, skill, incentive and courage on the part of individuals to bring the project to fruition. The racing achievements will undoubtedly become a notation in the record book, but the contributions to automotive technology should provide a lasting satisfaction to all those who participated in the program.

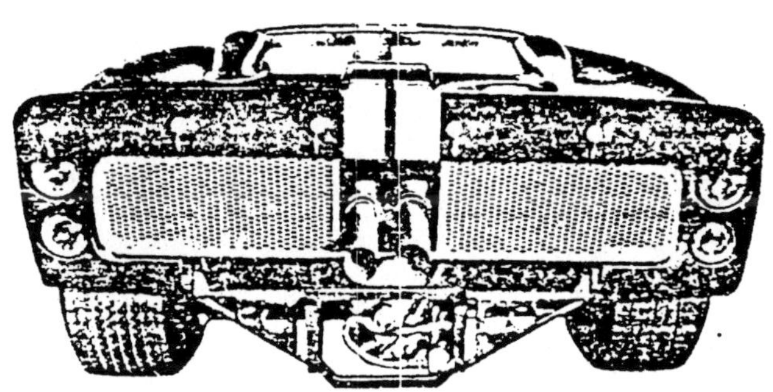

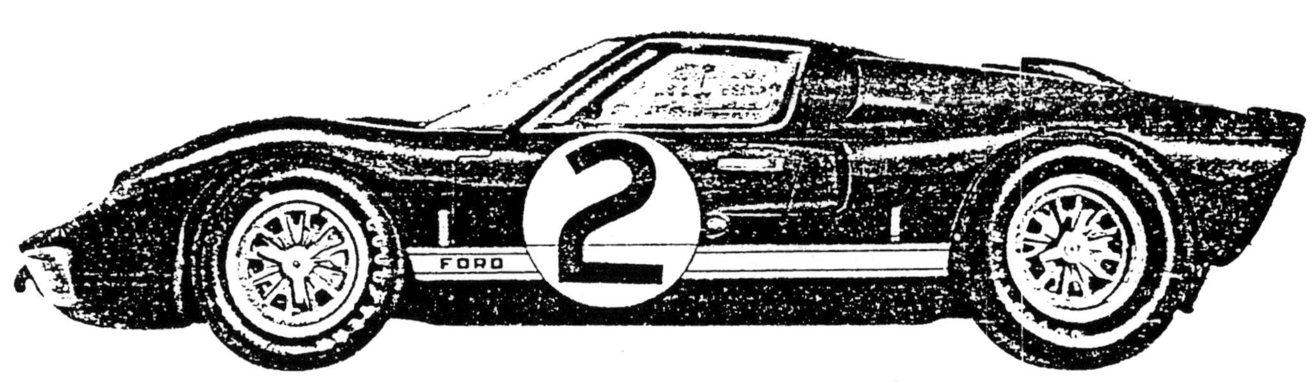

Mark II–427 GT Engine

INTRODUCTION

The Mark II Ford 427 cubic inch displacement GT engine used in the 1966 Le Mans race was a modified version of the 1966 High-Performance production 427. (Figure 1) Major changes from the standard version included the use of aluminum cylinder heads, "dry-sump" lubrication system, new carburetor, aluminum water pump, transistorized ignition, special flywheel, an aluminum front cover, and a magnesium oil pan.

These changes, along with various design refinements, were made to meet particular demands imposed by vehicle design, race regulations, endurance requirements, and vehicle performance characteristics.

Valuable background experience in developing engines for GT use had been accumulated since the fall of 1963 when the 255 cubic inch, pushrod "Indianapolis" engine was selected for the 1964 Le Mans race. This engine (Figure 2) was installed in a newly developed GT vehicle which was entered in the Nurburgring 1000 KM race in May, 1964, in order to test the vehicle and engine combination under actual racing conditions. Suspension problems forced the vehicle to retire in the fifteenth lap.

In June, three of the new GT vehicles powered by the 255 "Indianapolis" engine competed in the Le Mans race. One of the vehicles set a new lap record but various vehicle

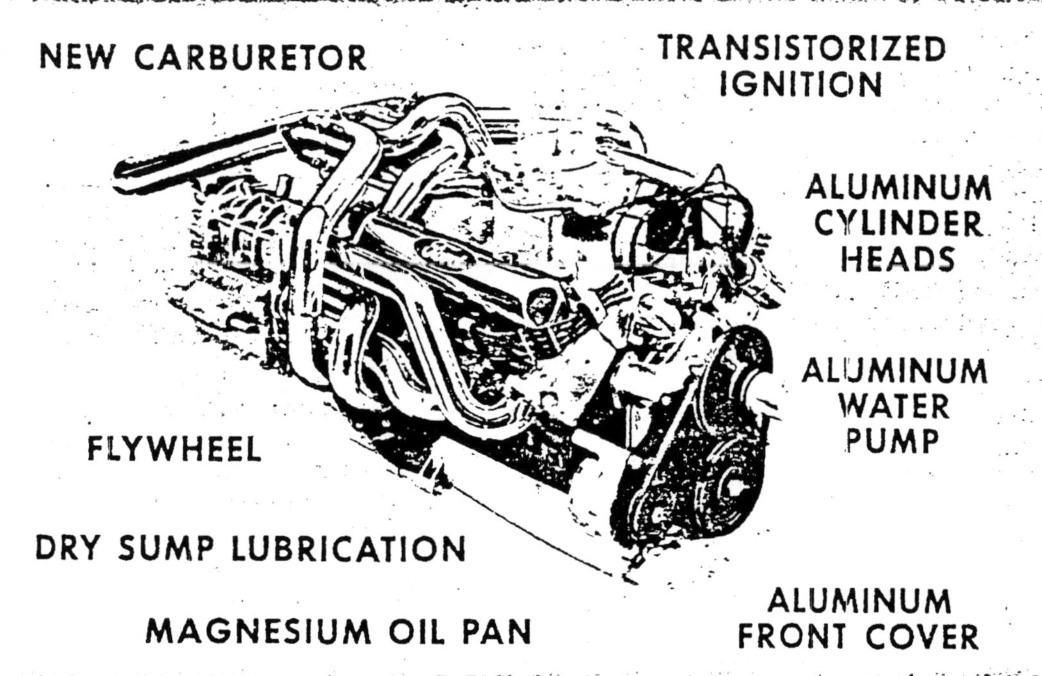

Figure 1

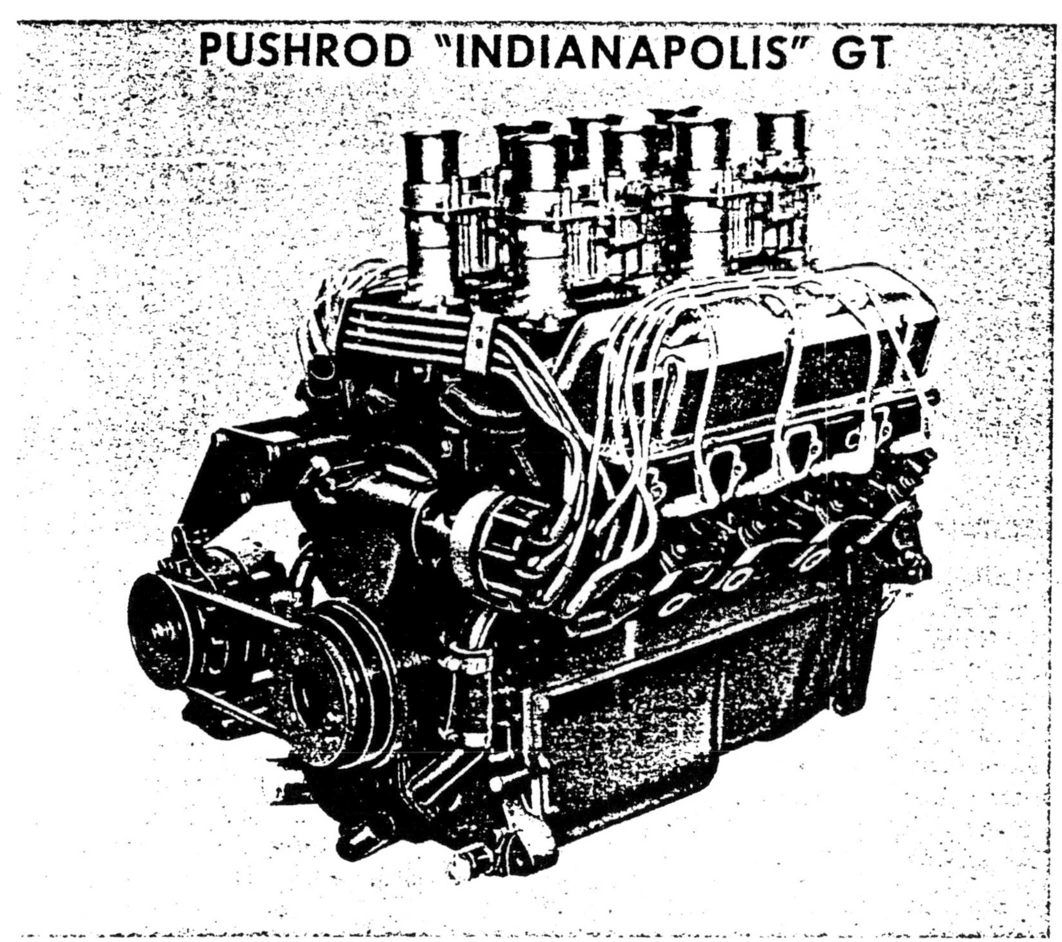

Figure 2

mechanical problems led to the retirement of all three entries by the fifteenth hour.

During 1964 a lightweight version of the 427 High-Performance engine had been developed for use in the Cobra. One of these engines, with slight modifications, (Figure 3) was installed in a GT vehicle undergoing tests at Riverside, California, early in 1965. The excellent performance of the vehicle prompted the decision to use the 427 cu. in. lightweight engine in the 1965 Le Mans race, primarily as a pacesetter. The durability level of the engine for this type of race was unknown, but performance was such that the vehicles would be capable of setting an unusually fast pace. By forcing the opposition to run beyond their capacity, the remaining 289 cu. in. powered vehicles would then have an excellent chance to finish in lead positions.

A new lap record was set during the race by one of the 427 cu. in. powered vehicles, but the strategy ended at that point, as all of the Ford entries retired early due to mechanical problems. The 427 cu. in. engines had performed extremely well during their short duration of running. The durability level of the engine, however, still remained unanswered.

On the basis of its excellent performance in '65, the 427 cu. in. engine was selected as the main powerplant for the 1966 Le Mans race. The 427 cu. in. stock car engine could readily meet the 450 horsepower objective. Our major effort during the year was directed toward improving the durability of the engine. To meet this objective, we planned to capitalize on the proven durability features of the production engine.

The major challenge that faced our engineers was to package the engine in the new vehicle and comply with necessary ground clearance, cooling requirements, oil capacity, intake and exhaust system clearance, as well as considering numerous details such as hot-fuel handling, driveability, servicing, maintenance, and instrumentation.

As with all of our high-performance and competition engine work, one of the most

Figure 3

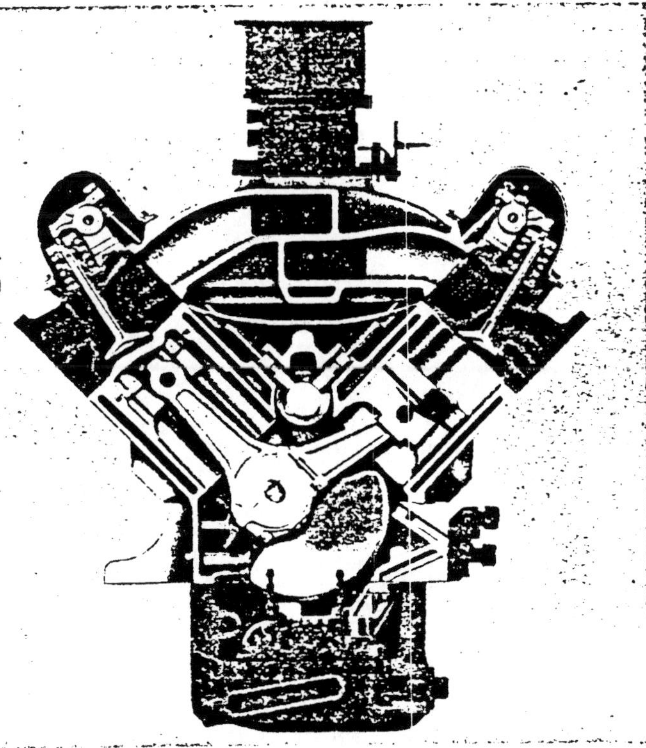

Figure 4

stringent requirements is to retain maximum use of existing production components. This cross-sectional view of the MK II 427 cu. in. engine (Figure 4) shows to some extent the degree of success we achieved, having retained the cylinder block, camshaft, crankshaft, connecting rods, pistons, valve train, and intake manifold.

DESCRIPTION

In order to promote a better understanding of the total engine engineering effort, the description of component and system changes is divided into four groups: detail modifications, unique components designed primarily to affect weight savings, a review of those parts that were changed for durability reasons, and the unique designs dictated by vehicle requirements.

DETAIL MODIFICATIONS

Piston

The standard production piston is an aluminum extrusion which affords maximum strength-to-weight ratio. The thread-type surface finish on the skirt is used for lubrication and scuff resistance. The piston is located on center, since "cold-start piston slap" is of no concern. The pin bores are honed for optimum geometry and piston-to-piston compatibility. Two slots are provided in the upper half of each pin hole at 45° from the centerline for added lubrication.

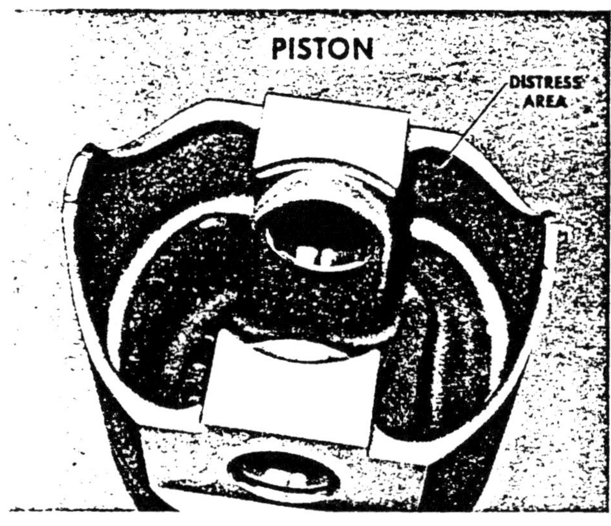

Figure 5

A number of incidents of skirt cracking at the large radius in the slipper occurred during dynamometer testing. (Figure 5) In order to correct this problem, a design modification, consisting of an offset radius to provide additional metal in the distressed area was made. The transfer of the stress to the pin boss resulted in chaotic engine failures instead of spasmodic piston failures. The failures were the result of the piston pin being pulled from the piston.

It was decided to use the standard production piston and polish the skirts in the distressed area to provide an extra measure of insurance.

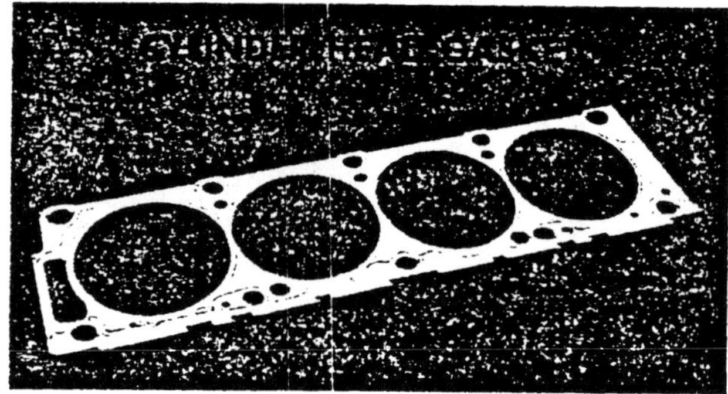

Figure 6

Cylinder Head Gasket

· The standard production cylinder head gasket is a stainless-steel beaded design, .016 thick, with a rolled-over flange on each bore. (Figure 6) The use of this gasket did not present any sealing problem with the aluminum cylinder head. It was necessary, however, to revise the bolt torquing procedure in order to prevent cracks in the cylinder head at the center bolt hole on the exhaust side. The new procedure consisted of torquing all cylinder head bolts in the conventional manner, but at 20 ft.-lb. intervals until the maximum value was obtained.

Valve Train Components

As indicated earlier, the standard production valve train components were retained.

The production nodular cast iron rocker arms were X-rayed prior to use to provide maximum insurance against inclusions and porosity, which could not readily be detected by the Magnaflux process. (Figure 7)

The rocker arm shaft is hardened in the rocker arm bearing area and is held in place by slotted malleable iron supports bolted to the cylinder head.

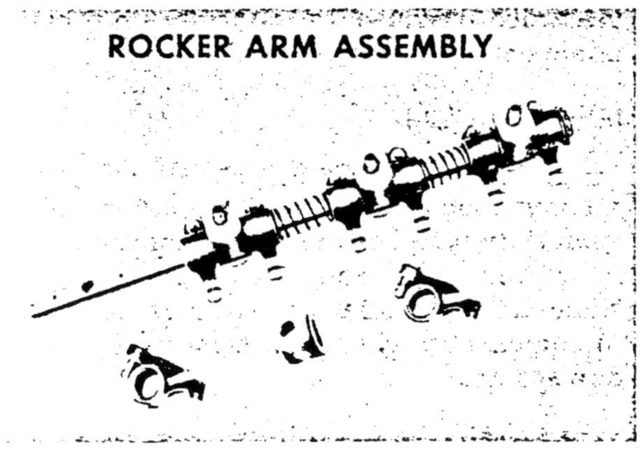

Figure 7

The pushrod is comprised of an electric-weld tubing with a ball bearing welded to the tappet end and a stamped case-hardened cup welded to the opposite end. One isolated push-rod failure was experienced during the course of vehicle tests and was attributed to an improperly heat-treated socket. To guard against further incidents of this nature, all pushrods were quality inspected.

Throughout the program, spasmodic valve spring failures were experienced. The failures always occurred in the lower closed coil approximately 1-1/2 coils from the end. In an effort to reduce the mechanical friction thought to be the source of failure, the production springs were cadmium-plated to provide a form of lubricant on each coil. Since this did not reduce the number of failures, the plating operation was discontinued.

When time literally ran out and no further changes could be incorporated to alleviate the problem, a concerted effort was made to train the pit crews in the art of changing valve springs with a minimum amount of time. Fortunately, no valve spring failures, which affected vehicle performance, were experienced during the race.

Crankshaft

A forged steel crankshaft with cross-drilled oil passages feeding the connecting rod bearings is standard on the 427 cu. in. production engine.

All of the "bobweight" is counterbalanced within the crankshaft counterweights. External counterbalancing has been avoided because of the excessive edge-loading it imposes on the front and rear main bearings. Main bearings are of the steel-backed over-plated copper-lead lining construction. The number three main bearing absorbs the crankshaft thrust loads.

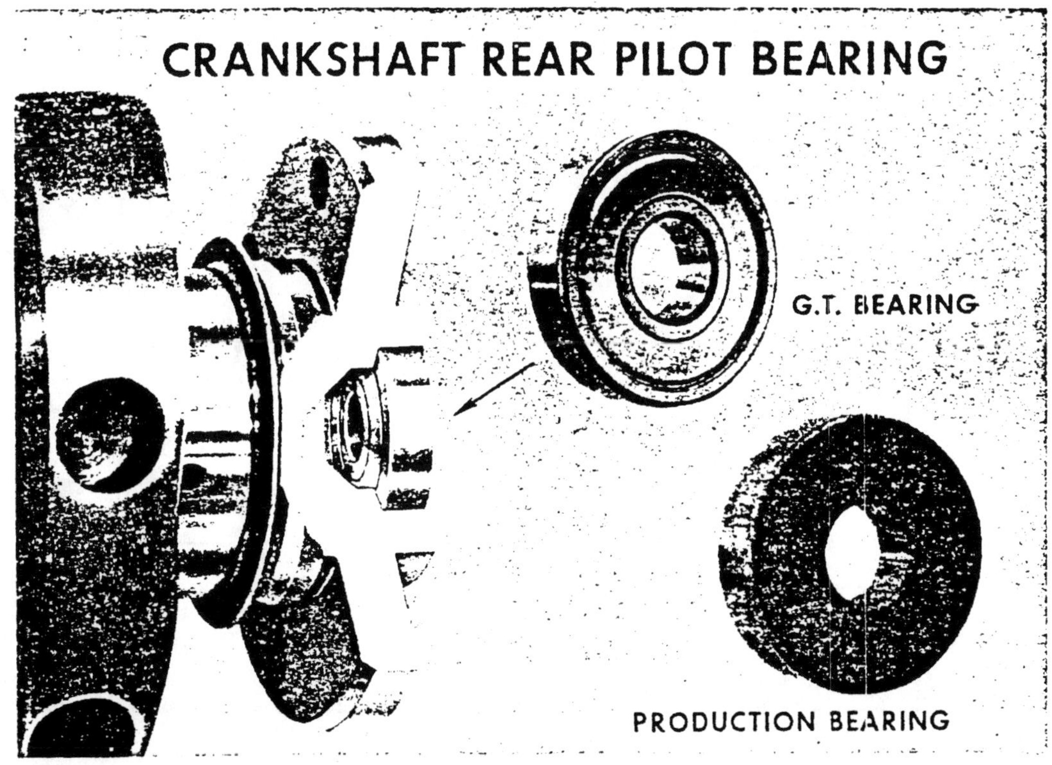

Figure 8

This crankshaft was used in the GT application with only one modification, the sintered-bronze clutch pilot bearing was replaced with a double-row sealed-ball bearing. (Figure 8) Early vehicle tests indicated a need for a sealed bearing to prevent the high-temperature resistant grease from leaking onto the flywheel face of the clutch, thus avoiding a slipping condition.

UNIQUE COMPONENTS FOR WEIGHT SAVINGS

The reduction of weight was based on the use of aluminum and magnesium castings, wherever feasible. The first 1965 experimental engines weighed 555 pounds. These engines were built with aluminum cylinder heads, water pump housing, crankshaft damper hub and a magnesium intake manifold. A twenty-five pound weight penalty was imposed on the 1966 engines through the requirement for an internal "dry-sump" lubrication system and a decision to use the standard production aluminum intake manifold. The magnesium manifold was discarded because of difficulty in obtaining homogenous castings.

Intake Manifold

The cast aluminum manifold (Figure 9) is a conventional "over-and-under" single four-venturi design. The optimum runner configurations and sizes had been obtained through static airflow studies of each individual runner in conjunction with each element of the induction system.

Figure 9

(Further information regarding the induction system is contained in Mr. A. O. Rominsky's paper No. 670067 on MK II — 427 Engine Induction System.)

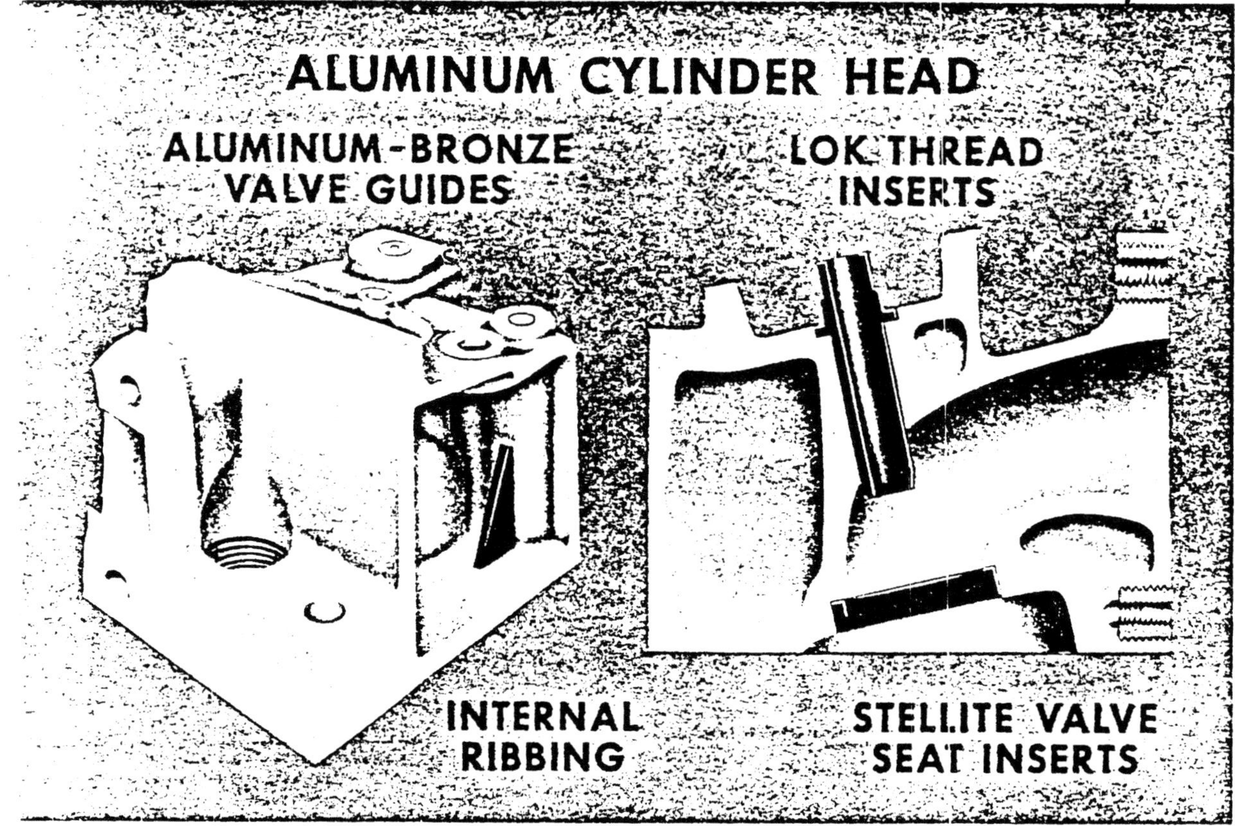

Figure 10

Cylinder Head

The aluminum cylinder heads were produced from modified plastic patterns obtained from existing production equipment. The heads contain internal ribbing and heavier sections consistent with design practices for aluminum. (Figure 10) Aluminum-bronze valve guides and stellite valve seat inserts were incorporated.

During the early stages of the experimental build program, .050" was added to the intake manifold mounting surfaces to eliminate head cracking during assembly torquing. Steel "Lok Thread" inserts were used in the exhaust header mounting holes to prevent mutilation of the threads during the installation of the exhaust headers in the vehicle.

Crankshaft Vibration Damper

The production crankshaft vibration damper consists of an alloy iron inertia ring mounted to an iron hub casting with an elastomer sleeve. The inertia member has sufficient mass to cope with the torsional amplitudes and frequencies induced by high rpm. The damper was redesigned (Figure 11) to incorporate a cast aluminum hub in order to reduce engine weight while retaining the same inertia member. The damper contains holes for the insertion of weights during mass balancing of the engine assembly.

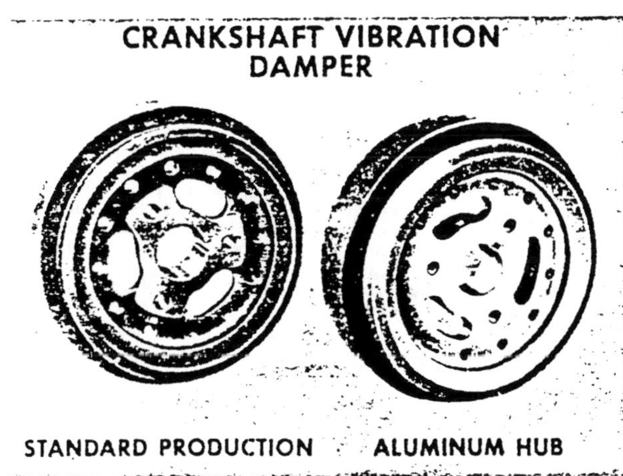

Figure 11

Water Pump

The water pump housing was cast in aluminum, utilizing the existing production equipment for the cast iron housing. (Figure 12) Because of the high water temperatures experienced during a test trip where the ambient temperatures were 90°F, a larger impeller was used to increase the capacity of the pump. Subsequent wind tunnel tests were conducted to finalize the pump capacity.

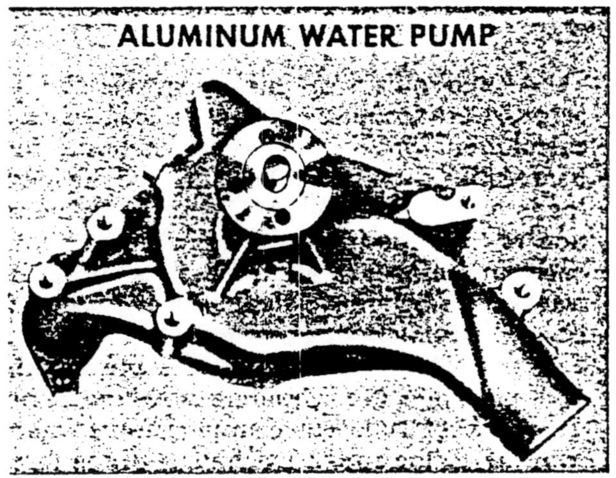

Figure 12

REVISED COMPONENTS FOR DURABILITY

As indicated earlier, the primary objective was to meet durability requirements. In line with this, the following production components were revised to achieve the objective.

Connecting Rod

The forged steel connecting rod is the type currently being used in the production engine. (Figure 13) The cap is located with respect to the rod half by two steel dowels which are contained in counter-bores at the parting line. The cap-screw has a 12-point head, washer face and undercut shank. The thread is a "tri-

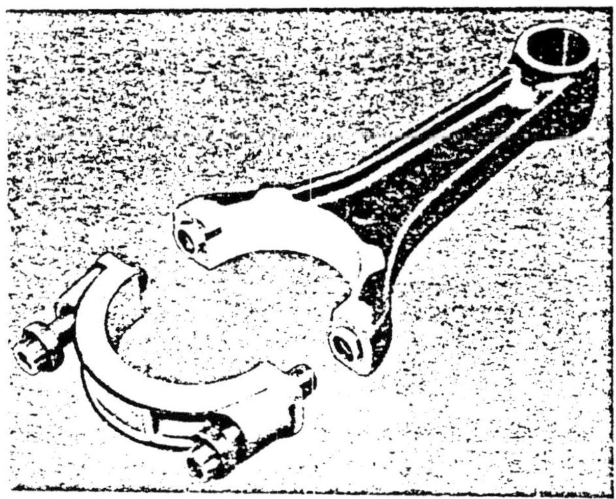

Figure 13

lobe" interference fit type designed to spread the load more uniformly along its full length. The cap screw is prelubricated with a graphite compound during assembly to minimize the high friction torque inherent in the interference fit thread.

At the outset of this program, the cap screws were torqued to 55-60 ft. lbs., which by calculation, provided sufficient clamping force between the cap and rod to withstand the inertia loads at maximum anticipated rpm. Examination of the rods from two engines, which failed early in the program indicated that the failure was due to bending fatigue of the bolt. Insufficient clamping force between the cap and the rod permitted movement at the parting line, thus placing a cyclic bending stress on the bolt. The lack of sufficient clamping force was attributed to the rough machining of the thread in the rod bolt hole. Some of the fixed torque load was expended in overcoming the additional friction caused by the poorly threaded hole. For the race, the connecting rod bolt holes were specially processed to improve the quality of the thread and the cap screws were installed by loading to a predetermined stretch.

Piston Rings

The first and second compression rings are the same as those used in the production engine. They are 1/16 thick, barrel-faced and chrome plated. The top ring is cast of high-tensile ductile iron and the second ring is standard piston ring iron.

In the early stages of the program, the production oil control ring expander was used. This expander had a locating tang at each end which was inserted into a hole in the piston oil ring groove. Several instances of broken tangs resulted from inherent loading. To eliminate this problem, the expander was redesigned to remove the tangs, and a standard butt-joint was used in conjunction with epoxy resin filled ends. (Figure 14)

Piston Pin

Production piston pins were used. However, production tolerances were such that the end play could vary between 0.000" to .024". During the test program, several piston pin retainers failed, which was attributed to the severe pounding of the piston pin against the retainer. The piston pins were then select-fitted to obtain 0.001/.005 end play. This did not resolve the problem. An analysis of subsequent failures indicated that those piston and connecting rod assemblies which were at or near the low limit were failing. This was attributed to negative end-play due to the bending of the piston pin boss of the piston. The end play was revised to .005/.010", and no further retainer failures were experienced. (Figure 15)

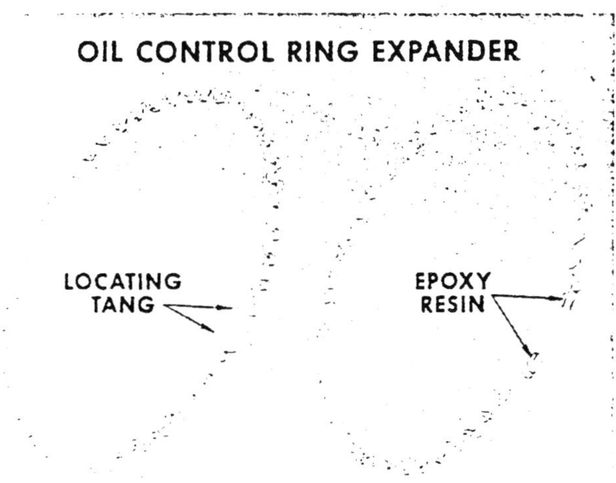

Figure 14

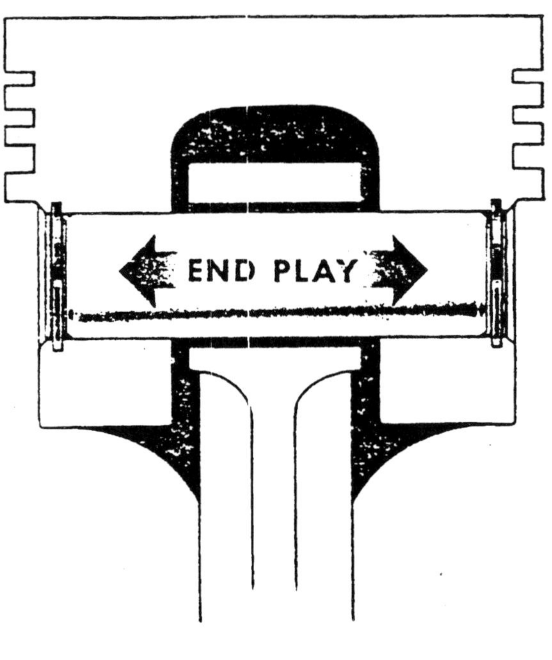

Figure 15

Valves

Intake and exhaust valves are of similar construction to those used in the production engine, except for smaller head diameters necessitated by the use of valve seat inserts in the aluminum cylinder heads. The lightweight valves have hollow stems which are flash-chrome-plated for scuff resistance. The 1.625 diameter exhaust valve is sodium cooled and is made from a 21-4N material with a hardened Silicrome #1 stem tip. The 2.06 diameter intake valve is made from Silicrome X-B material.

To improve durability, both valves were designed with tulip heads having a divergent cross section for constant stress. The stem blends into the head at a slight angle to eliminate the possibility of the diameter in the transition area being less than the stem diameter. (Figure 16)

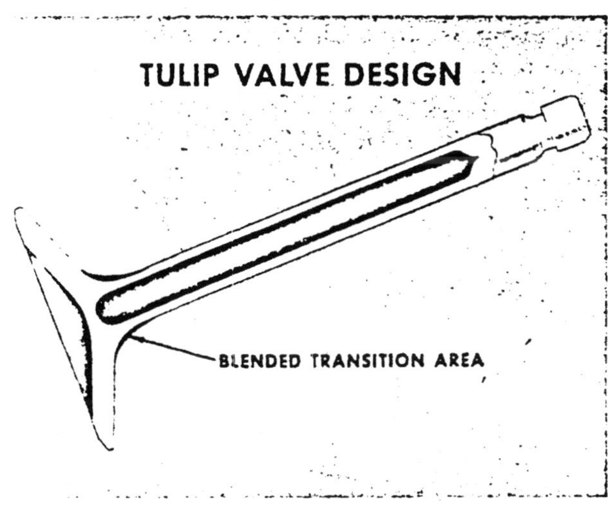

Figure 16

Valve Stem Seal

The production teflon valve stem seal, (Figure 17) which is pressed onto the upper end of the valve guide, was used. Several seals disintegrated during sustained high rpm operation. One seal failure almost resulted in an engine failure during testing at Daytona. Examination of the engine disclosed that the metal band of the failed valve stem seal had migrated to the lubrication system scavenge pump, causing it to seize. The failures were attributed to insufficient valve stem-to-seal clearance. This clearance was increased, and no further failures were experienced.

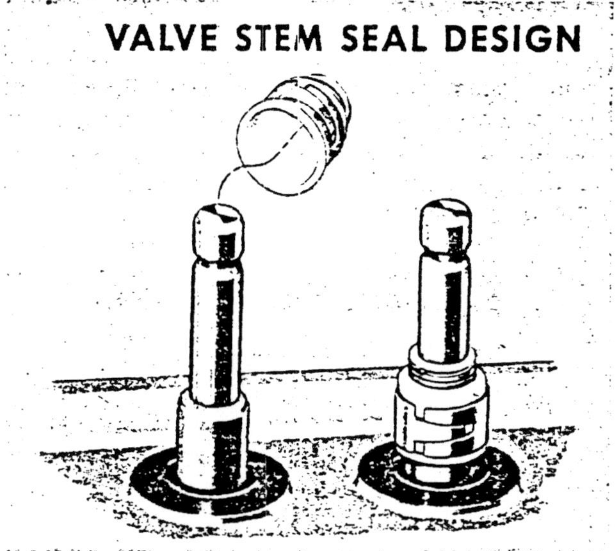

Figure 17

Rear Oil Seal

Early in the program, the engines were built with the conventional rope-type rear oil seal used in passenger car engines. This seal rides on a knurled journal behind an oil slinger machined into the crankshaft. A clutch slipping problem due to a leaky rear oil seal, was experienced during tests at Daytona. The rope seal was replaced by a two-piece rubber compound lip seal designed to fit into the existing groove. The knurl on the crankshaft journal was polished prior to the installation of the seal to prevent possible tearing of the lip. (Figure 18) No rear seal leaks were found after oil leak checks of the engines from races and dynamometer durability testing.

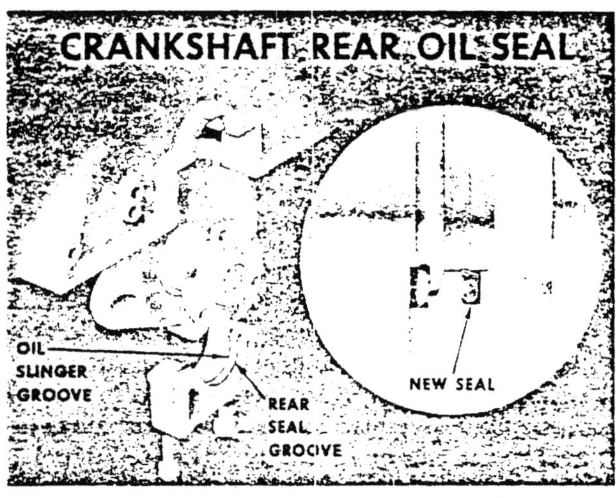

Figure 18

Electrical System

The Ford breakerless transistorized ignition system was used because of its excellent performance in the Indianapolis engine and during the previous two years in GT competition. A schematic diagram of this system is shown in Figure 19.

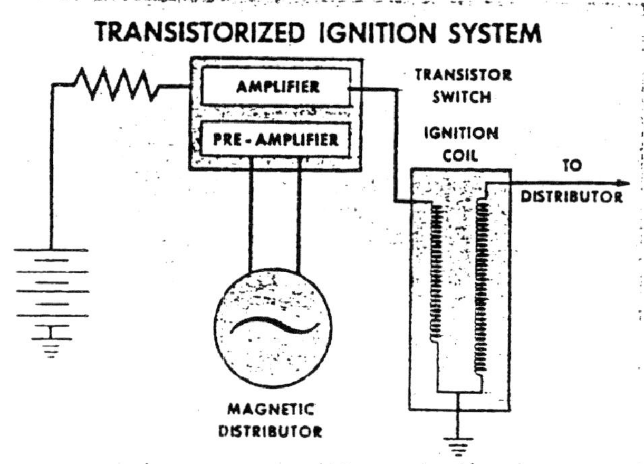

Figure 19

This system employs a variable reluctance for achieving voltage control in the triggering of the main ignition circuit. A concentric permanent magnet and coil, in conjunction with a tooth rotor, varies the reluctance of the magnetic circuit as it rotates. The proper voltage waveform is achieved to trigger the main ignition circuit. The main circuit is fully transistorized for compactness and high reliability. This circuit permits a spark coil to deliver an almost constant voltage throughout the entire high speed range.

UNIQUE COMPONENTS DUE TO VEHICLE REQUIREMENTS

The design of the vehicle dictated special requirements for the engine, clutch, lubrication system, and the cooling system.

Flywheel and Ring Gear Assembly

A new flywheel was designed to accommodate the double-disc clutch. It was machined from a steel casting with the standard production ring gear press-fitted to the flywheel and welded securely. Holes were drilled in the front side of the flywheel as required, for mass balancing the engine assembly. (Figure 20)

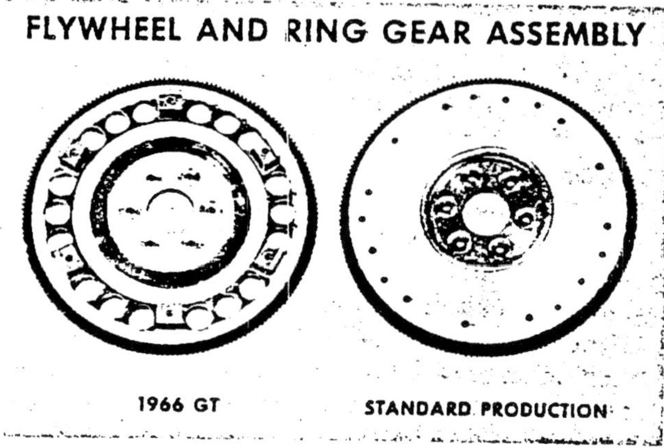

Figure 20

Front Cover

The sand cast aluminum front cover extends below the bottom of the cylinder block to accommodate the oil scavenge pump system. (Figure 21) It was extended as much as competition rules for ground clearance would permit. The front cover is sealed to the crankshaft in the conventional manner and a single gasket is provided to seal the cover to the cylinder block and oil pan.

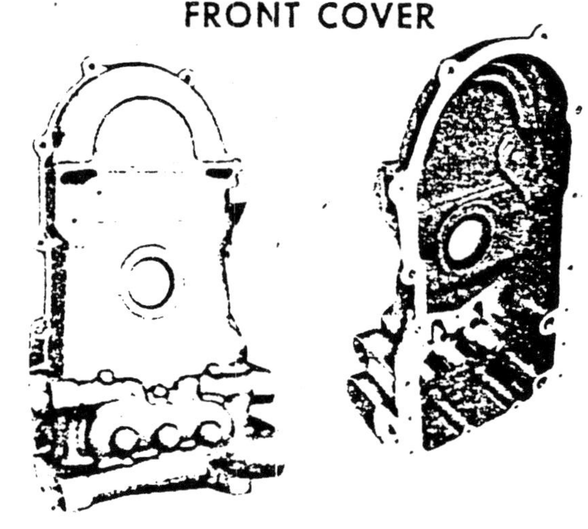

Figure 21

Oil Pan

The sand cast magnesium oil pan extends from the front cover to the rear of the cylinder block. (Figure 22) Provision is made on the inside of the oil pan for the installation of front and rear oil scavenge pickups. Drilled passages in the pan leading to the oil pick-ups, exit at the front of the pan and match the scavenge pump inlet holes in the front cover. The oil pan

Further information regarding the ignition system is contained in Mr. R. C. Hogle's paper No. 670068 on MK II — 427 Ignition and Electrical System.

is bolted to the front cover and cylinder block. Because of the rigidity of the casting as compared to the conventional stamped pan, it was possible to use a thin asbestos-rubber gasket. This permitted increased bolt torque to minimize the possibility of oil leakage. A cast lug was provided at each side of the pan for the installation of support bolts for the bell housing.

Figure 22

Lubrication System

The engine retains most of the components of the production lubrication system. Restrictions on adding oil at Le Mans required that we increase the capacity of the oil pan. It was not possible to do this in the GT car because of ground clearance limits and chassis interferences. The decision was made to install an oil scavenge system, or what is more commonly referred to as a "dry sump" system. This system also has the advantage of minimizing the whipping and aeration of the oil by the crankshaft. A diagrammatic sketch of the lubrication system is shown in Figure 23.

Oil is delivered by the pressure pump through a short drilled passage to an adaptor mounted externally on the left front skirt of the cylinder block. The standard 427 cu. in. engine production oil pump used in this application is bolted to an internally-cast pad inside of the oil pan mounting rail at the front of the engine. This pump is capable of delivering 20.5 gpm at 70 psi to satisfy lubrication requirements at 7000 rpm. The oil pump is driven through an intermediate hex driveshaft from the distributor. The pump contains sintered iron rotors sealed in an aluminum housing by a bolted-on cast plate.

From the adaptor, the oil passes through a remote mounted oil filter to an oil cooler. The oil filter, a screw-on throw-away type, is standard on the 427 cu. in. production engine. This filter differs from the passenger car filter

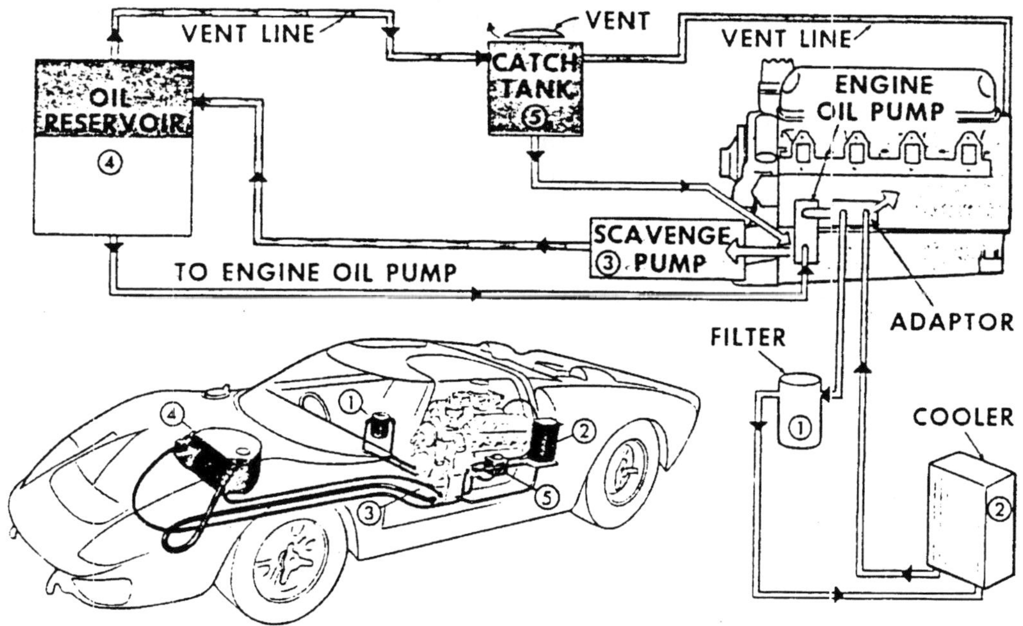

Figure 23

in that the water and sludge removing feature has been eliminated to hold the restriction to a minimum. The oil cooler is the air-cooled, single-pass type currently being used in stock car racing. It is mounted in the engine compartment adjacent to an air scoop on the side of the car.

The high-pressure oil from the cooler is then returned to the adaptor and into the main oil gallery. This gallery extends the full length of the block. Oil is routed to the bearings by a system of cross-drilled passages from the main gallery to an intersecting passage between the main and cam bearing bores. Full pressure oil is then taken off number two and number four cam bearings through passages to the block deck and into the cylinder head for lubrication of the overhead valve train. Overhead drain-back oil enters the oil pan through oil return holes at the front and rear of the heads.

The oil in the pan is then transferred to an oil reservoir located in the front of the car by means of the scavenge pump. The oil reservoir is baffled for de-aeration of the oil. This reservoir is vented to a "catch" tank which in turn is vented to the atmosphere. There is a drain line from the "catch" tank to the oil pan to drain off excess oil in the "catch" tank. The engine crankcase is also vented to the "catch" tank through the original crankcase road draft tube hole in the intake manifold. A return line from the bottom of the oil reservoir is connected to the suction side of the engine pressure pump.

Up to the time of the Sebring race, the lubrication system incorporated a dual-rotor, belt-driven scavenge pump externally mounted to the oil pan. (Figure 24) The oil was pumped via two external lines from the front and rear of the oil pan to the oil reservoir. The oil from the reservoir entered the pressure pump through a fitting in the oil pan leading to the inlet tube.

Although this system functioned satisfactorily at the outset of the program, a new internal scavenge pump system was designed. This work was done because there was some apprehension about the vulnerability of the pump and drive belt to possible damage by flying debris. A belt failure occurred during the Le Mans practice in April 1966 when the vehicle went off the road onto a sand bank. It was then decided to use the internal scavenge pump.

Figure 24

Several design parameters were established for the internal scavenge pump. It was required that the existing oil pressure pump be retained. Dual gear pumps were to be used for both front and rear oil scavenge pickup. A common outlet was required in order to provide a clean installation in the vehicle. The location of the oil pressure pump negated the possibility of using a design similar to that of the Indianapolis engine where dual gear pumps are mounted to the main bearing caps. The arrangement, which met all of the design objectives, was the gear-type pump incorporated in the front cover. There are three gears of which the center gear, chain driven from the crankshaft, drives the other two, thus duplicating two pumps.

The capacity of each pump was established at 40 gpm at 6000 rpm. The total scavenge capacity is approximately four times the capacity of the pressure system. This was required to maintain a dry sump as well as clear the crankcase of blowby gases. The two inlet holes to the pumps match the internal passages in the oil pan leading from the front and rear oil pick-up screen and cover assemblies. The objective of providing a common outlet was achieved by means of drilled passages in the front cover and the scavenge pump cover. The inlet to the oil pressure pump was located in the front cover adjacent to the scavenge pump outlet to simplify vehicle plumbing. This also provided oil pan accessibility without disturbing the remainder of the oil system. (Figures 25 and 26)

The outlet pressure from the scavenge pumps was estimated to be 10 psi at 6000 rpm.

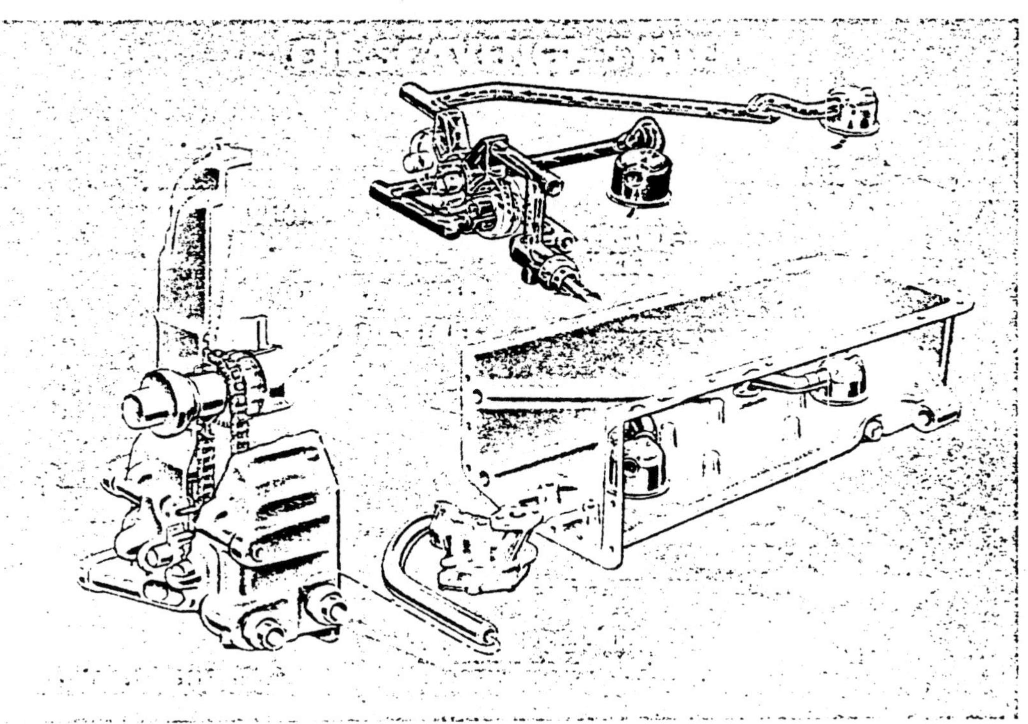

Figure 25

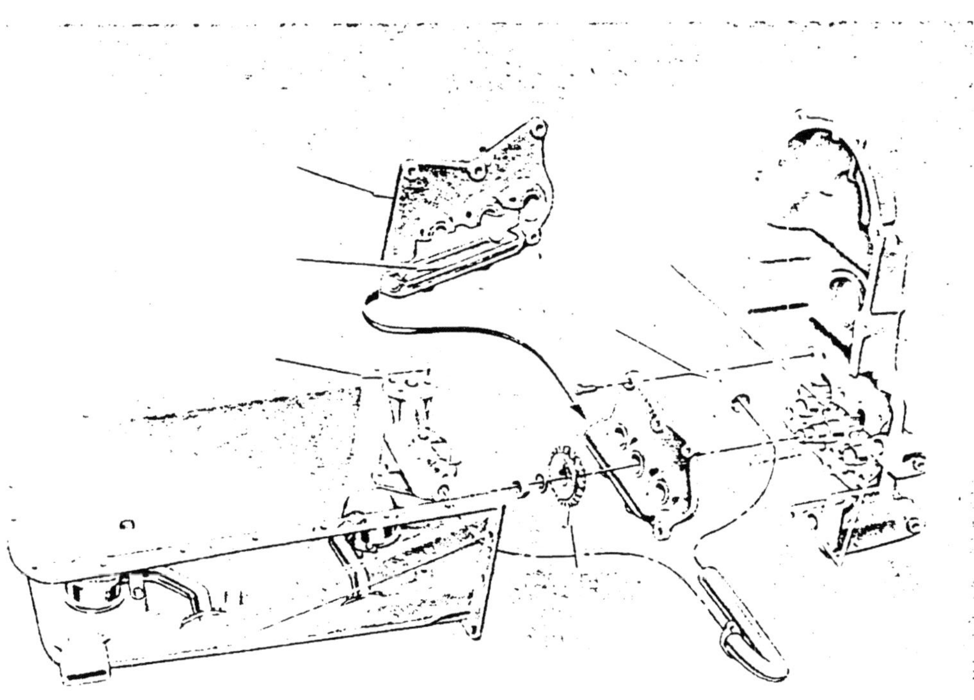

Figure 26

From test data obtained during a test trip at Sebring, it was found that the outlet pressure varied from 4 psi at 3500 rpm to 9 psi at 6200 rpm. From this information, the calculated horsepower required to drive the pump is 1/2 h.p. at 6200 rpm. This results in a load of approximately 5 lbs. on the 3/8 pitch single roller chain used to drive the pump.

The new system was successfully tested on the dynamometer using both the WOT high speed (6000/7000) cycling durability test, and the simulated " Le Mans" cycle test.

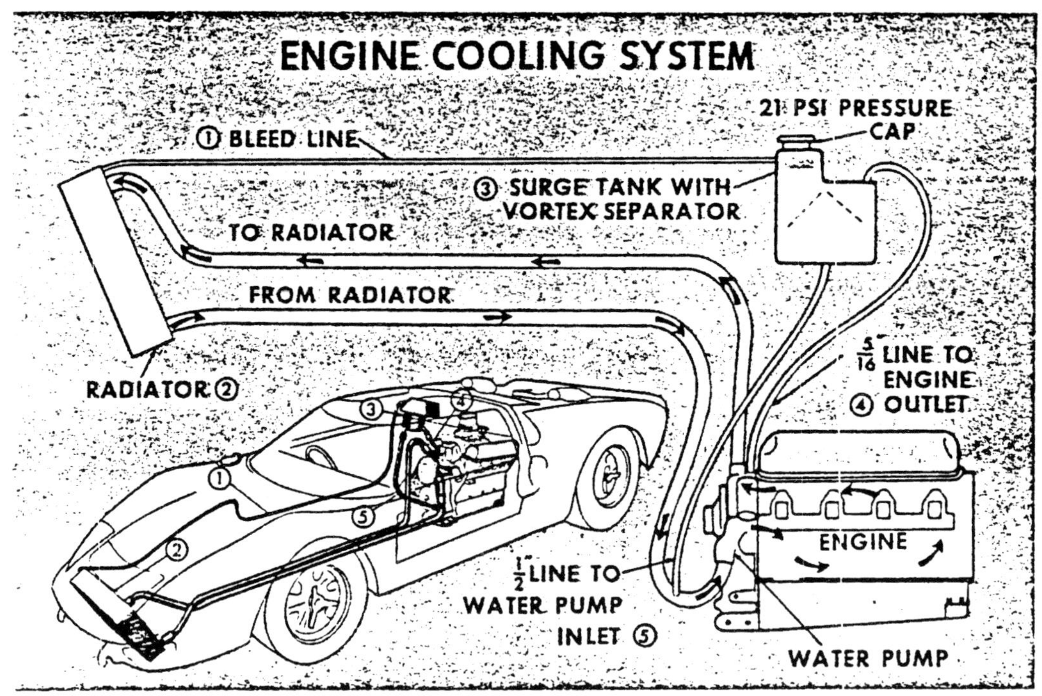

Figure 27

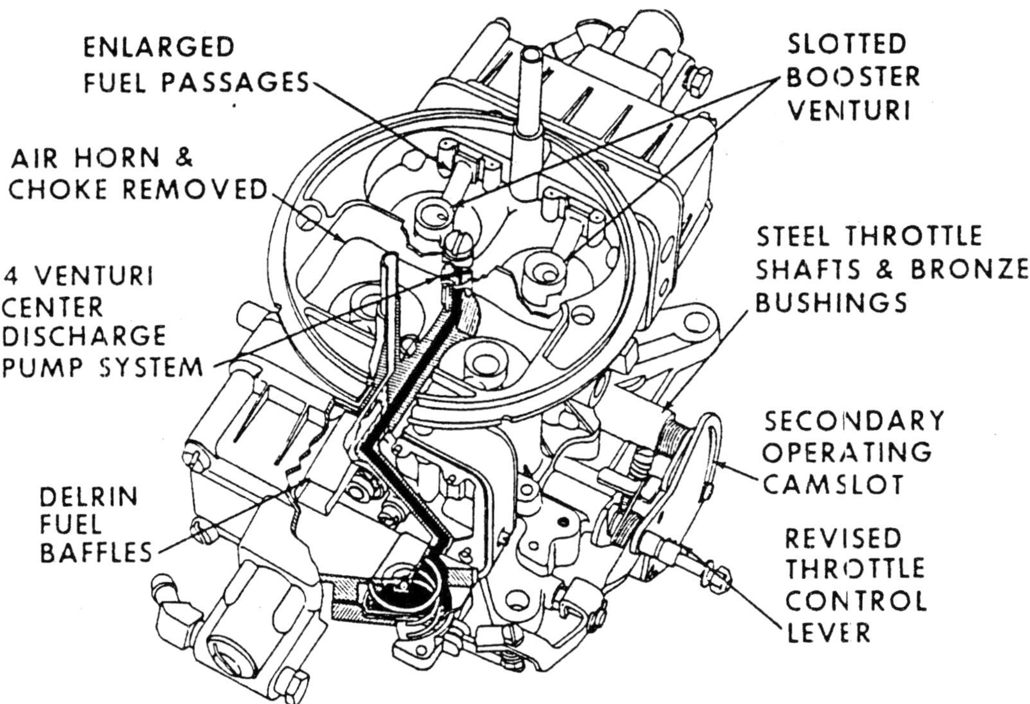

Figure 28

Cooling System

A diagrammatic sketch of the vehicle cooling system is shown in Figure 27. The production engine series-flow cooling system was retained.

The following modifications to the cooling system were adopted in order to stabilize the water temperatures at an acceptable level based on vehicle and wind tunnel tests:

- Larger surge tank equipped with a vortex separator to allow for expansion and deaeration of the water.
- Increasing the size of the line from the engine water outlet to the surge tank to obtain

acceptable de-aeration with minimum loss of cooling capacity.

- Increasing the size of the line from the surge tank to the pump inlet to insure positive pressure at the pump inlet regardless of pressure losses in coolant system.
- Installation of a 20 psi pressure cap on the surge tank to raise the air-to-boil temperatures of the system.
- Increasing the water pump capacity by installing a larger impeller.

Carburetor

Several modifications were made to the production 780 cfm carburetor to improve its operation in the GT car. (Figure 28) These modifications included the use of mechanically operated secondary throttles instead of vacuum operated secondaries to give the driver full throttle control. Steel throttle shafts and bronze bushings were added for durability and safety. A center-discharge accelerator pump was added to deliver an equal amount of fuel to each of the four venturi during acceleration. Delrin baffles were installed in the fuel bowls to eliminate spill-over on extreme acceleration and deceleration. The choke and the air horn were eliminated for unrestricted air flow. The booster venturi were slotted to improve fuel distribution. The fuel discharge passages were enlarged to meet fuel requirements at high rpm, and the throttle control lever was revised to accommodate the special throttle linkage required for the installation.

Fuel System

Figure 29 shows the complete fuel system in the vehicle. Three electric fuel pumps are used in this installation. Two of these pumps are connected in parallel and are used at all times unless the engine runs out of fuel. In such an emergency, the driver switches over to the third fuel pump. This enables the driver to return to the pit for refueling.

At the outset of the program, the standard production mechanical fuel pump was used in conjunction with one emergency electric pump. During vehicle testing, some mechanical fuel pump arm failures were experienced when running on the electric pump. Unloading of the arm caused a non-follow condition because of lack of fuel pressure. The mechanical fuel pump was then replaced by two electric pumps.

Exhaust System

Figure 30 shows the fabricated tubular header system that was designed to fit the vehicle installation. The 427 cu. in. engine has been quite responsive to this type of header system. The 32" length and 2-1/4" diameter of the headers, required to obtain optimum power at high rpm, were determined from dynamometer tests. Severe bends in the tubing appeared to have no effect on tuning as long as cross-sectional areas were maintained.

FUEL SYSTEM

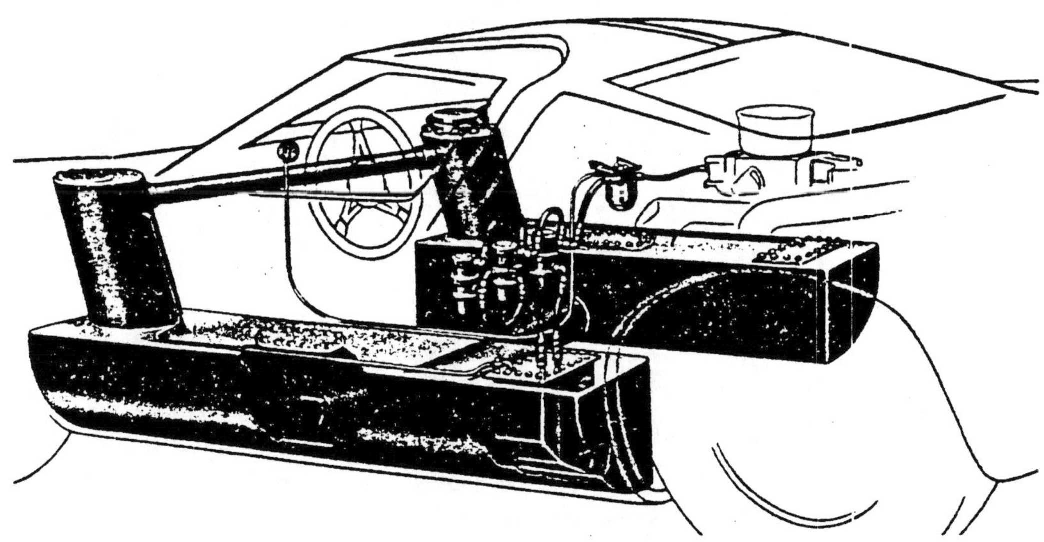

Figure 29

Figure 30

DYNAMOMETER TESTING

Dynamometer testing was predicated on the fact that while performance was important, durability was the prime consideration. With this in mind, it was decided to subject all new designs, revisions, etc., to two types of dynamometer durability testing. The first test was one which had been devised for checking the durability of the 427 cu. in. engine used in stock car racing. It consisted of cycling the engine between 6000 and 7000 rpm at wide-open-throttle for a period of six hours.

Upon the successful completion of this test, an exact duplicate of the engine was built and installed in a specially designed dynamometer setup. (Figure 31) This test simulated the actual running conditions prevailing during one lap of the Le Mans race with the driving pattern for the circuit duplicated relative to length of time in each gear, maximum rpm, etc.

The entire powertrain was cradled between two electric dynamometers which were coupled to the output shafts of the transaxle. The throttle, clutch mechanism, and gear shift selector were actuated by an air-valve solenoid system controlled electronically at a master panel.

Because of the limitation of the dynamometer equipment, some compromises were made. The final cycle was considered to be one which would test the durability of the powertrain to a degree equivalent to that anticipated in the race.

The duration of the test was established at 48 hours to provide a maximum safety factor.

(Further information regarding this installation is contained in the SAE paper No. 670071 "Laboratory Simulation — Mark II GT Powertrain".) .

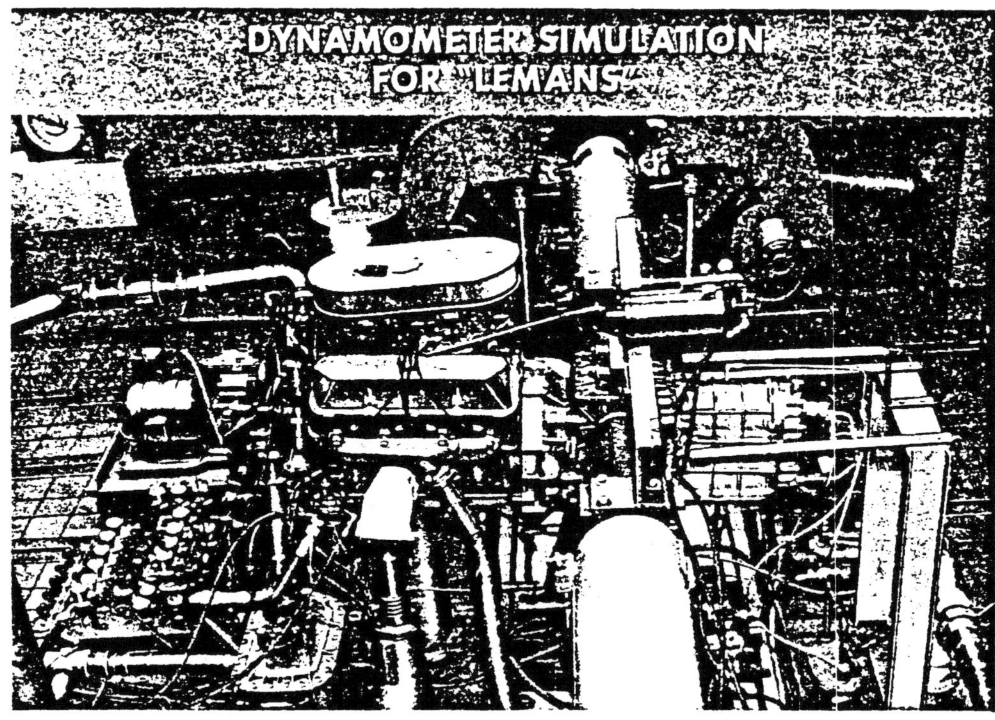

Figure 31

Each engine is given a four-hour dynamometer break-in and power run. An average of four hours of vehicle "shakedown" are required after engine installation in the vehicle. Five to nine hours of pre-race practice can be accumulated on each engine. This could result in considerable amount of running time prior to the race.

The simulation test was used extensively during the program. It proved to be an excellent instrument in exposing weaknesses of various components which were then redesigned and retested. The process was repeated as often as required on test engines until the durability objective of 48 hours was attained. It is interesting to note that the last experimental engine, which successfully completed the test, was compared with the engine from the winning car at Le Mans with respect to wear rate of the various components, general condition of the engine, etc., and no appreciable differences were found.

The performance phase of the dynamometer development of this engine was concentrated on obtaining 450 horsepower at 6200 rpm with a 10.5:1 compression ratio. Achieving this objective presented no problem. The 427 cu. in. push rod engine used in stock car competition had been refined to such an extent that it was capable of producing 525 h.p. at 6400 rpm with a 12.0:1 compression ratio.

Figure 32 shows the actual performance of the early GT engines using a moderate camshaft and having a 10.5:1 compression ratio compared to a representative production engine having a 12.0:1 compression ratio.

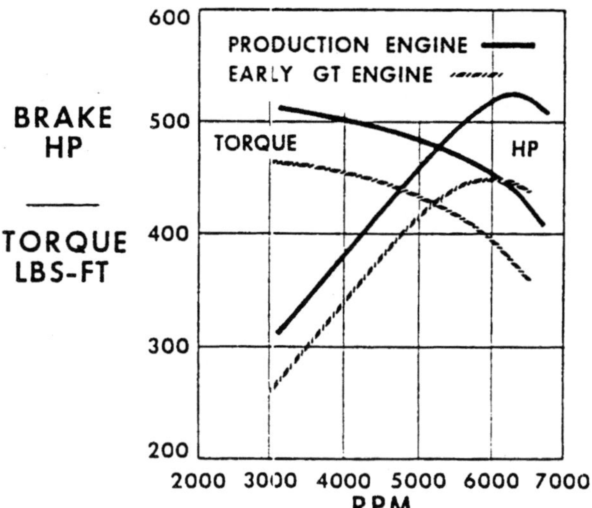

Figure 32

Detail refinements to the engine during the year resulted in progressive improvements in engine performance. Figure 33 shows four power curves for the GT engine, indicating the progress. The average power of all the engines used at Le Mans was 485 h.p. at 6400 rpm which easily surpassed our original objective. Except for the pre-Daytona engines, all of these runs were made with the same camshaft used in stock car competition.

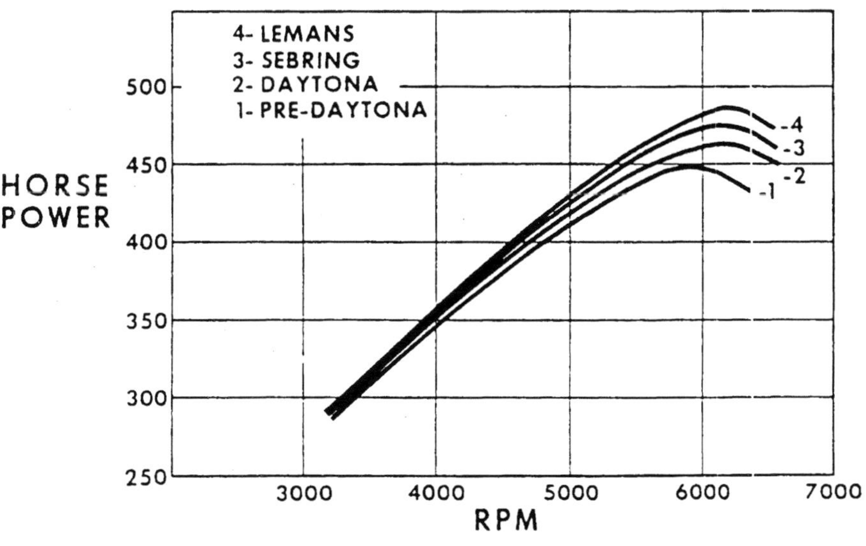

Figure 33

TRACK TESTING

Daytona International Speedway

This track was selected for testing because the maximum engine rpm obtainable on the back stretch could approach those expected on the Mulsanne Straight at Le Mans.

In order to evaluate the engine durability for a 24 hour period of continuous running, tests were conducted in August and December of 1965. The results of these tests revealed problems previously mentioned concerning the pushrod, piston skirts, valve springs, rear oil seals and valve stem seals.

Five cars were entered in the Daytona Continental with as many "fixes" as possible. The results of this race provided confirmation of soundness of these changes.

Sebring

A test was conducted in March at the track. Because of the many gear shift changes required to complete one circuit of this track, the associated high rpm would further evaluate the effectiveness of the "fixes" previously introduced in the Daytona Race.

The internal dry sump scavenge system was introduced in the Sebring Race because of its successful performance on dynamometer tests. No problems were experienced with this system during the race.

However, as previously mentioned, it was during this race that the serious shortcoming in connecting rod bolt fatigue life was revealed.

Riverside International Speedway

This track was used intermittently throughout the program primarily to "shakedown" the vehicle prior to a race or test.

Le Mans

At the practice in April, the vehicles were instrumented to record engine data. The results confirmed the accuracy of the theoretical data used to set up the Le Mans cycle on the dynamometer.

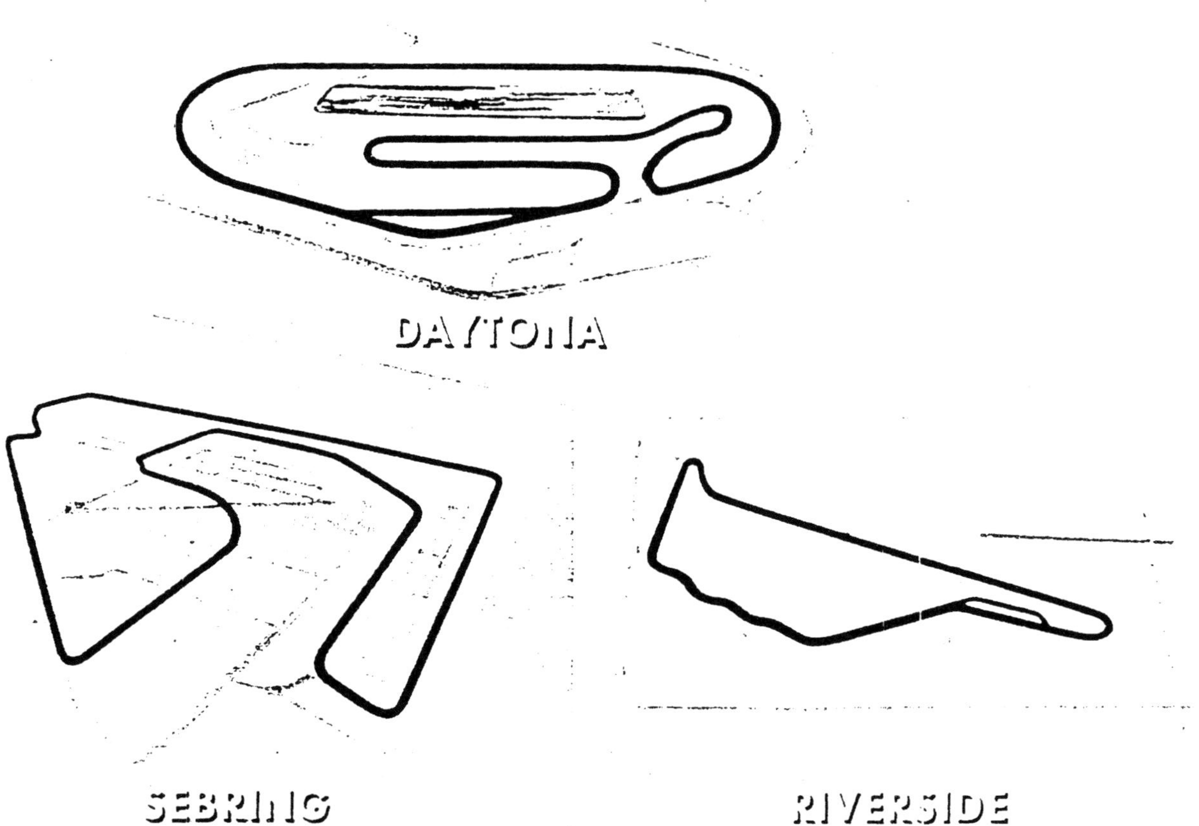

SUMMARY

The result of the 1966 Le Mans race speaks for itself in attesting to the success of the 427 cu. in. GT engine. This race is recognized to be the supreme test of durability in racing circles. To achieve our objective with an inexpensive, basically-production engine, equipped with a single four-venturi carburetor is a measure of the degree of engineering technology and manufacturing quality embodied in our production engines.

Because we were working with production designs, many of the engineering changes and design refinements necessary to achieve per-design refinements necessary to achieve performance or durability objectives can be incorporated directly into standard production engines. Accelerated schedules, common to projects of this nature, frequently lead to the development of more efficient techniques and methods, which then become a new standard for future production applications.

SPECIFICATIONS

ENGINE

Type	90° V-8 OHV
Bore	4.2346
Stroke	3.784
Displacement	427
BHP at Engine RPM	485 at 6200-6400
Torque at Engine RPM	474.5 at 3200-3600
Compression Ratio	10.5:1
Fuel System	Carburetion
Number Carburetors	One – 4V
Oil System	Dry Sump
Oil Pressure	80-84 PSI
Dry Weight (Less Air Cleaner, Clutch Assy., Exhaust Manifold)	580#
Firing Order	1-5-4-2-6-3-7-8
Ignition	Breakerless Transistorized
Maximum Spark Advance	38° at 4000
Alternator Rating	52 Amps.

PISTON AND PIN

Piston to Bore (Skirt)	.006-.007
Skirt Taper	.001
Top of Piston to Top of Block	.018
Piston Pin Material	SAE 5015 Steel
Piston Pin Clearance – (Piston)	.0007-.0009
Piston Pin Clearance – (Rod)	.0003-.0005
Piston Pin End Play	.005-.012
Piston Pin Offset	None
Piston Pin Diameter	.9751

PISTON RINGS

	TOP	SECOND	OIL CONTROL
Width	.062	.062	.125
End Gaps	.028	.028	.032
Groove Clearance	.0027	.0027	Snug in Groove

CONNECTING RODS

Material	SAE 1041 Steel
Length Center to Center	6.488
Weight	806 Grams
Clearance (Journal)	.002-.003
End Play (Two Rods)	.018-.028

CRANKSHAFT

Material	SAE 1046 Steel
Main Bearing Clearance	.002-.0031
End Play	.004-.008
Main Bearing Diameter	2.7488
Crankpin Journal Diameter	2.4384

CYLINDER BLOCK

Cylinder Bore Diameter (Range)	4.2328-4.2364
Thrust Bearing Face Runout	.001
Camshaft Bearing Clearance	.001-.003

CYLINDER HEAD

Volume CC with Valves & Spark Plug in Place	91
Valve Seat Material	High Cobalt Alloy Steel
Valve Seat Fit	.002-.004 Shrink Fit
Valve Guide Material	Aluminum Bronze SAE 68A
Valve Guide Fit	.0025-.0035 Press Fit

SCAVENGE OIL PUMP

Chain Drive	Single Strand Roller
Pitch	.375
Drive Ratio to Crankshaft	1/1

DISTRIBUTOR

Backlash	.002-.010
End Play	.004-.020
Initial Timing	8° at 700 RPM

SPECIFICATIONS — Continued

VALVE SYSTEM

	INTAKE		EXHAUST
Valves Per Cylinder	1		1
Gage Diameter	2.060		1.625
Lash (Hot)	.025-.027		.025-.027
Seat Width	.03		.055
Seat Angle	30°		45°
Stem Clearance		.001-.0024	
Camshaft Timing			
Opening at .100 Valve Lift	8°30' ATC		39°30' BBC
Closing at .100 Valve Lift	36°30' ABC		11°30' BTC
Duration		324°	
Overlap Theo.		96°	
Camshaft Drive Mechanism		Silent Chain	
Chain Pitch		.500	
Valve Lift — Zero Lash		.527	
Valve Spring Load	Valve Open		Valve Closed
Outer — Lbs with Damper	80-90		255-280
Rocker Arm Ratio		1.76:1	
Rocker Arm to R/A Shaft Clearance		.0015-.003	

UNIQUE BOLT TORQUE

Bolt and nut installation torque specifications with lubricated threads (preservative oil acceptable) except as noted by * or # or %.

	OPERATION	THD SIZE	INSTALLATION TORQUE
Bolt	— Oil Pan	5/16-18	20 Ft. Lb.
Bolt	— Rocker Arm Cover to Cylinder Head	5/16-18	10-12 Ft. Lb.
%Bolt	— Pressure Plate to Flywheel	5/16-18	15-20 Ft. Lb.
Stud	— Carburetor to Intake Manifold	5/16-18	Drive to Limit of Threads
Bolt	— Distributor Hold-Down Clamp	5/16-18	5-8 Ft. Lb.
%Nut	— Carburetor Mounting	5/16-24	120-156 In. Lb.
Cross Bolt	— Main Bearing Cap	3/8-16	38-42 Ft. Lb.
Bolt	— Intake Manifold	3/8-16	32-35 Ft. Lb.
Bolt	— Exhaust Manifold to Cylinder Head	3/8-16	12-18 Ft. Lb.
Bolt	— Flywheel to Crankshaft	7/16-20	75-85 Ft. Lb.
Bolt	— Main Bearing Cap	1/2-13	95-105 Ft. Lb.
Bolt	— Cylinder Head	1/2-13	100 Ft. Lb. (Cold)**
Plug	— Oil Pan Drain — Safety Wired	1/2-20	15-20 Ft. Lb.
Bolt	— Crank Damper to Crankshaft	5/8-18	70-90 Ft. Lb.
Oil Filter	— Oil Filter	3/4-16	95-105 Ft. Lb.
Screw	— (Attach Timing Pointer to Front Cover)	10-24	2-4 Ft. Lb.
*Spark Plug		18 MM	15-25 Ft. Lb
#Bolt	— Connecting Rod	7/16-20	.0055-.006 Elongation
Bolt	— Cam Sprocket to Camshaft	7/16-14	40-45 Ft. Lb.

*Use Graphite on Threads

#Moly Kote on Threads and Under Bolt Head

%Dry

**No Hot Torque Required

670067

Mark II — 427 GT Engine Induction System

A. O. Rominsky
Engine and Foundry Div., Ford Motor Co.

FACTORS ATTRIBUTABLE to Ford's success at Le Mans, France in 1966 are many. The factor we would like to single out for discussion is the high efficiency air induction system of the 427 GT engine.

Contrary to the basic design parameters of passenger car engine induction systems, which consider part-throttle transitions and fuel economy, induction systems for high speed, high-output competitive engines are quite the opposite. These engines demand maximum airflow capacity, and equal balance of air/fuel mixture to each cylinder to insure high efficiency throughout the engine speed range under full-throttle conditions.

Recognition of this need, its resolution, and the application of resultant techniques to the Ford 427 engine since its inception in 1963, are the basis for this paper.

INDUCTION SYSTEM FLOWSTAND

The development of the 427 induction systems was dependent upon the use of an airflow measuring device which had been used predominately for research purposes by the automotive industry for many years. The instrument measures steady-state airflow through complete induction systems or individual components under a regulated pressure drop (Fig. 1). Air is drawn through the component being evaluated by a constant-speed, constant-flow vacuum pump. The air flows through the components, an orifice tank, and a system of control valves before exhausting to atmosphere. A manometer is used to measure pressure drops, and a calibrated inclinometer measures steady-state airflow in cubic feet per minute.

By selecting various orifice combinations, airflow can be controlled from 1 - 1140 cfm with approximately 0.1-10 in. Hg pressure drop across the component being evaluated for airflow.

For our purpose, induction system airflow evaluations are measured at a pressure drop of 5 in. Hg. This value established a baseline for comparison of all flowstand development and has no significance with regard to pressure drops in an engine.

EARLY INDUCTION SYSTEM DEVELOPMENT

Based on the premise that a downward-flow intake manifold and large intake ports were required to produce increased performance, a high-riser intake manifold and cylinder head

ABSTRACT

Among the factors attributable to Ford's success at Le Mans, France in 1966 is the high efficiency air induction system of the 427 GT engine. Contrary to the basic design parameters of passenger car engine induction systems, which consider part-throttle transitions and fuel economy, induction systems for high speed, high-output competitive engines are quite the opposite. These engines demand maximum airflow capacity, and equal balance of air/fuel mixture to each cylinder to insure high efficiency throughout the engine speed range under full-throttle conditions.

Recognition of this need, its resolution, and the application of resultant techniques to the Ford 427 engine since its inception in 1963, are the basis for this paper. The GT induction system is the product of a new development technique that involved the extended use of an induction flowstand, as well as extensive studies of various types of manifolds and cylinder heads. During the course of this work, successive improvements in modeling and pattern making have combined to refine the art to a high degree of sophistication.

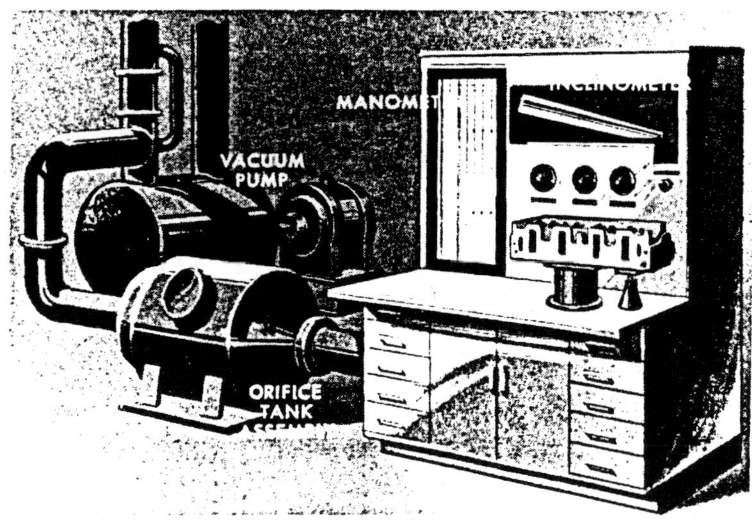

Fig. 1 - Induction system flowstand

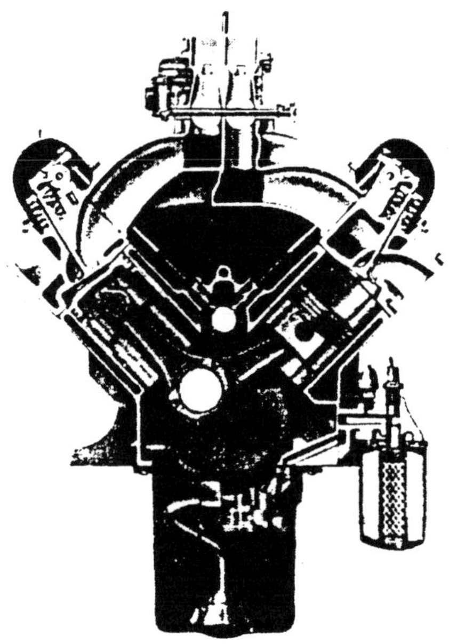

Fig. 2 - High-riser induction system, 427 engine

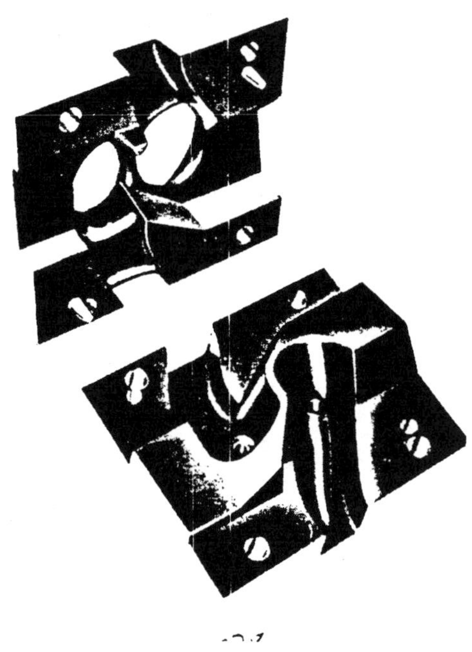

Fig. 3 - Plastic cylinder head flowbox

design was introduced with the 427 engine in 1963 (Fig. 2). At that time, special consideration was given to breaking away from the traditional method of designing cylinder heads and intake manifolds. Normally, drafting layouts were made which stressed constant or gradually changing areas throughout the intake manifold runners, and smooth diverging cross-sectional areas from the intake port entrance to the port throat. Manifold design variations, generated by this method, would not guarantee a significant performance improvement after new manifold castings were made, machined, and tested on an engine.

In analyzing this problem, it was thought that the ports and runners of the intake manifold and the cylinder head could possibly be utilized as development tools to achieve maximum induction system airflow and efficiency in a more reliable and less time-consuming fashion. To implement the approach, the induction system flowstand was brought into use as a development tool.

CYLINDER HEAD - To study the airflow characteristics in the intake and exhaust ports of the high-riser cylinder head, a plastic cylinder head flowbox was constructed which contained an intake port, exhaust port, and a combustion chamber built from drafting outlines (Fig. 3). This flowbox was made in two pieces and split along the normal parting lines to enable making internal revisions with clay or by grinding. The flowbox was mounted to the induction system test stand on a flanged cylindrical tube of the same ID as the actual cylinder bore (Fig. 4). Clay was used at the air entrance to the flowbox to provide a smooth, nonturbulent airflow into the system.

Fig. 4 - Cylinder head flowbox mounted on flowstand, intake port

Fig. 5 - Cylinder head intake port modifications

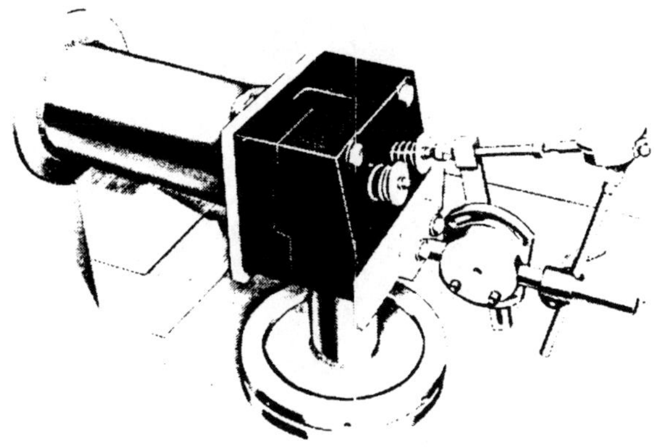

Fig. 6 - Cylinder head flowbox mounted on flowstand, exhaust port

Fig. 7 - Rubber core of high-riser cylinder head contours

A dial indicator and adjusting screw were installed above the intake valve to accurately control the valve at various opening positions. This enabled plotting the airflow curve throughout the valve opening and closing cycle for any camshaft.

Clay modifications to the intake port floor and throat effected an airflow improvement (Fig. 5). The port floor was raised, or humped, in the path of the airstream where the air turns from the port into the throat. This redirected the air to the roof of the port near the throat and resulted in an appreciable increase in airflow.

The throat and seat of the intake port were found to be critical at high air velocities. Various valve seat angles and throat diameters built into the port throat area resulted in minor airflow improvements. The addition of a venturi-type ring adjacent to the valve seat increased the airflow substantially.

The exhaust port was investigated in a similar manner. The cylinder head flowbox was remounted on the flowstand (Fig. 6) with the exhaust port held against a tube mounted on the inlet of the flowstand. The cylindrical bore tube remained mounted on the flowbox to direct the air through the combustion chamber and out the exhaust port, similar to the piston moving upward on the exhaust stroke in a running engine. The exhaust port floor was modified in conjunction with minor revisions to the port roof. Modification to the combustion chamber afforded little improvement in airflow; therefore, it was left unchanged.

Having completed development of the entire flowbox, a rubber compound was poured into the cylinder head flowbox and allowed to harden. This hardened core was used for dimensioning the port contour. A layout was made to determine the external shape of the cylinder head (Fig. 7). When these new cylinder heads were cast, machined, and checked, airflow capacity was slightly less than the flowbox model. This loss was due to the drafting limitations in transferring the rubber core contours to the layouts.

Fig. 8 - Four-venturi high-riser intake manifold

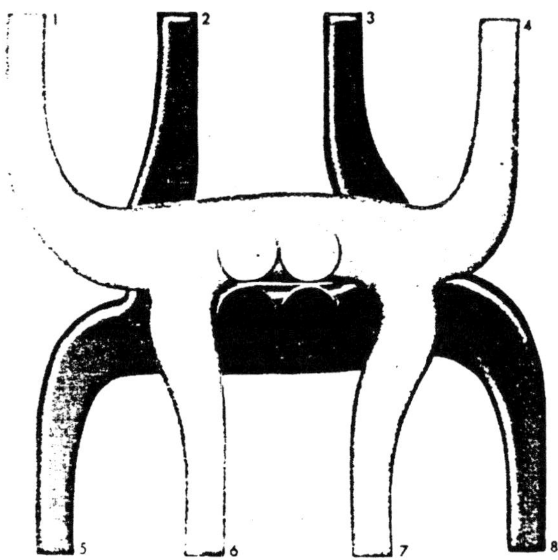

Fig. 9 - Four-venturi high-riser intake manifold runner system

INTAKE MANIFOLD - Our attention was then turned from the cylinder head to the intake manifold. It was required that the manifold be a four-venturi over-and-under design which is used extensively on passenger car engines (Fig. 8). It consists of two dependent sets of intake manifold runners which are connected to their respective plenum chambers (Fig. 9). One set of runners is at intake port level, and the other slightly above this level. This arrangement is necessary, since each set of runners feeds cylinders on both banks of the engine and must therefore cross each other.

This four-venturi manifold design separates the successive firing pulses by alternating the engine firing order from one runner system to the other. Therefore, each cylinder draws from separate runner systems of the manifold.

The cylinder head flowbox was placed on the flowstand in the intake port flow position, and the manifold was attached. Each runner of the intake manifold was checked separately to measure the cubic feet per minute of airflow at various intake valve openings ranging from 0.050 - 0.500

Fig. 10 - High-riser cylinder head and intake manifold mounted on flowstand

in. (Fig. 10). The test results indicated the degree of unbalance that existed from one runner to the next. To correct this unbalance condition, and to increase total airflow in all runners, the intake manifold was physically divided into three sections; one cut through the lower plenum area, and the second cut through the upper plenum area (Figs. 11 and 12). To determine if the sectioning had any effect on airflow, the intake manifold was reassembled and all seams were covered with clay to prevent air leaks. The flow values were found to be exactly the same, indicating that the amount of area lost by cutting had no effect on overall airflow.

It was now possible to open the manifold and to make various contour changes in the manifold runners by adding clay to the passages or by grinding away metal. It was found that the airflow of each runner could be increased and that the runners could be balanced to flow equal amounts of air.

Wood pattern equipment was then made to simulate the cavities in the developed manifold. From these patterns, changes were made in the casting equipment, new castings were produced and machined, and rechecked on the airflow stand.

As was the case with the cylinder head castings, flowstand results indicated a slight loss of airflow as compared to the sectioned manifold (Fig. 13). This was attributed to the fact that an absolute duplication of the clayed or ground contours in the sectioned manifold was difficult to transfer to wood patterns. However, airflow of this modified induction system, as compared to the original cylinder head and manifold, showed a marked increase.

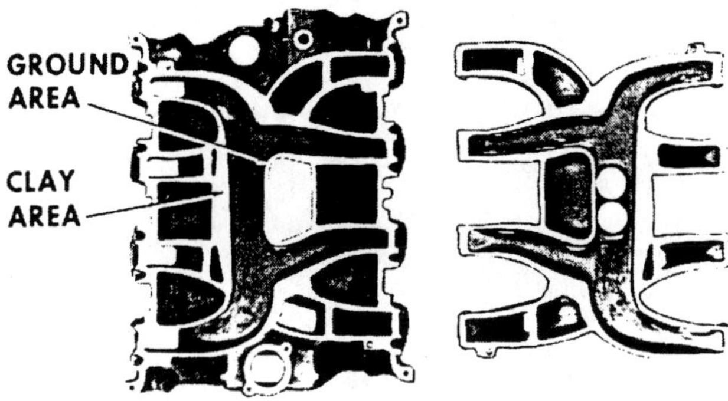

Fig. 11 - High-riser intake manifold sectioned through lower plenum

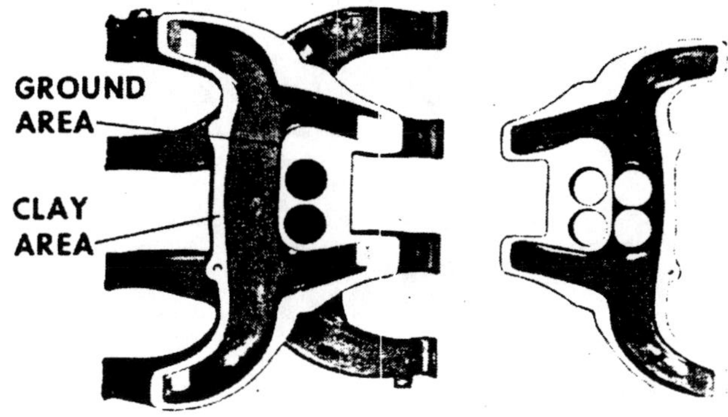

Fig. 12 - High-riser intake manifold sectioned through upper plenum

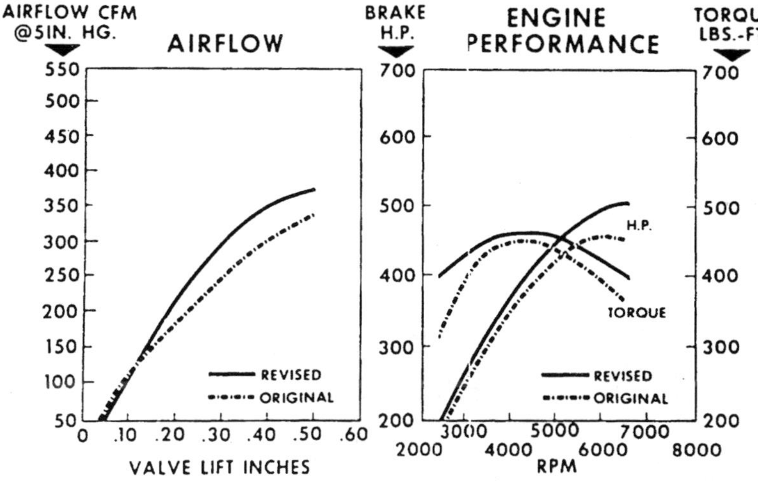

Fig. 13 - Induction system comparisons, high-riser intake manifold and cylinder head

Later airflow studies confirmed that some contours in a manifold runner can be so critical that the addition, movement, or removal of 0.010 in. clay thickness in an area can change the airflow by 5%.

The revised cylinder head, coupled with the new intake manifold, was installed on a dynamometer engine to measure the actual variation in performance attributed to the design changes. The results indicated an increase in horsepower.

MEDIUM RISER INDUCTION SYSTEM DEVELOPMENT

Further development of the intake port and manifold design clearly indicated that the high-riser concept alone was not essential to increased performance (Fig. 14). Induction airflow studies verified that reducing the cylinder head intake port closure, and lowering the intake manifold carburetor mounting pad to facilitate engine packaging and manufacturing did not affect engine output. This engine-induction system combination became known as the medium-riser engine.

CYLINDER HEAD - Using the cast iron high-riser cylinder head as a flowbox, the intake port roof line was reshaped by adding clay, and the port opening was reduced 1/2 in. with-

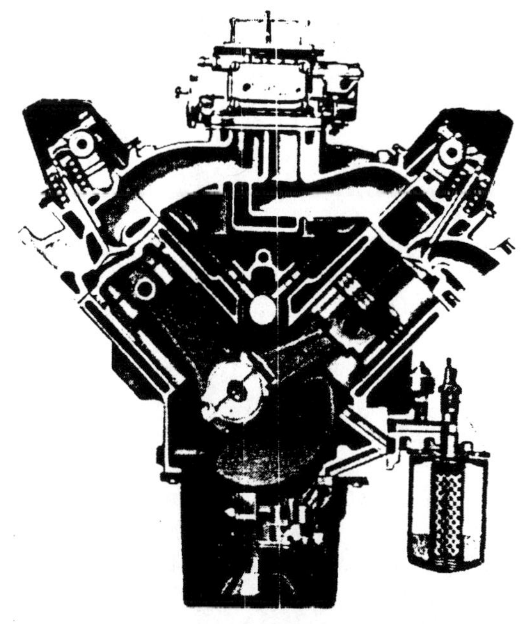

Fig. 14 - Medium-riser induction system, 427 engine

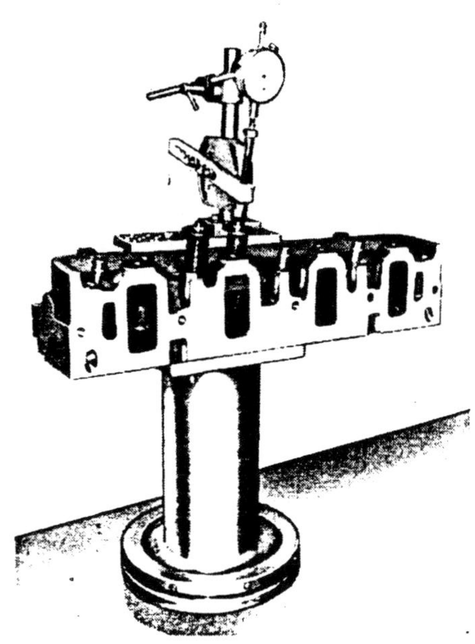

Fig. 15 - High-riser cylinder head on flowstand for intake port modifications

Fig. 16 - Rubber core of medium-riser cylinder head contours

out sacrifice of airflow (Fig. 15). All attempts to raise or lower the floor line at the port entrance resulted in airflow losses. This 1/2 in. reduction in the port roof line resulted in an intake port entrance which was smaller in area than the regular passenger car version of this engine, but it flowed as much air as the high-riser intake port. The exhaust port remained unchanged, since it was not increased in height when the high-riser cylinder head was first designed.

From the redeveloped cylinder head model, rubber and plastic models of the intake port were made to complete new casting equipment for the revised cylinder head castings (Fig. 16). Subsequent airflow checks of these new castings revealed the exact flow curve exhibited by the high-riser cylinder head.

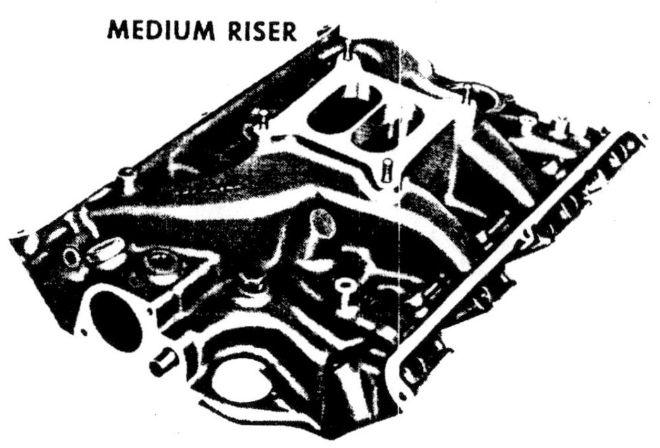

Fig. 17 - Four-venturi medium-riser intake manifold

Fig. 18 - Fiberglass flowboxes of manifold runner system

FOUR-VENTURI MEDIUM-RISER MANIFOLD - The program to design and develop a new, lower four-venturi intake manifold to mate with the medium-riser cylinder head provided the opportunity to advance the technique. A design layout was first generated which defined the cylinder head port locations, the new carburetor pad location, the existing pushrod locations, and the manifold runners (Fig. 17). It was required that the runners be similar in design but lower in silhouette than those developed for the high-riser manifold.

Wood runner coresticks were made, and two fiberglass flowboxes were made from the wood coresticks, one flowbox for the upper manifold runner, and the other for the lower runner system (Fig. 18). These fiberglass flowboxes were then ground and clayed in the same manner as the sectioned casting of the high-riser manifold to obtain maximum airflow and balance between all cylinders. Once airflow comparable to the high-riser manifold was obtained, a plastic epoxy resin was poured into the flowboxes and, upon hardening, the plastic coresticks were removed (Fig. 19). To compensate for casting shrinkage, the plastic coresticks were sectioned at constant cross-sectional areas so as not to affect the final airflow capacity. Actual casting corebox equipment could now be made directly from the plastic coresticks, assuring exact duplication in the manifold casting.

Fig. 19 - Plastic coresticks of upper and lower manifold runner system

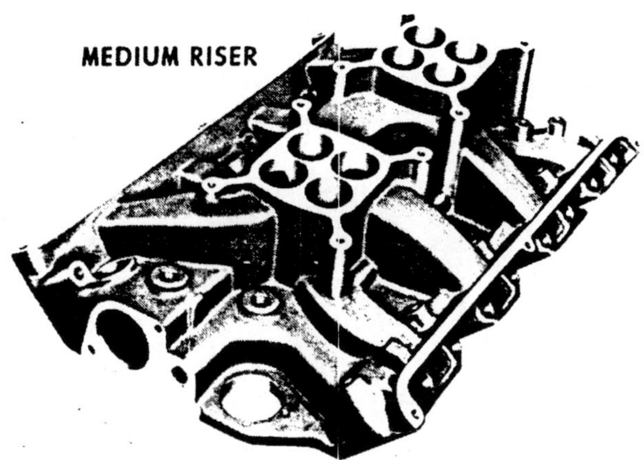

Fig. 20 - Dual four-venturi medium-riser intake manifold

To complete the design, a layout was made to provide normal metal thickness around the coresticks, to determine attaching locations of external components, and to provide machining dimensions. The finished intake manifold was rechecked on the flowstand. Airflow was almost identical to the original fiberglass flowboxes. The runner coreboxes were made directly from the plastic coresticks, and near-perfect duplication of the runner shape and airflow capacity was realized.

The success of this advancement in port and runner development technique was proved when the new cylinder head and intake manifold were installed on a dynamometer engine which had been "baselined" with the high-riser induction system. The new medium-riser engine power curve was within 1 hp of the high-riser system, which was well within limits of our objective.

EIGHT-VENTURI MEDIUM-RISER MANIFOLD - Following the same procedures used to develop the four-venturi manifold, a new dual four-venturi intake manifold was developed (Fig. 20). The manifold runners were essentially the same as the four-venturi intake manifold and were divided into an upper and lower system to even out firing pulses.

The four riser bores for the front carburetor were located as far forward on the manifold as possible without interfering with the distributor. In order to position these bores as close to the center of the front four runners as possible, the carburetors were rotated 180 deg, positioning the secondary venturies forward and the primaries to the rear of the manifold. The rear four riser bores were also located as far forward as possible to orient them near the center of the rear four manifold runners, thus allowing minimum clearance between carburetors.

This riser bore and manifold runner configuration shortened the runner lengths considerably and provided an almost straight-line airflow path (Fig. 21). One primary and one secondary carburetor venturi were now located diagonally opposite each other, feeding one cylinder on each bank. Cylinders 1 and 6, 2 and 5, 3 and 8, and 4 and 7 each used a different pair of carburetor venturies.

This type of straight-flow design also minimized fuel

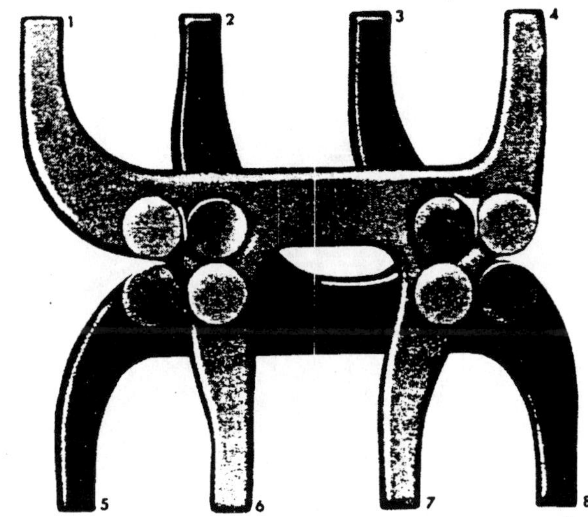

Fig. 21 - Dual four-venturi medium-riser intake manifold runner system, SOHC engine

distribution problems. The fuel distribution was further improved and the airflow increased by the incorporation of balancing passages and a plenum located between the front and rear carburetor riser bores. This permitted each runner to receive an additional boost of air/fuel mixture from the carburetor located on the opposite end of the manifold. Vertical vanes were oriented around the riser bore pillars to aid in directing the mixture from the riser bore through the plenum and to the runners. These vanes were also necessary to obtain balance between runners with maximum possible flow.

The resulting airflow figures of the new dual four-venturi intake manifold and cylinder head combination indicated a substantial increase in breathing capacity as compared with the single four-venturi manifold (Fig. 22). Corresponding improvements were experienced in performance.

GT INDUCTION SYSTEM

The medium-riser induction system was adapted to the Ford GT engine. The four-venturi intake manifold design

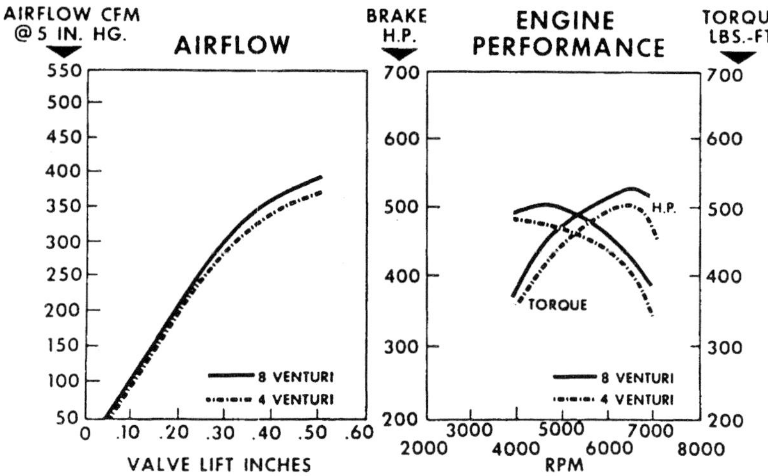

Fig. 22 - Induction system comparisons, medium-riser cylinder and intake manifold

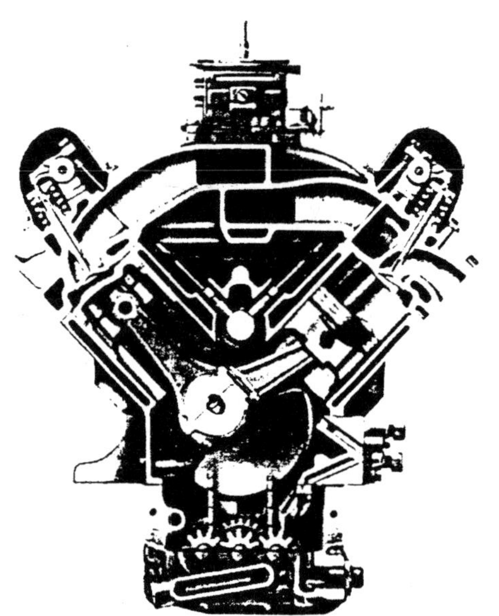

Fig. 23 - Medium-riser induction system, 427 GT engine

was adopted in its entirety, but the cylinder head design was slightly modified to enable the heads to be cast in aluminum rather than cast iron for reasons of weight reduction (Fig. 23).

In conjunction with the material change, valve seat inserts and valve guides were incorporated in the cylinder head. To provide clearance between intake and exhaust seat inserts and retain a common valve centerline, the valve head diameters were reduced from 2.19 to 2.09 in. for the intake valve, and from 1.72 to 1.65 in. for the exhaust valve. The intake valve seat inserts maintained the same venturi configuration as the iron cylinder head, even though the throat diameter was reduced by 0.100 in.

Airflow checks of the intake and exhaust ports of the aluminum cylinder head showed a slight reduction in airflow, which was also reflected as a slight loss in performance on the dynamometer (Fig. 24).

SOHC INDUCTION SYSTEM

Continuing demands for performance improvements led to another engine project -- the development of the 427

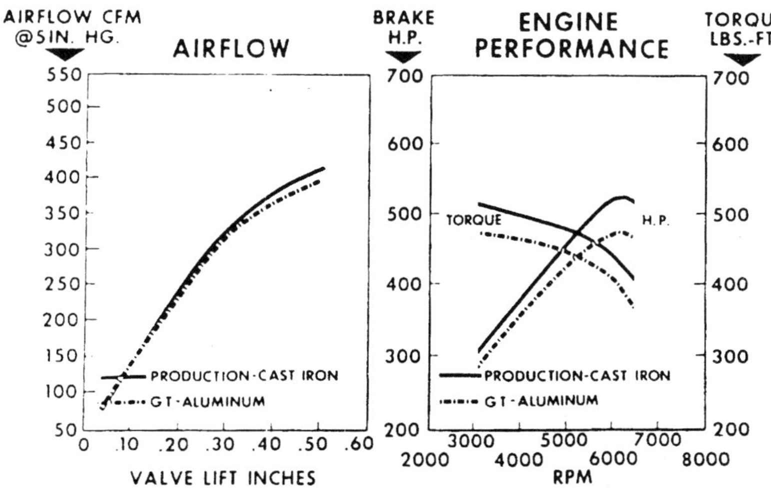

Fig. 24 - Induction system comparisons, GT cylinder head

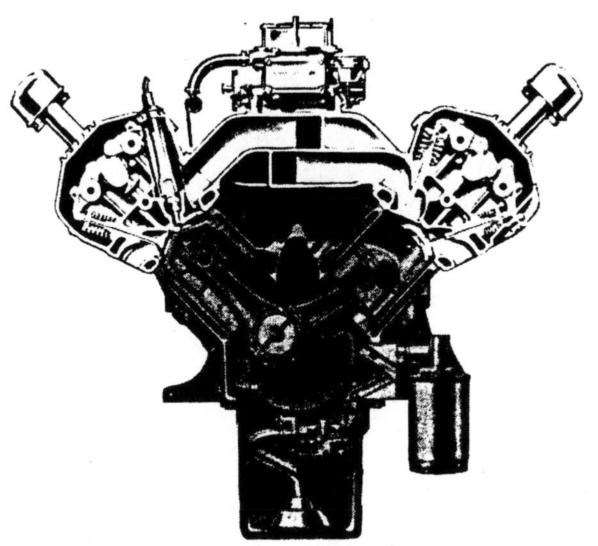

Fig. 25 - Induction system, 427 SOHC engine

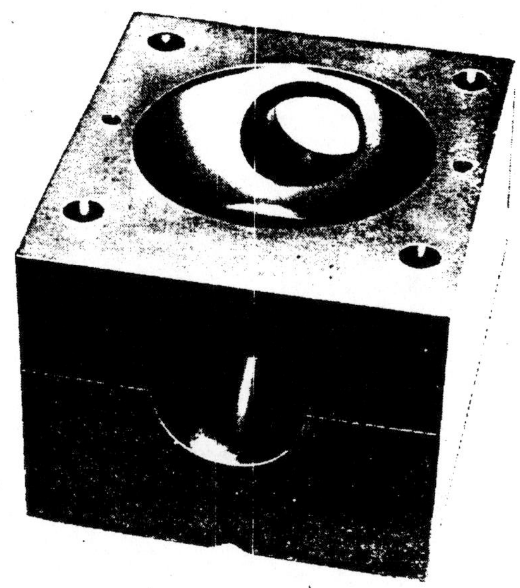

Fig. 27 - SOHC cylinder head flowbox

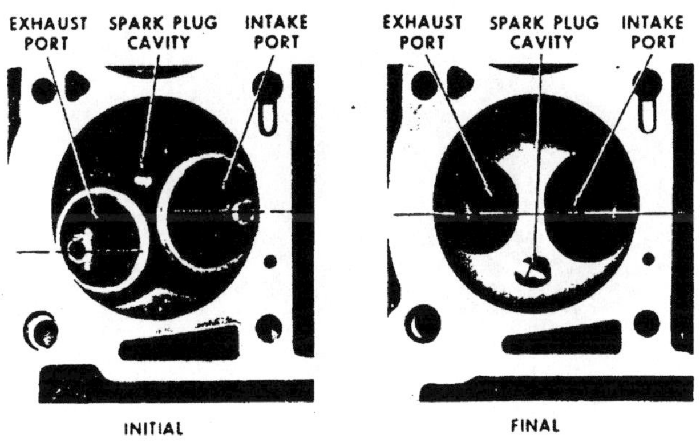

Fig. 26 - Combustion chamber designs, SOHC cylinder head

SOHC engine (Fig. 25). This engine was a natural evolutionary step from the proven concepts of the 256 DOHC engine and the 427 pushrod engine. To accommodate the new 427 SOHC engine, another induction system was developed.

CYLINDER HEAD - The initial cylinder head design used large intake and exhaust valves (Fig. 26). The exhaust valve was recessed and offset to one side of the combustion chamber to aid in more centrally locating the spark plug to provide equidistant flame travel. The intake valve was located on the combustion chamber centerline to provide the least valve shrouding possible for optimum breathing. The cast combustion chamber, which was a combination polyspherical-kidney shape, had to be qualified by machining for clearance between the piston and the chamber roof.

Airflow studies were conducted to establish baseline airflow data of this cylinder head design and to determine whether improvements could be realized through further contour development. Revisions in the combustion chamber were directed at modifying the exhaust valve cavity recess. More gradual slopes improved airflow slightly. By gradually filling in the cavity and raising the exhaust valve toward a spherical roof line, a considerable increase in airflow was gained. Further chamber development evolved a fully machined hemispherical combustion chamber design. In this chamber design, the exhaust valve was now seated on the chamber roof but relocated directly on a transverse centerline opposite the intake valve in the chamber. This necessitated relocating the spark plug cavity to a location further from the geometric center of the chamber. It was felt that the increased airflow more than offset the less desirable spark plug location. Another advantage to the completely machined chamber was that the compression ratio could be closely governed.

Once the new combustion chamber was established, a plastic cylinder head flowbox was made to incorporate the new chamber exhaust port design along with the original intake port (Fig. 27). The intake port and exhaust port both had round cross sections from the port entrance to the valve seat for more efficient flow. Rectangular cross sections are used in pushrod engine port designs because of space limitations imposed by the pushrod location and valve positioning in the combustion chamber.

Modifications to intake port floor and roof resulted in airflow improvements (Fig. 28). A hump on the port floor above the port throat area redirected the air toward the roof of the port which utilized more of the effective throat curtain area and improved airflow efficiency. The roof of the intake port was moved outward by grinding directly above the throat area. The 30 deg intake valve seat was replaced by a radius to eliminate the sharp edges of an angular valve seat. Turbulence was thereby reduced in this area. Valve sealing was also improved by providing a positive-line contact on the valve seat.

Development of the exhaust port was accomplished in the same manner as development of the intake port. Exhaust port airflow responded to both a hump on the floor and a modified roof contour. The final design of the exhaust

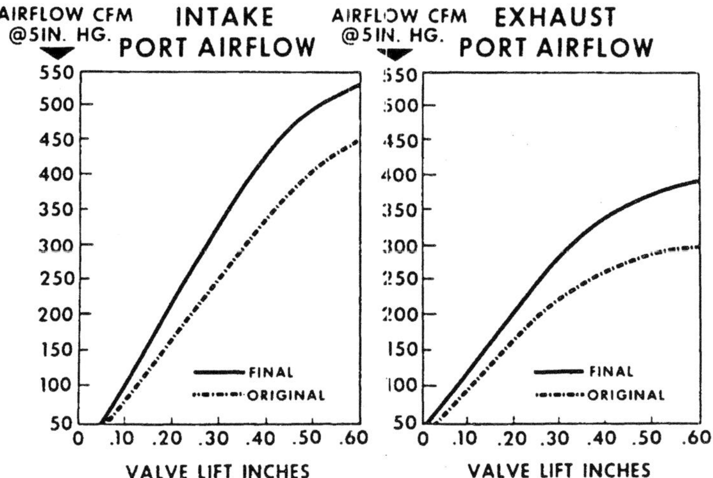

Fig. 29 - Induction system comparisons, 427 SOHC cylinder head

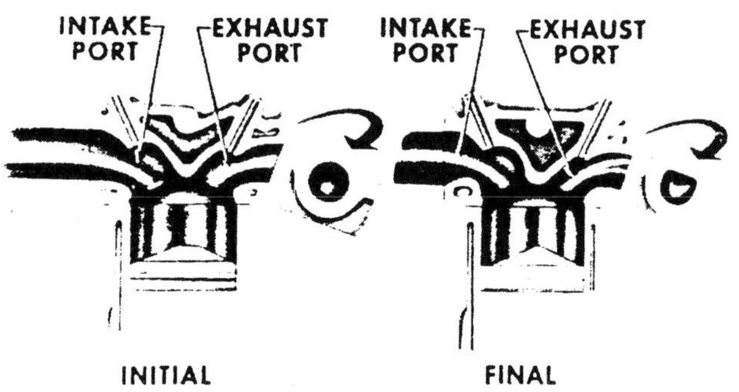

Fig. 28 - Port contour development, 427 SOHC cylinder head

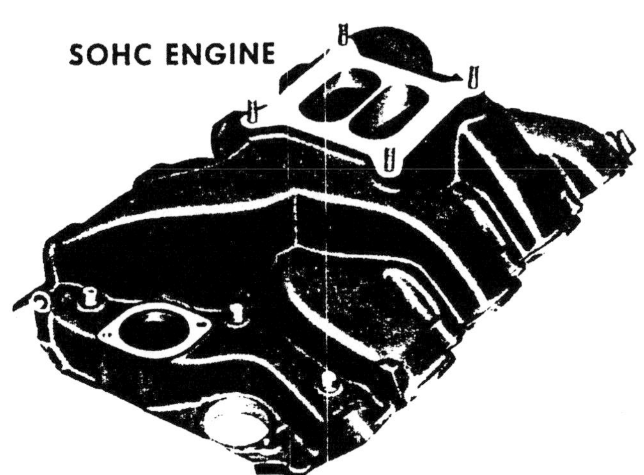

Fig. 30 - Four-venturi intake manifold, SOHC engine

port exit showed an airflow preference for a flat roof and rounded floor; whereas, maximum airflow is normally obtained using round cross sections. This design accommodates the radical change in direction from the top of the combustion chamber to the exhaust system. The rounded section at the bottom of the port exit was essential for good airflow when the valve was slightly off the seat and during the lower valve lift portion of the opening cycle. The flattened roof was necessary for maximum flow when the valve approached the wide-open position. The resulting design of the exhaust port exit of the cylinder head was a horizontal "D" shape.

The exhaust valve seat was changed from 45 to 30 deg, and the seat was modified from a 45 deg chamfer to a seat with a radius and throat similar to the intake side of the cylinder head.

The resulting airflow differential between the original cylinder head design and the airflow-developed final cylinder head is shown in Fig. 29.

FOUR-VENTURI INTAKE MANIFOLD - The four-venturi intake manifold runner and plenum designs followed those of the pushrod engine. The initial intake manifold design was accomplished without the benefit of airflow studies. Later manifold development, using fiberglass flowboxes patterned after the original manifold design, balanced the airflow of all the runners and increased the total airflow capability of the manifold. The manifold runners were rectangular in shape throughout most of the distance from the plenum to the cylinder head mounting face. At the mounting face, the cross section became circular to conform to the intake port shape. The four-venturi manifold design presently used on the single overhead cam engine utilizes circular sections throughout (Fig. 30).

Due to the high air velocities experienced in this four-venturi manifold design, fuel separation was found to be a problem. The four manifold outboard runners had a fairly direct path to the cylinders, while the other four runners had abrupt angles of more than 90 deg (Fig. 31). Airflow of all eight manifold runners had previously been balanced by contour development, but the engine experienced lean mixtures in the four cylinders fed by the inboard runners. As the air and fuel mixture descends from the carburetor through the riser bores and turns into the plenum, the velocities increase to the point where inertia drives the fuel particles to the ends of the manifold rather than to the inboard cylinders.

An airflow check, using the new cylinder head and four-venturi intake manifold combination showed an increase in

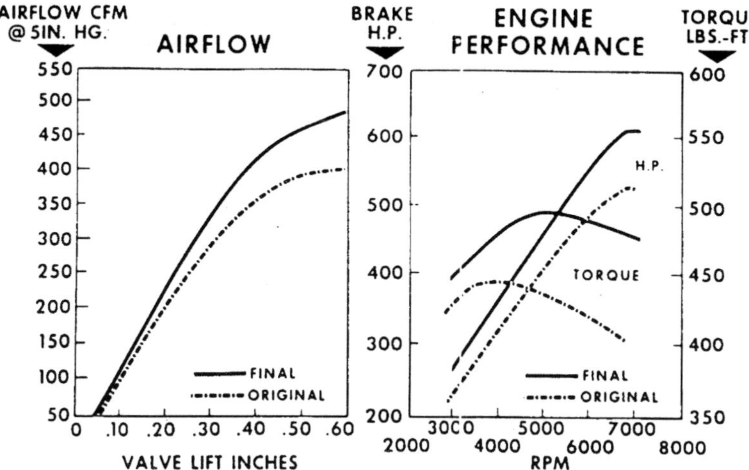

Fig. 32 - Induction system comparisons, SOHC cylinder head with four-venturi intake manifold

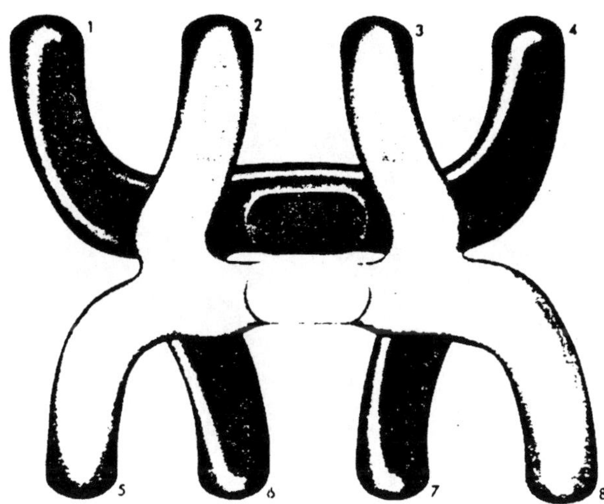

Fig. 31 - Four-venturi intake manifold runner system, SOHC engine

Fig. 33 - Dual four-venturi intake manifold, SOHC engine

airflow of 20% as compared with the original components. Dynamometer power curves also showed an increase of 20% in peak horsepower (Fig. 32).

EIGHT-VENTURI INTAKE MANIFOLD - A dual, four-venturi intake manifold was designed upon completion of the single four-venturi intake manifold development program (Fig. 33). The runner, plenum, balance passage and riser bore configuration were almost identical to the manifold designed for the pushrod engine. The intake runner cross section changed from rectangular to round near the runner exit to conform with the cylinder head port entrance design. All of the internal passages were also somewhat larger in cross section, primarily because of the increased flow requirement of the single overhead cam cylinder head port.

The airflow capacity curve for the dual four-venturi manifold showed gradual improvement, as compared with the single four-venturi manifold throughout the entire valve-opening cycle (Fig. 34). This indicated the potential performance advantage of the eight-venturi over the four-venturi engine package.

Induction system comparisons of the four-venturi versus eight-venturi intake manifolds conducted on a dynamometer engine showed an improvement in performance throughout the speed range of this multiple carburetion system. It is evident that an increase in airflow is inherent in this type of manifold design.

PRESENT INDUCTION SYSTEM DEVELOPMENT TECHNIQUES

The development technique for induction systems has evolved to the state currently being practiced. This technique has been simplified to the point where both design time and labor have been reduced appreciably, but most important was that the duplication accuracy has been improved significantly.

CYLINDER HEAD DEVELOPMENT - The next evolutionary step in the development of the cylinder head was concentrated mainly on the design of the intake port valve seat and throat. The objective was to improve design compatibility with manufacturing for the venturi throat, while main-

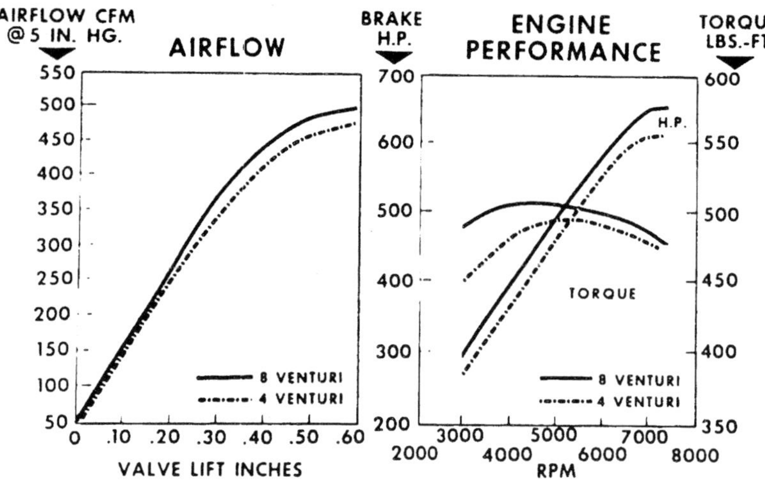

Fig. 34 - Induction system comparisons, SOHC engine

Fig. 36 - Valve seat cutting tool with contour template

Fig. 35 - Cylinder head flowbox with rubber cores of intake and exhaust ports

taining a high degree of airflow inherent in the present port design.

To facilitate cylinder head development, a rubber impression was made of the intake port, exhaust port and combustion chamber (Fig. 35). A plaster flowbox was made from this rubber impression which closely simulated the cast iron cylinder head. This plaster flowbox was easier to handle when removing material to make design changes.

Chamfers machined on the valve seats produced edges which had undesirable effects on airflow. As a result, a rounded-type of valve seat was considered. To make relatively accurate valve seat variations that would represent machined valve seats, a valve seat cutting tool was devised which used the valve guide as a pilot (Fig. 36). The opposite end of the tool contains a slot to hold the various valve seat contour templates and a large washer and nut to lock the templates in place. Rotating these templates in the plaster flowbox generated various valve seat shapes and contours. After each cut and flow test, the valve seat area was rebuilt with clay to its original state for additional evaluations.

Determining a final valve seat radius presented a problem, since each radius seemed to complement only one valve position.

Before the final seat contour could be established, a throat diameter also had to be determined. Having arrived at the most desirable throat diameter, the final valve seat became a developed curve which favored the low valve lift positions near the combustion chamber end of the seat. The curve progressively rolled down to the throat area which added to airflow at the higher valve lift positions.

With the final seat contour and throat diameter established, the valve seat cutting tool was mounted on a comparator to multiply the image 20 times its size. This enabled the contour to be accurately scaled and the dimensions to be established for designing production cutting tools.

Airflow through the intake port of both the high-riser and medium-riser cylinder heads responded well through the valve-opening cycle from the time the valve left its seat to a 0.500 in. valve position. Airflow at an increased valve-opening position, such as 0.550 or 0.600 in. afforded no appreciable improvement in airflow. This discouraged

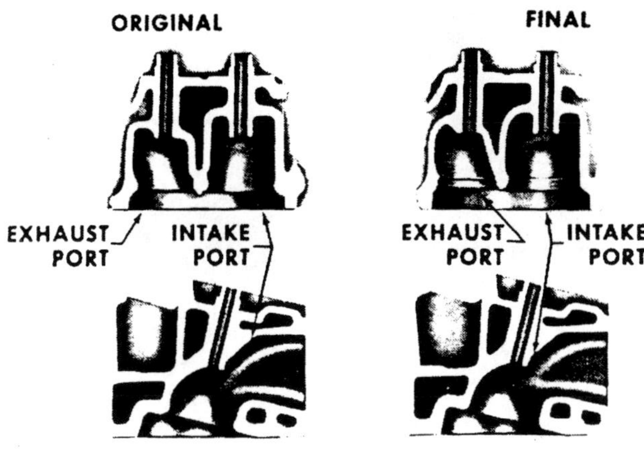

Fig. 37 - Cylinder head modifications

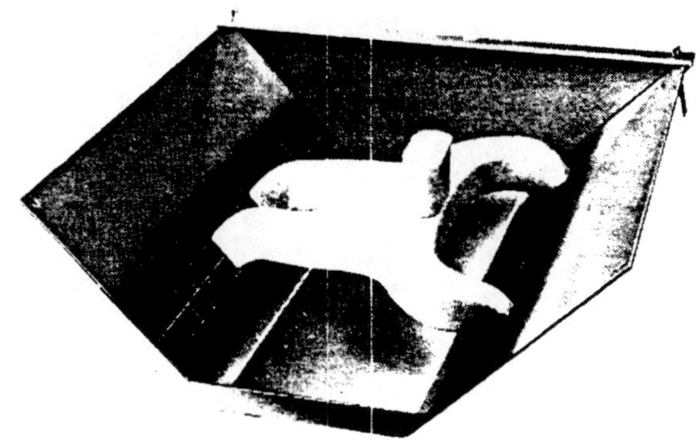

Fig. 38 - Styrofoam corestick model in cradle

use of high lift camshafts to gain additional performance improvements.

Further development to improve port airflow resulted in moving the port side wall nearest to the combustion chamber center outward approximately 0.100 in. (Fig. 37). This increased the cross-sectional area at this point and caused the intake port to be receptive to increased valve lifts exceeding 0.600 in. which was beyond the valve lift contemplated for use on this engine.

The shifted side wall of the intake port created a problem in casting the water passage between the intake and exhaust ports. To enable adequate cooling between these ports, the exhaust port was modified by applying clay on the exhaust port side wall nearest the intake port to represent the shift required to reestablish the coolant passage. Since this resulted in a loss of exhaust port airflow, the port wall, roof and floor were modified until a combination was found that recovered the lost airflow.

The plaster flowbox was then used to make rubber impressions for engineering layouts. The rubber impression allows the port core to be easily sectioned in any area or plane for an actual view of that area. Laying the sectioned piece of rubber on a copying machine provides an immediate reproduction of the section being interpreted.

Simulating the rubber impression technique, a plastic core model is made in the plaster flowbox. This plastic core model is used as an aid in making the coreboxes for the cylinder head castings.

The correlation of cylinder heads cast, using this technique, closely duplicates the airflow capacity of the original plaster flowboxes.

INTAKE MANIFOLD DEVELOPMENT - The intake manifold design and development technique was also made more sophisticated and accurate. The initial steps for intake manifold development now consists of defining the locations of the cylinder head port, the carburetor pad, and the pushrod, in sketch form, since these elements run through the intake manifold. The sketch also roughly defines the flow path centerlines of the proposed manifold runners.

The next step is the fabrication of a cradle, which consists of two metal plates fastened together to form a 90 deg wedge to simulate the cylinder head on the engine. Each plate has scribed layout lines which represent the cylinder head intake port openings. A third plate is then added across the top of the side plates at the height established for the carburetor mounting pad. This top plate has machined holes or slots which represent the riser bores.

Using the cradle and the locating sketches, a styrofoam model of the upper and lower manifold runner and plenum corestick is made (Fig. 38). The model is carefully formed to extend from the scribed cylinder head port openings on the cradle side plates, through the plenum, and to the top of the riser bores at the carburetor mounting pad in the top plate. Consideration is also given to provide adequate wall clearance between coresticks for casting.

The styrofoam corestick of the lower runner is then placed in the cradle. Screw holes are drilled through the side cradle plates to fasten the corestick in place. An approximate parting line is then established on the corestick, and metal shims are inserted in the styrofoam corestick at this point. Metal end-plates are then installed in the cradle and clay is applied to seal all joints. Next, a plaster mixture is poured into the cradle to the level of the shim stock parting line. Once the plaster hardens, the shim stock is removed from the styrofoam, and a parting agent is applied to all the exposed plaster surfaces (Fig. 39). Another plaster mixture is then poured into the cradle to the top of the riser bores and carburetor pad. When this has hardened, the plaster sections are opened and the styrofoam runner model is removed. The two plaster casts now form an intake manifold flowbox of the proposed lower runner section (Fig. 40). This sequence is repeated using the upper styrofoam runner core model. Both flowboxes are now ready for preliminary flowstand evaluation.

With one of the flowboxes attached to a cylinder head on the airflow test stand, individual runners are checked to determine airflow range and degree of unbalanced airflow.

The intake manifold runners are developed while mounted on a cylinder head to insure proper direction of airflow. The ultimate result is a complementary induction system (Fig. 41). Manifolds can be airflowed and developed independently. An appreciable drop in airflow, however,

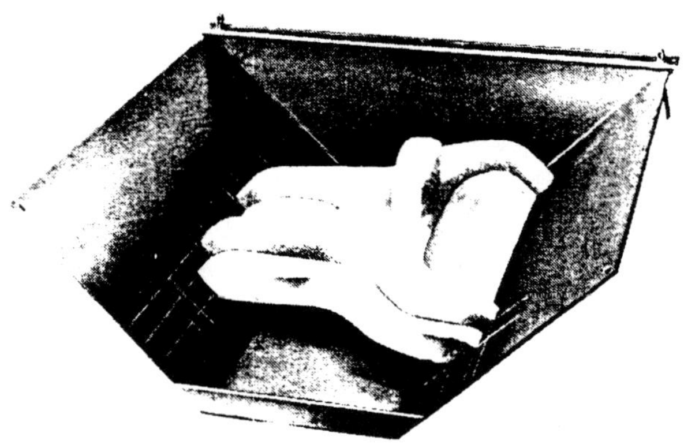

Fig. 39 - Lower plaster flowbox poured in cradle

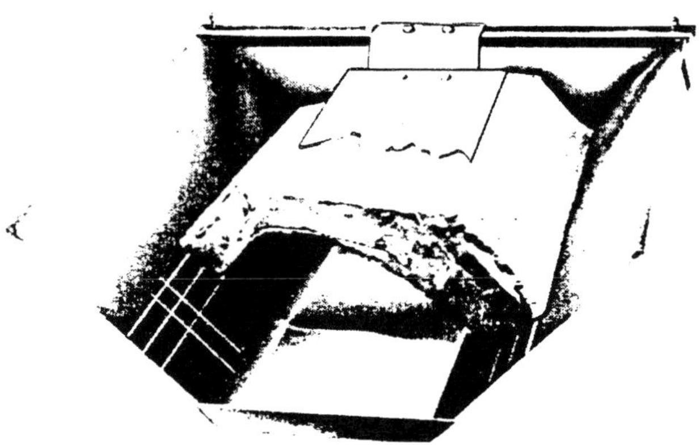

Fig. 40 - Upper plaster flowbox poured in cradle

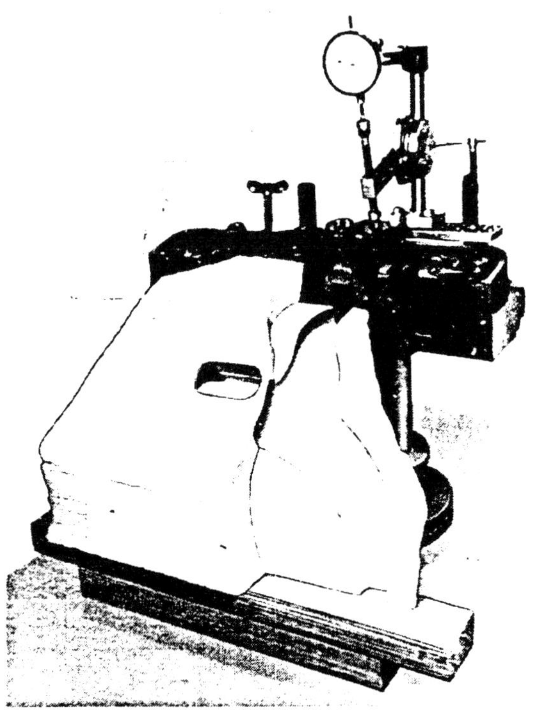

Fig. 41 - Plaster runner flowbox mounted to cylinder head on flowstand

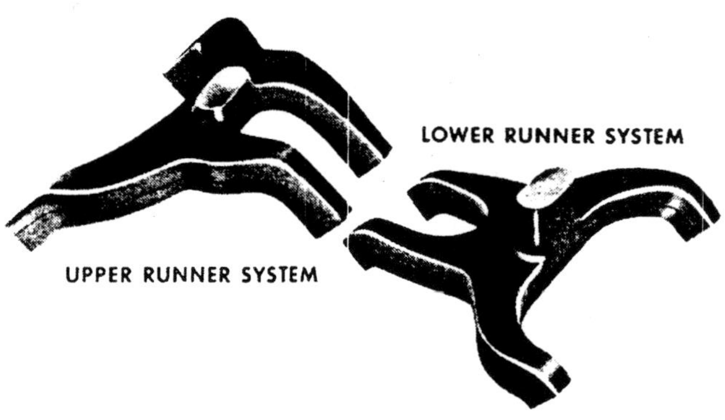

Fig. 42 - Secondary plastic coresticks

may result when coupled to a cylinder head. On occasion, port and manifold design limitations require that humps be added to the runner floor to improve airflow direction for increased efficiency.

Each runner of the plaster flowbox is then modified by grinding or adding clay to achieve maximum airflow and equal airflow between cylinders. Optimum plenum width and height must also be established to obtain increased flows. Particular attention must be taken to properly develop the areas where the runners connect to the plenum. Another critical area is the transition from the riser bores to the plenum. In most instances, a 1/2 in. radius produces the maximum airflow.

Once this stage of refinement is attained, additional clay is added to various areas in the manifold runners to determine if a reduction in size would maintain airflow volume while increasing flow velocity.

The next operation consists of pouring a plastic mixture into the plaster flowbox. After the plastic sets, the plaster flowboxes are parted and the plastic coresticks are removed. These coresticks are then nested into the metal cradle to determine if the proper metal thickness is still available between coresticks. Fiberglass flowboxes are then formed around these secondary plastic coresticks (Fig. 42). These flowboxes may contain slight modifications to the parting lines to facilitate casting. Once the fiberglass flowboxes are completed, they are rechecked for airflow and reworked, as necessary, to correct flows.

The primary plastic coresticks are then poured using the reworked fiberglass flowboxes (Fig. 43). These coresticks are used to construct the production casting equipment.

PORT AND RUNNER POLISHING - Another aspect of induction system development, which was investigated, was the polishing of ports and intake manifold runners. This tedious and time-consuming effort was found to be of no actual benefit to induction system airflow. Polishing has in some cases been found to be detrimental to airflow capacity. As mentioned earlier, some areas in the port and runner are so sensitive to variations in contour that the removal of the slightest amount of metal can redirect airflow and result in performance losses. In cases where polishing did have a beneficial effect, the port casting was found to be origin-

Fig. 43 - Fiberglass flowboxes with primary plastic coresticks

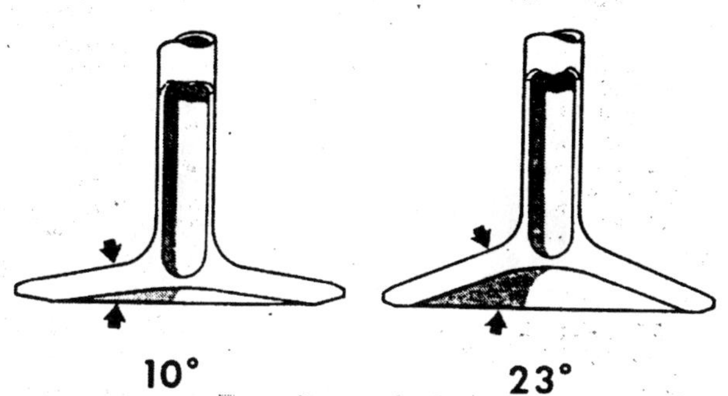

Fig. 45 - Modified valve underhead angle for improved durability

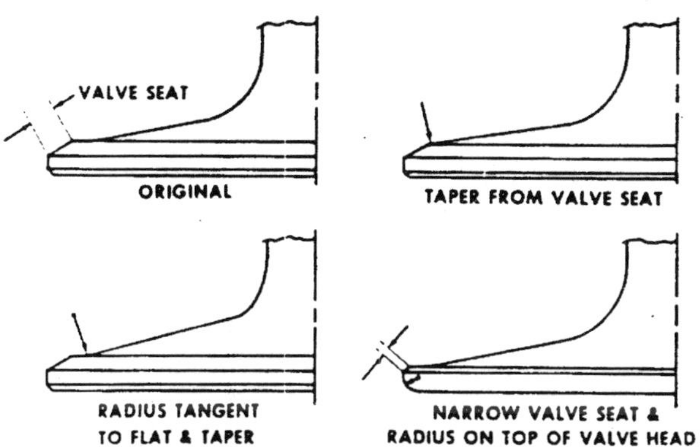

Fig. 44 - Intake valve designs

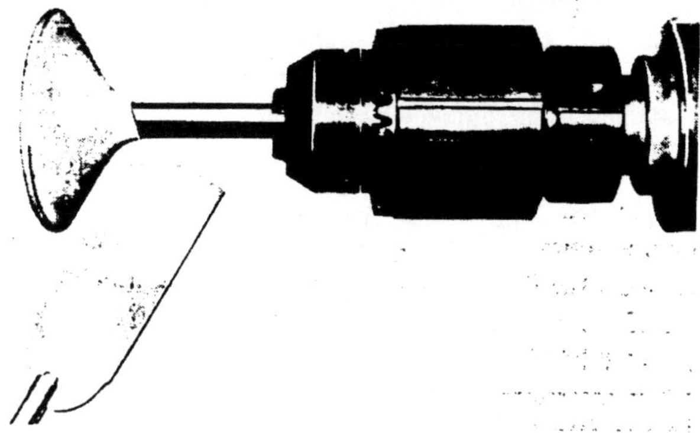

Fig. 46 - Valve head shaping

ally undersize, and the improvement was actually the result of correcting cross-sectional area. Of course, slight metal protrusions or casting fins which sometimes are the result of faulty cores should be removed by lightly grinding to eliminate the possibility of inducing turbulence.

VALVE DESIGN

Optimum valve design, to complement the intake port and exhaust port, is essential for maximum airflow. The valves initially used in the wedge combustion chamber of the 427 engine were similar in design to those found in a passenger car engine. The intake valve had a 30 deg valve seat angle and a 10 deg underhead angle, while the exhaust valve had a 45 deg valve seat angle and a 20 deg underhead angle. In the course of developing the intake and exhaust ports, various valve head configurations were evaluated (Fig. 44). Starting with the existing intake valve, several modifications were made. The underhead of the valve, adjacent to the seat, was first reworked to taper directly from the edge of the valve seat. Another modification was the addition of a slight radius, tangent to the flat and taper, before moving toward the stem at a 10 deg angle. Airflow favored the original valve design with the short flat.

Variations in valve seat angle had little effect on airflow in the various port and chamber configurations, although a 30 deg exhaust seat showed a definite improvement on the 427 single overhead cam engine. The valve seat width was then reduced, resulting in a definite increase in airflow. On the opposite side of the valve head, the radius replacing the chamfer had no effect on airflow.

The combination of the preceding data was the basis of the valve head design used on the early 427 cu in. engines (Fig. 45). The hollow stem of the valve was incorporated primarily as a means of reducing weight, thus increasing the valve toss or no-follow speed. The head was designed with a constant-stress head section for maximum head flexibility and minimum weight. Also, this valve was forged rather than cast for optimum strength and incorporated a swirl-polished underhead for removal of grind marks, as well as to provide turbulence-free airflow.

In later valve designs, the valve underhead angle was revised from 10 to 23 deg for durability at increased engine rpm. Studies indicated an airflow loss when using the 23 deg angle valve, but the airflow sacrifice was offset by increased valve durability and permitted higher engine speeds for longer periods of time. This valve design was used in the 427 GT engine and is currently being used in production.

Earlier airflow studies of the 23 versus 10 deg angle valves with the SOHC hemispherical combustion chamber design responded in the opposite manner. Increasing the angle on the intake valve progressively increased airflow. These results prompted a thorough investigation of valve head design.

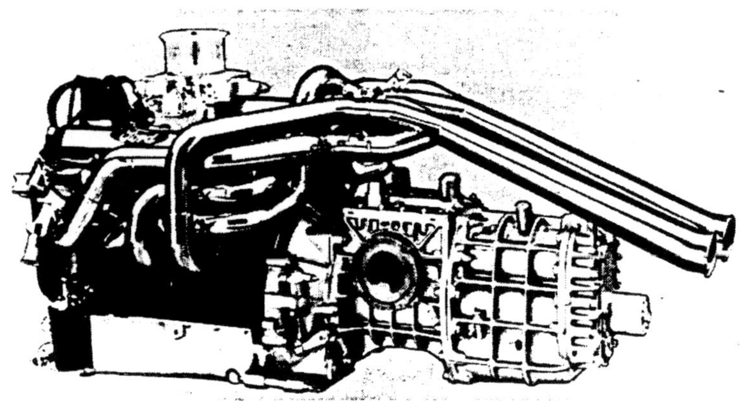

Fig. 47 - 427 GT engine

An early cast valve design with a nearly flat head was ground until the valve head-to-stem angle was 90 deg. This valve was then mounted in a motor-driven chuck and clay was added to or removed from the head to form various head shapes and angles (Fig. 46). As indicated previously, the increased valve angle improved airflow when using a hemispherical combustion chamber. With the wedge combustion chamber, the 90 deg valve provided maximum airflow. This type of valve, however, has very high stress levels through the valve head section.

It is interesting to note that many intake port designs will actually flow more air with the intake valve opened between 0.500 and 0.600 in. than a port with the valve removed. This is the type of complementary design desired for optimum airflow.

FUEL DISTRIBUTION

Fuel distribution was the only problem area encountered on engines equipped with four-venturi manifolds, which were developed using the airflow technique. This problem is caused by inadvertently designing fuel-separating features in the induction system even though the airflow is equally balanced throughout the manifold. The degree of unbalanced fuel distribution can be approximately determined in a running engine through the use of thermocoupled spark plugs, temperature-sensitive valve alloys, and exhaust gas analysis. In most instances, the problem can be minimized by making carburetor modifications in jetting, notching carburetor booster venturies to redirect fuel dispersions, or adding deflectors or pins. If the problem persists and cannot be alleviated by these means, a mechanical type of deflector can be introduced somewhere inside the manifold. Balance passages, notches and ribs may also be added.

INTAKE MANIFOLD AND EXHAUST MANIFOLD TUNING

Tuning an intake manifold consists of lengthening or shortening the intake manifold runners to vary the magnitude and rpm of the horsepower and torque peak (Fig. 47). Dynamometer development work demonstrated that the shortest possible runner lengths produced higher horsepower at higher speeds. This is practical for certain applications, but the prime objective is to obtain optimum horsepower throughout the higher engine rpm range.

Exhaust manifold tuning has, primarily, the same effect on engine performance as intake manifold tuning. Short pipe lengths favor high rpm power peaks, and longer lengths increase torque in the mid-range speeds. Exhaust system designs of different vehicles may vary drastically due to different chassis designs and space availability in the engine compartment. Optimum pipe diameter must first be established in order to realize the full benefit of exhaust system tuning. To determine the optimum pipe length for the exhaust system, pipes of an appropriate diameter are attached to an engine. While the engine is running in the rpm range in which the tuning is to be most effective, the true pipe length is determined. The pipes may be bent to fit the engine compartment, and all four pipes on one side of the engine must terminate in one plane. Severe bends in the tubing are permissible and have no effect on tuning, providing the proper cross-sectional areas are maintained. The pipes may then be connected to a common collector and then routed through a larger diameter pipe toward the rear or side of the vehicle to be expelled to atmosphere. The 427 cu in. engines have been responsive to the tuning of fabricated tubular header systems. The most successful systems produce higher horsepower throughout a wide speed range.

CONCLUSION

The GT induction system is the product of a new development technique that involved the extended use of an induction flowstand, as well as extensive studies of various types of manifolds and cylinder heads. During the course of this work, successive improvements in modeling and pattern making have combined to refine the art to a high degree of sophistication.

A large measure of credit for the development of tools and techniques, such as this, must be given to the challenge and impetus supplied by the Ford Motor Co.'s participation in the highly competitive sport of automobile racing.

670068

Mark II–GT Ignition and Electrical System

Robert C. Hogle
FORD MOTOR COMPANY

There is no handbook of vehicle electrical system design or no secret formula which can serve as a guide in the development of electrical components for competitive events. The basic approach to the extent possible, therefore, was to use production components or components for which we had considerable background experience.

The design group assigned to the MK II project consisted of engineers with experience on components of their responsibility. The members of this group were asked to limit their considerations to failure mechanisms which were actually causing part failures, rather than to speculate on problems more imaginary than real. Failure analysis data were examined on components used in prior competitive events. Vibration was the major cause of failure; parts which were defective prior to installation or damaged at installation were a secondary cause of failure. Corrective design action was taken at the outset to eliminate these types of failures. Corrections were verified during dynamometer engine and driveline tests, as well as in vehicle durability tests.

The following excerpts, from the general regulations for the 24-hour race, have a bearing on electrical system performance and reliability.

- The vehicle must be equipped with a self-starter that may only be operated by the driver from his seat.
- The engine must be stopped during all pit stops.
- The starting motor, generator or battery may not be replaced as complete assemblies.

  Recharging of batteries by outside means is prohibited.
- The windshield wiper, self-starter, statutory lighting equipment and warning devices must be in working order throughout the race.
- Spare parts or tools must be on board the vehicle or in the working pit.
- The driver may not take spares or tools from the pit nor may these be brought to him.

From these few regulations, you can realize the importance of a reliable electrical system. Our total objective was to simply reduce the chance of electrical failure to an absolute minimum.

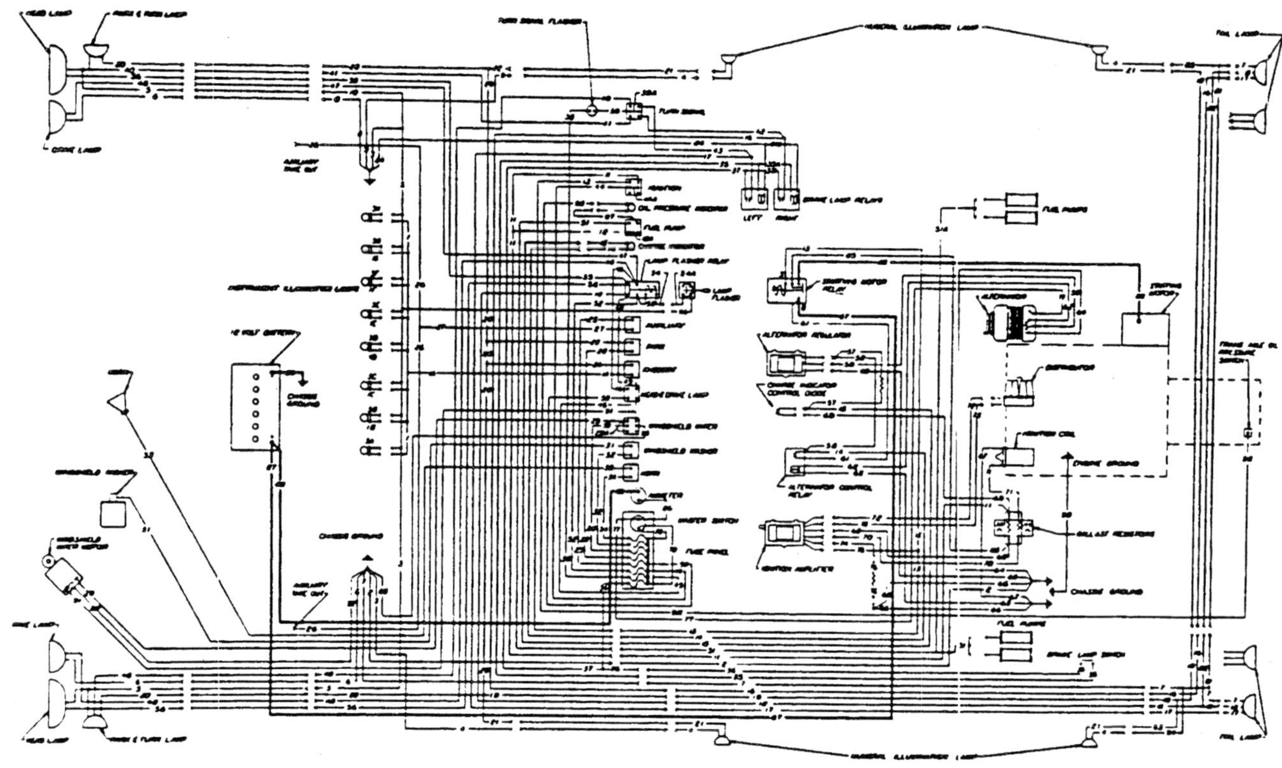

Figure 1 — Vehicle Circuit Diagram

WIRING SYSTEM

Wiring is a flexible component which is relatively easy to change and re-route. More often than not, it is kicked around until the last moment, which usually results in too many compromises for the best service.

The first task in developing an efficient wiring system is to determine the location of all electrical components with respect to their functional, environmental and servicing requirements. The engine electrical component locations were fixed by the basic engine design, and no changes could be made to the standard production locations and mountings. The transistorized ignition amplifier and voltage regulator were located behind the left seat, near the floor pan, adjacent to the alternator field relay. All control switches, fuses and lighting control relays were located on or behind the instrument panel.

A schematic diagram of the final vehicle wiring is shown in Figure 1. From the diagram you would think it was a fairly complex electrical system. A detailed review of the system, however, would show that it contains only the basic sub-systems found in any normal vehicle used on the public highways.

Figure 2 shows the physical components in their relative vehicle position. In spite of a relatively light electrical load in some of the circuits, the wire size chosen for the minimum acceptable tensile strength was #14 AWG with heavy-wall, chlorosulfonated polyethylene insulation, SAE Type HTS. This wire has a maximum service temperature of 275° F. and is the recommended SAE practice for heavy duty trucks. The insulation of this wire is thermo-setting and therefore, does not melt when subjected to the heat produced by an overloaded circuit. Adjacent wires in a harness using this insulation are not damaged when an overload occurs.

As a servicing convenience, electrical components were equipped with multiple disconnects wherever possible. Important polarity-sensitive components, such as the voltage regulator and ignition amplifier, could then be removed and replaced readily with no risk of hook-up error.

Determining a safe routing for wiring harness assemblies is an important consideration in maintaining the security of the wiring system. The shortest route results in significant weight savings and in the least voltage drop.

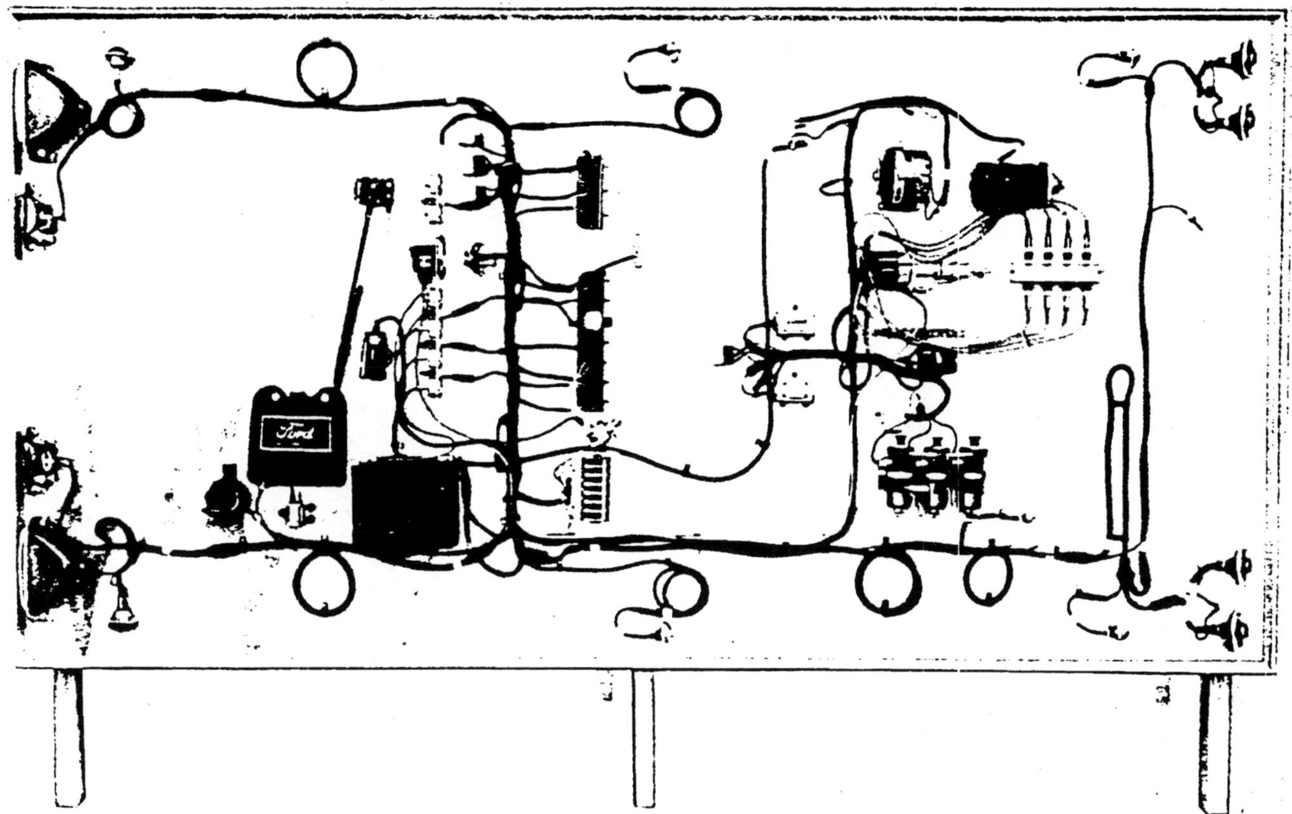

Figure 2 – Display Buck

Care must be taken to avoid congestion, damage by hot engine components, moving parts of the vehicle and other such potential trouble makers. Full utilization was made of the available protection afforded by existing channels in the vehicle structure.

Two harnesses, which take considerable abuse, are the front and rear body lighting harnesses. The front and rear body sections were made removable for ease of service; this required the use of high tensile strength harnesses with waterproof quick disconnects shown in Figure 3. The required tensile strength was obtained by extruding a neoprene jacket over the wiring harness. It is also required that these harnesses be provided with strain relief to prevent damage if the body sections were removed without first separating the disconnects manually.

All switches used were heavy truck or aircraft-types. A master disconnect switch in the primary circuit was used for safety purposes at pit stops and when the vehicle was being transported or stored.

GENERATING AND STARTING SYSTEM

Electrical loads for the MK II application were planned to be all continuous except the starting motor, horn, windshield washer and turn signals. Differences in day-night loads were not considered, since the entire race might be run under adverse weather conditions requiring lights and windshield wiper.

An alternator with a rating of 52 amperes was chosen; it exceeded the full-load requirements and no deficits had to be made up by the battery, except for pit stops.

The alternator and transistorized voltage regulator used were basically production units modified to withstand extreme vibration. The regulator incorporates 2 transistors, 3 diodes, several condensers and resistors and no moving parts whatsoever; the limiting factor in its application was that the ambient temperature must not exceed 200° F. The transistorized voltage regulator was a discrete component assembly in which all the components were locally potted to the circuit board.

Critical components, such as the alternator diodes, were selected for high reverse voltage and low-leakage characteristics. The rectifier

Figure 3 – Wiring Disconnects

assembly, shown in Figure 4, was potted to eliminate the possibility of relative motion between any of the components. The field coil was epoxied to the pole pieces to minimize the effect of vibration and rotational stresses.

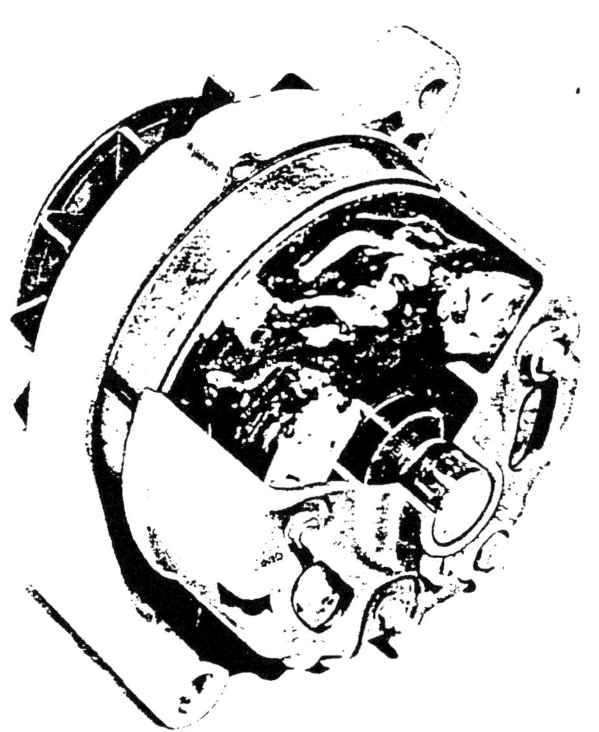

Figure 4 – Alternator

Figure 5 shows the 53-ampere hour SAE Group 2SMB battery with 9 plates per cell selected for this application. This is a standard multipiece battery sealed with hard epoxy in place of the normal soft asphalt seal. The plates are anchor-bonded to the bridges with an epoxy.

The starting motor has only one function to perform. It must crank the engine at a sufficiently high rpm to insure a start. In the interest of starting system reliability, it is essential to limit cranking to the shortest possible time. The starting motor must be matched with the battery and battery cables to insure high cranking rpm. The principle concern was, of course, to hot-crank a loaded engine, purge it, and accomplish a quick start.

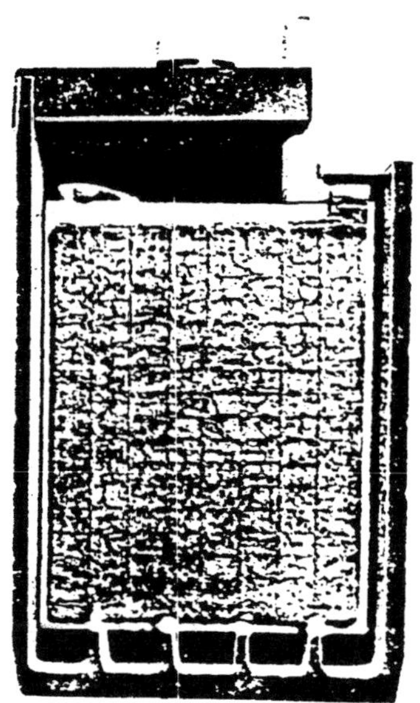

Figure 5 – Battery

The starting motor for this application was a 4.5" diameter, positive engagement production unit with minor modifications, as shown in Figure 6. The field coils, connecting links and brush shunts were bonded to the frame with silicone rubber. The actuating coil aluminum sleeve was replaced with brass to avoid possible fracture and interference with starter drive engagement.

ELECTRIC FUEL PUMPS

The electric fuel pumps shown in Figure 7, were production units mounted directly above the fuel tank to minimize the possibility of vapor lock. The inlet and outlet ports were enlarged to reduce flow losses.

Two pumps were used in the main fuel system with a third pump in reserve. The pumps were connected with common manifolds, and the selection of pumps was done electrically. No provisions for back flow is required, since each pump has two check valves preventing reverse flow in a non-operating pump.

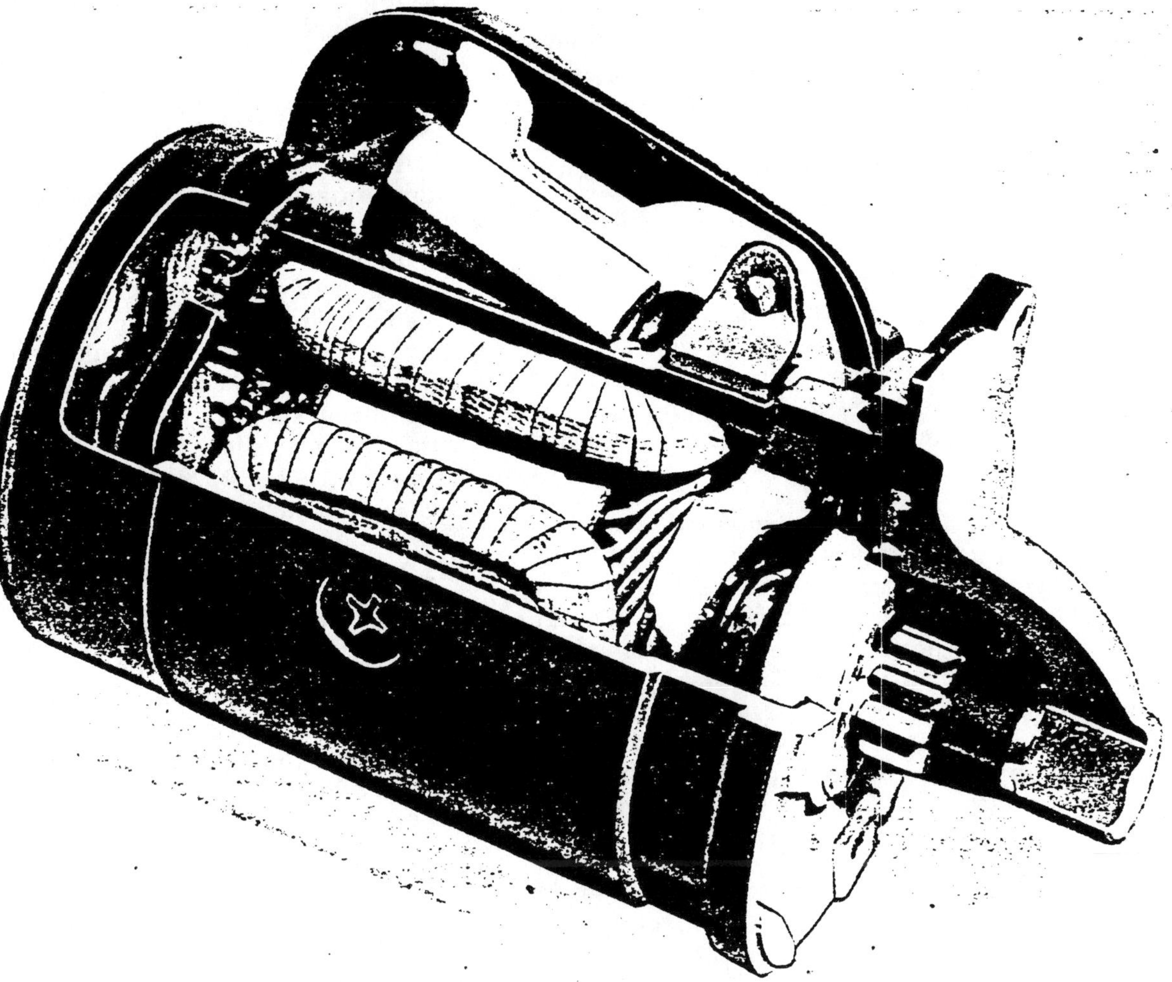

Figure 6 – Starting Motor

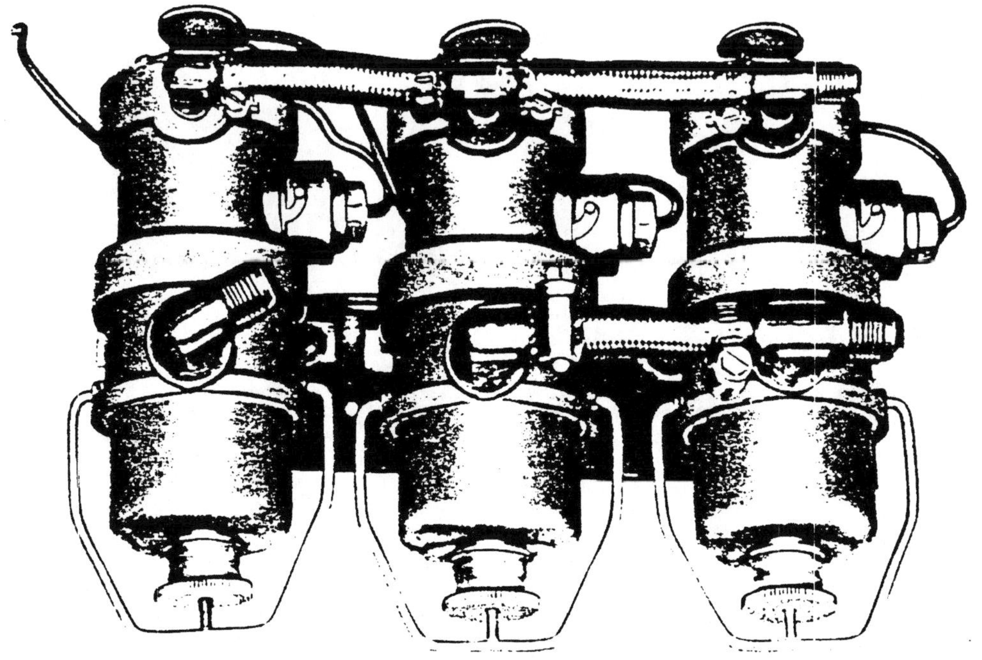

Figure 7 – Fuel Pumps

WINDSHIELD WIPER SYSTEM

The windshield wiper components (Figure 8) were modified to our specifications from units used on Boeing 707-type aircraft. The system used consisted of a D.C. motor, oscillating gear box and a flexible shaft, 15" long, connecting the motor and gear box.

Normal wiping speeds for aircraft applications are in the order of 240 wipes per minute through an angle of 60 degrees. The D.C. motor as applied to aircraft 28 volt systems operates at 11,000 rpm. Since the MK II has a 12 volt electrical system, the resulting wiping speeds were between 105 and 115 wipes per minute.

The wipe angle was increased from the normal aircraft application of 60 degrees to 110 degrees. The arm and blade were also a modification of equipment designed for aircraft use. The main portion of the arm was stainless steel tubing. A spring-loaded blade pressure adjustment was incorporated in the section that mounted to the gear box output shaft. The blade was an anti-windlift design modified to withstand vehicle design speeds, and was adjustable with respect to the arm.

Windlift was non-existent at any vehicle speed. Blade pressure was in the order of 30 ounces, which is slightly higher than that specified for passenger cars. The arm and blade assembly were given a black-mat finish to reduce glare.

LIGHTING SYSTEM

The head lamps used were high intensity, iodine-quartz units used on some European vehicles. These lamps have a single filament; no dimming is required since there is not oncoming traffic. The passing lamps, sometimes called driving lights, were also iodine-quartz. For nighttime operation, all four lamps are normally on, and the headlamps are turned off momentarily by means of a steering column-mounted flick switch, to signal a vehicle being overtaken. For daytime operation, the flick switch turns the driving lamps on to indicate a passing situation. All other lighting on the vehicle was of the normal incandescent-type, except for the heavy duty bulbs; these were developed jointly by the SAE and the industry for long-life heavy truck use. The most troublesome lights on the vehicle proved to be the side number identification and vehicle recognition lamps which were often wiped off on straw bales.

IGNITION SYSTEM

The decision to use a breakerless transistor ignition system was based on its highly satisfactory performance. This dates back to the

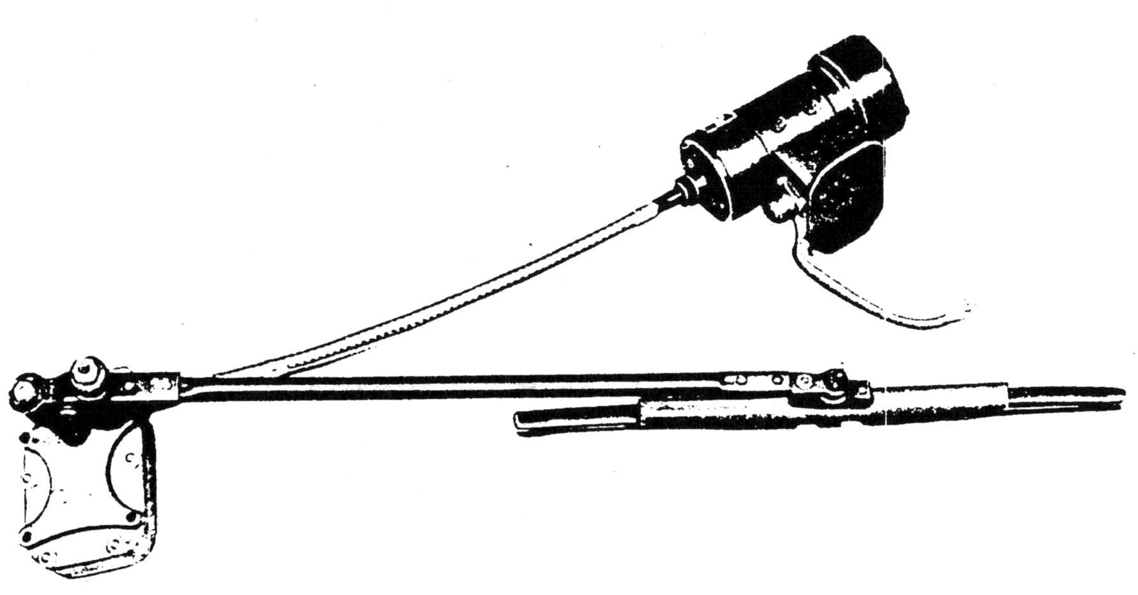

Figure 8 — Windshield Wiper

Figure 9 – Distributor Assembly

initial breakerless application on the 4.2 liter push rod engine used at Indianapolis in 1963.

There are two salient features of this system which are particularly advantageous for a racing engine:

1. Elimination of distributor contact bounce at high speeds which results in consistently high voltage to the spark plugs and no engine misfiring.

2. Elimination of distributor contact rubbing block wear resulting in no losses of engine timing.

These features enable the engine to finish the race in a "just-tuned" condition.

The breakerless ignition distributor incorporates a variable reluctance magnetic pickup in place of the breaker points. The distributors used on the Mark II and GT 40 are identical and differ from the original 4.2 liter application only in minor details. The principle change involves the addition of a centrifugal spark advance mechanism.

Figure 9 shows the distributor assembly incorporating a mechanical tachometer drive and the magnetic pickup. All other parts of this distributor are of a standard production design.

Figure 10 — Distributor Assembly — Less Cap and Rotor

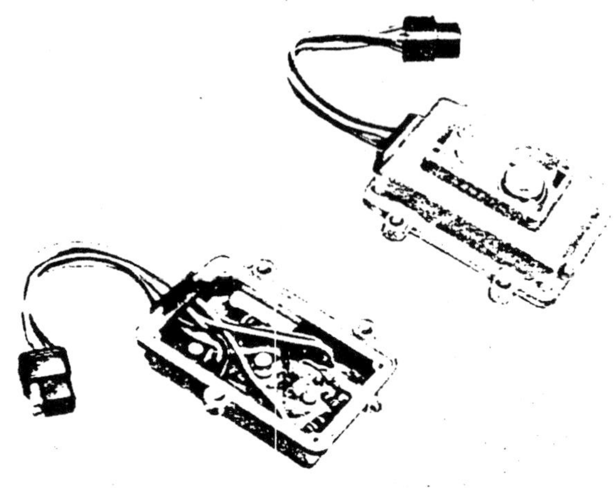

Figure 11 -- Ignition Amplifier

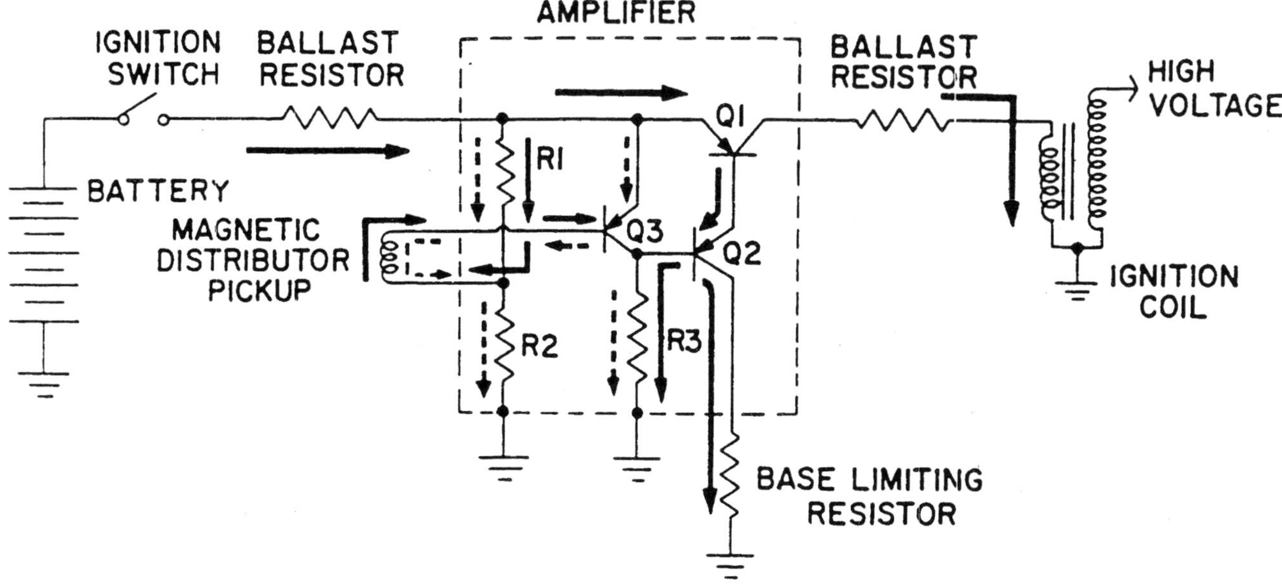

Figure 12 — Functional Circuit Diagram — Ignition Amplifier

Figure 10 shows the distributor with the cap and rotor removed. The eight-fingered metallic rotor and twin-coil stator form the key parts of the variable reluctance magnetic pickup. This arrangement produces eight pulses per revolution of the distributor shaft. The magnet is cast into the rotor assembly.

Figure 11 shows the electronic package, or amplifier. The heat sink housing is a modified production transistor regulator casting. The circuit utilizes discrete components interconnected on a printed circuit board. The power transistors are mounted directly to the heat sink and enclosed in the "dog house." The completed assembly is fully potted with epoxy resin to reduce the possibility of vibration damage and to improve the heat transfer characteristics of the internally mounted components.

Figure 12 is a functional schematic of the system.

The electrical signal from the magnetic distributor is referenced in the amplifier by the bias level set by resistors R1 and R2. This level is established so that with zero signal from the magnetic distributor, the input transistor (Q3) is turned ON, thereby clamping transistors Q2 and Q1 OFF. (The current flow is shown by the dotted line.) A negative signal from the distributor provides additional "ON" drive to Q3; however, since Q3 is fully saturated — that is, turned ON in its static state — this additional drive has no effect.

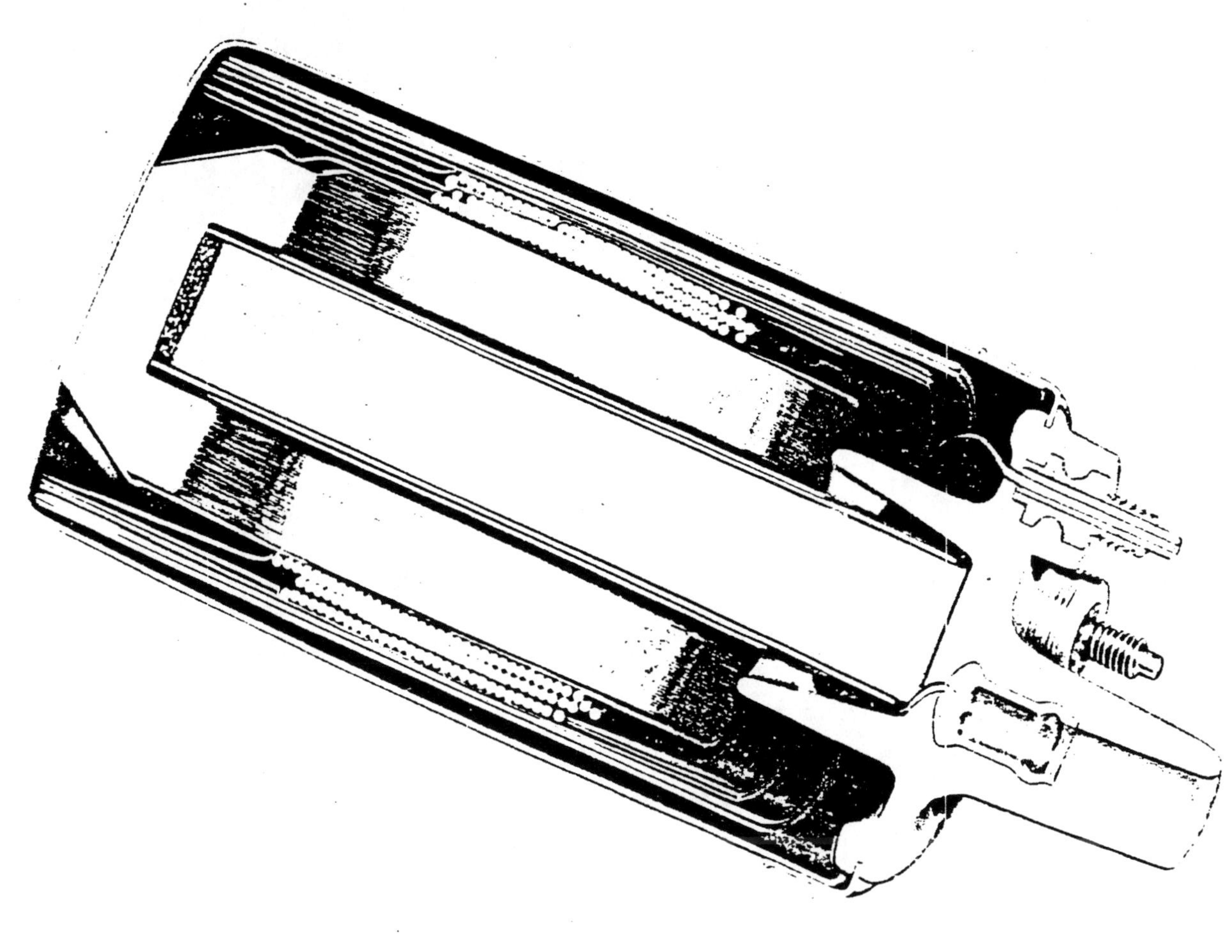

Figure 13 — Ignition Coil

When the positive output voltage from the distributor exceeds the bias level set by R1 and R2, the input transistor (Q3) turns "OFF"; this allows current to flow from the base of Q2 through resistor R3, thus turning Q2 "ON" and subsequently turning Q1 "ON." Current will now flow in the primary of the ignition coil. (The current flow when the amplifier is ON is shown by the solid line.)

The amplifier is turned "OFF," and high voltage is generated in the ignition coil by the rapidly collapsing magnetic field when the distributor output falls below the bias level set by R1 and R2.

The ignition coil, shown in Figure 13, is basically a production transistor unit. Like the amplifier, it has been fully potted with epoxy resin to guard against vibration damage and any loss in dielectric properties.

The high tension ignition wires from the distributor to the spark plugs and coil are special silicone rubber with a steel core. Silicone boots are fitted over the spark plugs, distributor and ignition coil terminals. The high temperature characteristics and increased flexibility of the silicone rubber, make this material especially desirable for the Mark II application.

The spark plugs, shown in Figure 14, were of the same design and construction as our standard gap automotive plugs except for the higher IMEP rating — cold heat range — required for use in a racing engine of this type.

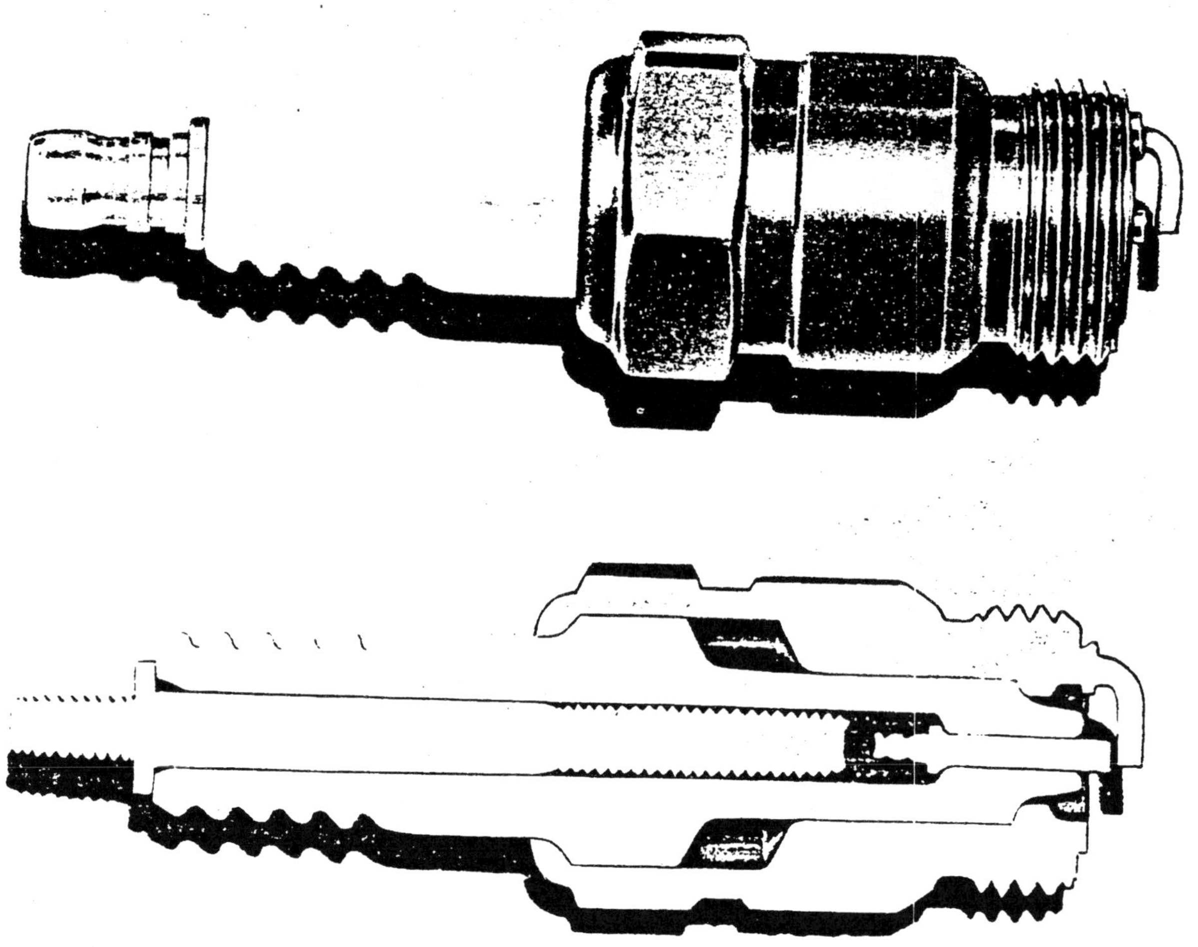

Figure 14 – Spark Plugs

670069

Mark II–GT Transaxles

H. L. Gregorich
and C. D. Jones
FORD MOTOR COMPANY

ABSTRACT

Mark II GT Transaxles

This paper provides a description of the Ford Mark II GT transaxle as used in the 1966 Le Mans race. The functional requirements, as established by a simulated Le Mans dynamometer test cycle, are summarized through integration of the engine torque and speed data and compared with gear and bearing design parameters to reflect their capacities in hours of Le Mans usage. The areas where the development work was particularly important are highlighted in the discussion.

MARK II GT TRANSAXLES*

While the specific subject of this paper, in a sense, did not exist as little as two years ago all of the mechanics for its functional needs were available. All of the mechanics, in fact, have been well developed for several years; the arrangement, however, is not in common use in this country and its application is treated as new. Interest in the subject is believed to stem, principally, from the demonstrated performance of the Mark II GT prototypes in the 1966 Le Mans Grand Prix race. Such interests most probably relate to the particular arrangement, mechanical action, functional capability and reliability, of the transaxle components, as well as the development work that brought them to their present state. This paper begins with a chronology of the subject's brief history. Following the chronology, a description and discussion of the particular type of transaxle used at Le Mans is given. The functional requirements for the transaxle, as established by a simulated Le Mans dynamom-

* The term transaxle is a contraction of the transmission and axle names into a single word. The precise subject refers to the transaxles which were in the 1966 Le Mans Mark II GT prototypes.

eter test cycle, are summarized through integration of the engine torque and speed data and compared with gear and bearing design parameters to reflect their capacities in hours of Le Mans usage. The discussion of development work is limited to those problems occurring within the transaxle; particularly those areas where the development work came to the rescue of the design concepts.

Chronology of the Transaxle

Following the 1964 Le Mans race, Ford Motor Company began the design of a heavier and more powerful experimental GT prototype car. The requirements of this new car excluded the use of any available proprietary transaxle; the contemplated use of a V-8 427 CID engine required heavier and stronger transmission and axle components than those that were currently packaged for race applications in either domestic or foreign cars.

In December of 1963, Ford had introduced a newly-developed four-speed manual transmission for the 427-4V passenger car applications. The ruggedness of these new four-speed transmissions led to their being considered for the GT transaxle application. By January 1965, a transaxle with conventional clutch, synchro-mesh four-speed transmission and axle with hypoid gearing and locking differential was being considered, at least as an alternate design.

In early 1965, prototype hardware for each of two transaxle designs was fabricated and assembled for vehicle evaluation. One of these designs featured a torque converter coupled to a two-speed transmission with Colotti-type gear engagement. The transmission was directly geared to the axle pinion gear shaft and powered the axles through a locking differential. The second design used the conventional clutch and four-speed transmission, with axle and locker components interchangeable with those of the first design.

Track performance of the initial experimental prototype vehicles was so satisfactory that two experimental cars were entered in the 1965 Le Mans Grand Prix race. Each of the GT vehicles was equipped with similar transaxle units; their design featured the conventional four-speed synchro-mesh transmission. Although the initial prototypes had low durability characteristics, Phil Hill, driving one of the Mark II cars, set a new lap record of 141.1 mph early in the race. This modest accomplishment undoubtedly created the incentive as well as established the objectives for the program.

Through mid-year of 1965, development work continued on the transaxles. By November 1965, several evaluation tests, including durability runs under race conditions, had permitted further evaluation of gear ratios and study of individual components for the need of modifications. Toward the close of 1965, two transaxle designs had been developed and were considered race-ready. One was the four-speed manual shift transaxle, and the other was a fully-automatic power shift transaxle.

The power shift transaxle axle was developed primarily for the J-Car, a prototype lighter than the Mark II's. Although the automatic-type transaxle was run at Daytona in a Mark II, the development work on the light car had not progressed sufficiently to be ready for the June 18, 1966 Le Mans race, so the unit did not receive further competitive evaluation.

The design with conventional clutch, transmission, etc., went on to make race history in the Mark II GT prototypes at the Daytona International, Sebring Grand Prix and Le Mans Grand Prix races and is, of course, the subject of this paper.

Transaxle Arrangement

The general arrangement of the transaxle is shown in Figure 1. In these sectional views, three principal construction areas may be identified.

SECTIONS THROUGH MANUAL TRANSAXLE

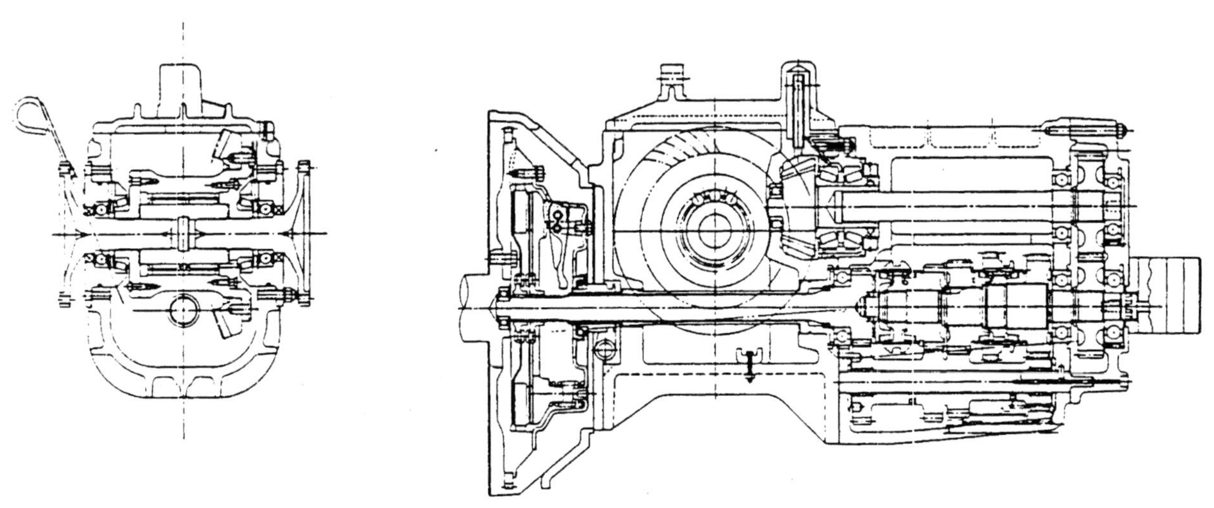

Figure 1

At the left side in the longitudinal section is the flywheel housing, which encloses a two-disc clutch, and provides support for the clutch actuating mechanism. This housing is mounted to the rear face of the engine block.

The differential housing is immediately behind the flywheel housing and is attached to it (a two-piece construction is used to provide flexibility in design without loss of too much rigidity). The differential section, in addition to encasing the locking differential and ring and pinion gears, provides support for the pinion bearings and differential side bearings. An access opening at the top of the case is provided with a cover with two integral mounting attachments to serve in conjunction with a supporting cross member as the rear mounts for the engine.

The transmission case, in turn, is mounted to the axle housing. This component, besides housing the transmission proper, has provisions for attaching, on the right, the shift mechanism and, at the rear, the transfer drive gear cover. A small gear pump, mounted on the rear face of the transfer drive cover and driven from the transmission, is used for maintaining flow of lubricants to and from an external cooler.

Structurally the flywheel housing, axle case and top cover provide support for the complete transaxle assembly, as well as carrying the reaction loads due to input shaft, and the axle shaft loads. All of the principal housings are of sand-cast magnesium alloy to minimize weight. Each of the cast housings was designed with sufficient wall strength and external ribbing so that no serious structural problems were experienced in the development tests. Problems related to deflections and interference fits did appear; discussions of such problems are included in the development discussion on the particular components affected.

The transaxle arrangement provides for power flow from the rear-mounted engine through a dual disc clutch to the transmission. From the transmission, power flows through a pair of transfer gears mounted at the rear of the transmission case, to the pinion gear, ring gear and locking differential. Stub shafts, splined to the locking differential, drive the rear wheels through Rzeppa-type jointed shafts.

Functional Requirements

The functional requirements of the transaxle and its components are, to a great extent, dictated by the engine torque characteristics, vehicle weight, weight distribution and wheel size. They are further established by the terrain and layout of the race course, and the anticipated vehicle speed and engine, as well as the number of laps to be completed.

For the Mark II GT transaxle application, the vehicle weight was 2850 lbs. (less the driver) with a weight distribution which provided sufficient traction to induce wheel-slip torques at 2500 ft. lbs. The engine of the Mark II developed up to 485 brake horsepower with available torque up to 460 ft. lbs. While the torque capacity and gear ratio requirements for the transaxle unit fall within or near the limits of those in use in current passenger cars, the high operating engine speed (up to 6800 rpm intermittently) requires special lubrication and cooling features to maintain satisfactory operating temperatures.

The layout and topography of the Le Mans race course requires vehicles with exceptional accelerating and braking characteristics, as well as high speed stability. The transaxle contributes to these requirements by its ability to respond quickly to any required gear change, either up or down shifts, and to retain gear engagement under all torque and speed conditions. Approximately 9000 gear shifts are required during the 24 hours of the race; many of the shifts, because of the high speed, require the synchronizers to perform three to four times the work required in conventional passenger car use. The anticipated transmission gear usage is itemized in Table 1; i. e., the approximate time spent in each gear whether in drive, coast or cruise operation. The vehicle is in the drive and cruise position 55% of the time. Except for cruise, this time is spent accelerating in the various transmission gear engagements. The coast time (amounting to 30%) is with the clutch engaged resulting in engine braking. The time spent in changing gears amounts to 15% of the total time or 3.6 hours.

G.T. TRANSMISSION GEAR USAGE

(LE MANS CYCLING - 24 HRS.)

GEAR USE	RATIO	PERCENT OF TIME	HOURS
1ST GEAR DRIVE	2.32	4.8	1.15
1ST GEAR COAST		5.8	1.39
2ND GEAR DRIVE	1.54	10.3	2.47
2ND GEAR COAST		3.2	.77
3RD GEAR DRIVE	1.19	12.3	2.95
3RD GEAR COAST		8.4	2.02
4TH GEAR DRIVE	1.00	12.1	2.91
4TH GEAR COAST		12.1	2.91
4TH GEAR CRUISE		16.0	3.84
CHANGING GEARS		15.0	3.60

Table 1

The operation of the transmission when downshifting requires that the clutch discs must be capable of running at speeds of at least 10,000 rpm. Further, all transmission operation in gear is at 100% of available engine torques (except in first gear). This usage, with variable torque and speed, determines the load-carrying requirements of the transfer drive gears, bearings, differential locker, and ring and pinion gears.

Design Analysis

To establish quantitative values of torque, speed and usage for the design analysis, estimates of their respective averages were taken as 427 ft. lbs., 6000 rpm, and 3 hours in 1st gear, 6 hours in 2nd gear, 6 hours in 3rd gear and 9 hours in direct drive. While these estimates proved adequate for the initial design and development work, a verification of their accuracy and possible identification of areas having design improvement potential was needed. As a consequence, the engine-applied torque and speed data were integrated relative to time for each gear application. The results of these latter calculations are presented in this paper.

The transmission section of the transaxle provides for the gear ratio changes needed to utilize the engine horsepower effectively. There are four forward drive ratios, with values as shown in Table 1. As used in the race, their application was limited to acceleration by upshifting through each gear in sequence; holding in fourth gear at cruising speed for a short time; and then braking with the engine by downshifting through the gears in sequence. A typical shift pattern is shown in Figure 2; in this graph the engine speed is plotted against time. It should be noted that the engine speed versus time is nearly a straight line for each gear application; to simplify gear tooth fatigue stress and life analysis, the speed is approximated by a straight line.

The available engine torque plotted against the engine speed in rpm is shown in Figure 3. As in the graph of the engine speed, the torque data curve is approximated by straight-line segments; in this instance, three. Further, the engine torque applied in low gear is limited by the rear wheel tractive effort and is assumed to have a constant value near 400 ft. lbs. until the ratio of tractive effort to available engine torque becomes greater than the over-all gear reduction.

The bases for the speed and torque curves shown in Figures 2 and 3 are graphs run in the dynamometer from a simulated Le Mans test cycle. Four such cycles are equivalent to one lap over the Le Mans course at a competitive speed.

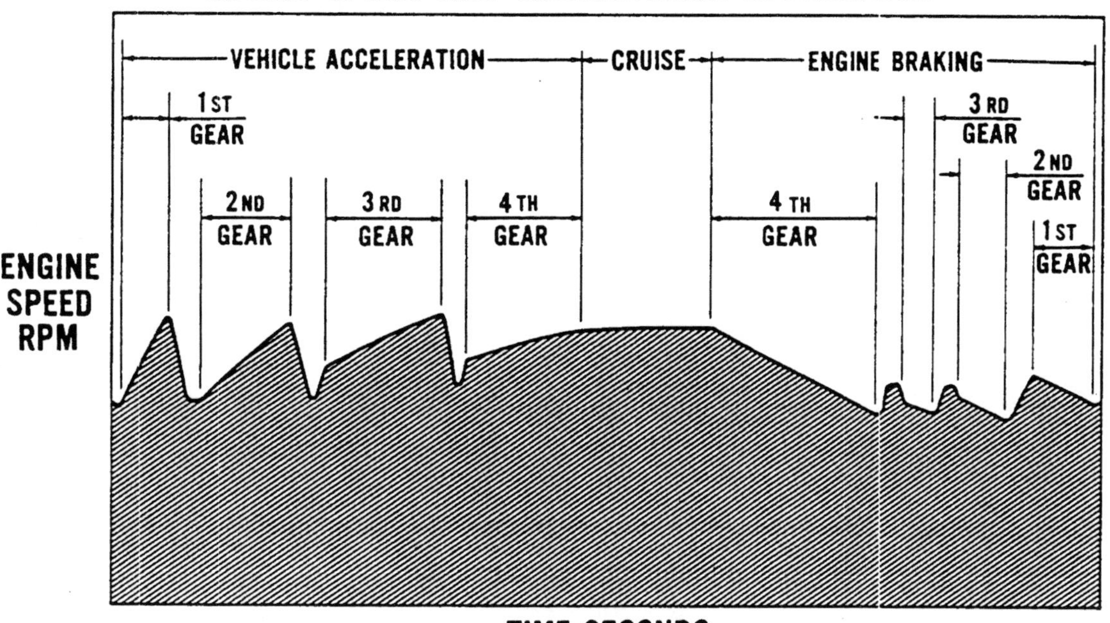

Figure 2

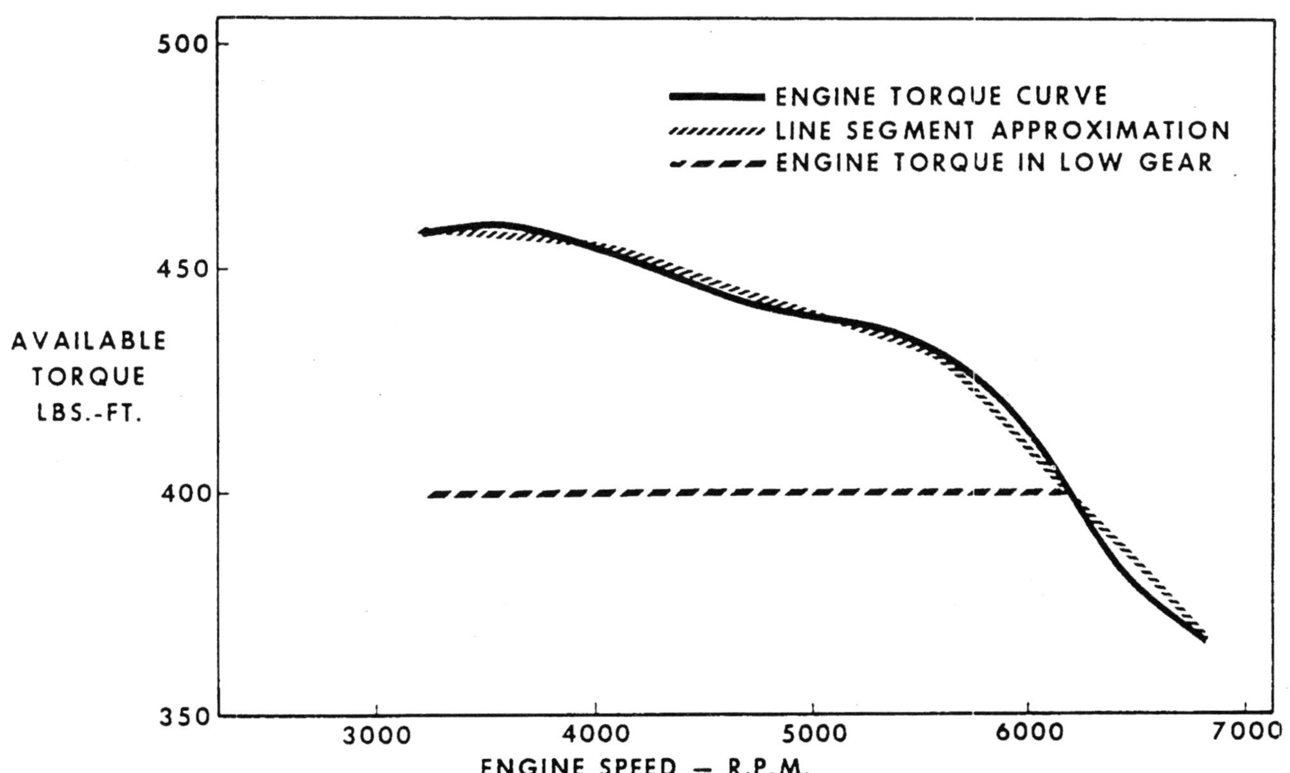

Figure 3

The formula for maximum gear tooth bending stress, as published in an AGMA report, (1)* is given as:

$$S = \frac{3 \pi T_G}{f Z N_G X} \tag{1}$$

where S = Maximum Stress, psi
T_G = Torque of Gear, Pounds Inches
f = Face Width of Gear, Inches
Z = Length of Contact, Inches
N_G = Number of Gear Teeth
X = Strength Factor, Inches

If the torque T_G is taken as $T_G = 12$, then the value for stress per ft. lb. of torque is

$$G = \frac{3 \pi 12}{f Z N_G X} \tag{2}$$

Also included in this report (1) is an S-N curve for carburized, hardened and shot-peened gears, which for our purpose is represented as:

$$\Gamma = 19 \times 10^{64} S^{-12} \tag{3}$$

where Γ = Expected fatigue life, number of cycles
S = Maximum working stress, psi

The engine speed graph with transmission gears under load, as shown in Figure 2, is readily approximated by a series of straight-line segments. All of the segments may be represented analytically by a function linear in time of the form.

$$n(t) = C + Dt \tag{4}$$

where t is restricted to the interval of time, $O \leq t \leq T$ and unless stated otherwise this range is to be assumed, and where

C = the initial engine speed, rpm

D = acceleration, rpm per sec.

t = time in seconds

T = interval of time a gear is in use, seconds.

For the engine torque graph in Figure 3, the line segment approximations can, of course, be represented by a function linear in engine rpm such as:

$$\tau(n) = E + Fn$$

where n has the range, $n(O) \leq n \leq n(T)$ and where

τ = engine torque, ft. lbs. (5)

---

* Numbers in parentheses designate references at the end of paper.

n = engine speed, rpm

E = initial torque, ft. lbs.

F = change in torque, ft. lbs. per rpm

Having developed the engine speed as linear to time and the engine torque linear to speed, it follows that the torque is also linear in time. Its expression as a function of time may be derived by substituting values for speed from equation (4) into equation (5); the result is represented by:

$$\tau(t) = A(B-t) \qquad (6)$$

where A & B are constants determined by C, D, E and F in the above substitution.

Returning to equation (1) and making use of equations (2) and (6), we find that the maximum stress as a function of time is derived as:

$$S(t) = G\,\tau(t) = GA\,(B-t) \qquad (7)$$

Introducing this expression for stress in equation (3) gives:

$$\Gamma(t) = 19 \times 10^{64} \left[GA\,(B-t)\right]^{-12} \qquad (8)$$

Equation (8) gives the fatigue life in cycles as a function of time; this may be reduced to life in minutes by:

$$L(t) = \frac{\Gamma(t)\,R}{n(t)} \qquad (9)$$

where R is the ratio of the engine speed to the speed of the particular gear.

To establish the expected fatigue life for the total service of a gear, the life function L (t) is integrated over one cycle which is representative of the gear's service. The integration is made through use of Miner's ratio for cumulative fatigue damage [(2)]; the composite life is expressed as:

$$L = \frac{T}{\int_{0}^{T} \frac{dt}{L(t)}} \qquad (10)$$

where L is given in minutes of composite fatigue life.

Results of the transmission gear life calculations are shown in Table 2; the life in minutes for each gear has been converted to hours of Le Mans service life. The life for each gear as calculated reflects the drive loads only. With the engine braking torques being only 30 to 40%

of the drive torques, the S-N curve, as given by equation (3), is adequate to predict the fatigue life satisfactorily without further allowance for reverse bending stresses from the braking torques. Also, with the transmission in first or low gear, the rear wheel traction torque limits the use of available engine torque to approximately 400 ft. lbs.; this value of torque was assumed constant in calculating the fatigue life for low gear service.

In vehicle operation, rapid release of the clutch throw-out mechanism induces impact loading on the powertrain, which may be several times the normal engine torque. To prevent gear tooth damage from shock loads, extra strength above that required for normal fatigue life was provided as required. The gear tooth bending stresses obtained from equation (1) by using the impact torque value, in place of that for the nominal engine torque, should be under 250,000 psi to avoid premature gear tooth failures.

The fatigue life expectancy of the axle ring and pinion gears and the B_{10} life expectancy of the bearings is established by use of a more general expression. A generalized form is developed by writing equation (3) as:

$$\Gamma(t) = K \times 10^{X} \, (G\,\tau(t) \times R \times E)^{J} \tag{11}$$

where $\Gamma(t)$ = Expected fatigue life in number of cycles for prevailing conditions at time t.

$\tau(t)$ = Engine torque in ft. lbs.

R = Ratio of engine speed to component speed

E = Mechanical efficiency factor

G, J, K and X are constants for a particular consideration.

With n(t) and $\tau(t)$ as given by equations (4) and (6) respectively, the general expression for composite life becomes:

$$\text{Life} = \frac{T \times R \times K \times 10^{X}}{(G \times R \times E)^{J} \sum_{i} \Delta i} \tag{12}$$

where $\Delta i = A_i^{J} \int_{T_{i-1}}^{T_i} (C_i + D_i t)(B_i - t)^{J} \, dt$

$$= A_i^{J} \left\{ \frac{\left[(C_i + D_i T_{i-1})(B_i - T_{i-1})^{J+1} - (C_i + D_i T_i)(B_i - T_i)^{J+1}\right]}{J+1} + D_i \frac{\left[(B_i - T_{i-1})^{J+2} - (B_i - T_i)^{J+2}\right]}{(J+1)\,(J+2)} \right\} \tag{13}$$

and $T = \sum_i (T_i - T_{i-1})$ (14)

These expressions give the expected life in minutes and are applicable if the engine speed is constant or linear to time, and if the torque is not constant but otherwise linear in the interval of time. However, if the torque is constant with a value of A_k ft. lbs. on any interval of time, such as $(T_k - T_{k-1})$, then for this k^{th} interval of time the expression

$$\left[A_k^J \left[\frac{(C_k + D_k T_{k-1})^2 - (C_k + D_k T_k)^2}{2D_k} \right] \right] \quad (15)$$

must be substituted for the more general expression given in equation (13).

In case the speed also is constant at C_k rpm on this interval of time, the expression:

$$\left[A_k^J C_k (T_k - T_{k-1}) \right] \quad (16)$$

is then used in equation (13).

The formula for life as shown in (12) was applied to the transfer gears, as well as the axle pinion and ring gears to determine their Le Mans service life. The composite fatigue life values for these parts are also included in Table 2.

GEAR FATIGUE LIFE

LE MANS SERVICE

TRANSMISSION	NUMBER OF TEETH	USAGE – % TIME DRIVE	USAGE – % TIME BRAKE	LIFE HOURS
1ST GEAR	32	4.8	5.8	37
CLUSTER	15			53
2ND GEAR	27	10.3	3.2	96
CLUSTER	19			80
3RD GEAR	24	12.3	8.4	124
CLUSTER	22			99
MAINDRIVE	23	27.4	17.4	76
CLUSTER	25			61
TRANSFER DRIVE				
DRIVER	30	55.5	29.5	1240
DRIVEN	26			1370
AXLE				
PINION	11	55.5	29.5	470
RING	34			470

Table 2

For the bearing B_{10} life in hours of Le Mans service, the calculations generally follow manufacturers' recommended methods. One exception is for all ball bearing applications, the "equivalent radial load" is taken as being proportional to the radial load; i.e., the conventional Y_2 factor[3] remains constant. Also, in the pinion bearing load analysis, the gear thrust load moment reaction loads are taken at the pilot and rear tapered bearings. The bearing B_{10} life in hours of Le Mans service for each of the various ball and roller bearings, as determined by application of formula (12), are shown in Table 3.

BEARING B-10 LIFE
LE MANS CYCLE SERVICE

LOCATION	TYPE	USAGE – % TIME DRIVE	USAGE – % TIME BRAKE	B-10 LIFE HOURS
CLUTCH				
CRANKSHAFT POCKET	B	27.4	17.4	5000+
TRANSMISSION				
INPUT SHAFT	B	27.4	17.4	69
INPUT SHAFT POCKET	B	27.4	17.4	330
CLUSTER GEAR FRONT	N	27.4	17.4	263
CLUSTER GEAR REAR	N	27.4	17.4	91
OUTPUT SHAFT FRONT	B	55.5	29.5	72
TRANSFER DRIVE				
OUTPUT SHAFT REAR	B	55.5	29.5	88
COUNTERSHAFT FRONT	B	55.5	29.5	453
COUNTERSHAFT REAR	B	55.5	29.5	221
AXLE				
PINION GEAR PILOT	R	55.5	29.5	45
PINION GEAR SHAFT FRONT	TR	55.5	29.5	46
PINION GEAR SHAFT REAR	TR	55.5	29.5	3060
RING GEAR SIDE	TR	55.5	29.5	54
DIFFERENTIAL SIDE	TR	55.5	29.5	476

Table 3

The synchronizer work requirements are best defined by establishing a comparison with known test data on production-type four-speed transmission synchronizers. The number of shifts per synchronizer in Le Mans service is weighted by a work factor established as the energy transfer ratio between Le Mans cycle use and the established test data. The equivalent number of shift cycles for each synchronizer are shown in Table 4. These may be compared to a test average of over 7000 shifts per synchronizer for passenger car applications.

SYNCHRONIZER SHIFT CYCLE REQUIREMENTS

SYNCHRONIZER LOCATION	TYPE OF SHIFT	ENERGY TRANSFER PER SHIFT LBS-FT	WORK FACTOR	NUMBER OF SHIFTS	EQUIVALENT NUMBER OF SHIFTS
1ST GEAR	2-1	1265	3.0	1125	3375
	3-1	2625	4.1	375	1435
2ND GEAR	1-2	1987	4.6	1400	5040
	3-2	749	1.3	1125	1460
3RD GEAR	2-3	1345	3.0	1400	4200
	4-3	864	2.0	1400	2800
4TH GEAR	3-4	1050	2.4	1400	3360

Table 4

The frequent use of transmission gears imposes severe loading upon the synchronizer components, particularly on the gear cone surface and the mating blocker ring cone surface. Energy transfers of up to 2600 lbs. ft. (with over 70% of the shifts above 1000 lbs. ft.) without synchronizer malfunction require that the surfaces of both elements be correctly processed. In particular, the character of surface finish on the gear cone and the micro-structure of the blocker ring need to be maintained near an optimum.

The differential drive is a proprietary* item. This locking device consists of two independent (except for the outer race) two-way roll clutches, each one splined to an output shaft. In operation, the unit is not subjected to an excessive number of cycle loads; therefore, analysis is limited to the static load stresses. Published formulae[(4)] for stresses of similar loadings are applicable and may be used to evaluate the maximum tensile, compressive and Hertz stresses. For a wheel torque of 1250 ft. lbs. (limited by the tractive effort), the tensile stress in the outer race has a range of 35,000 to 75,000 psi, depending upon the locking angle of engagement. Under the same conditions, the inner race Hertz stress ranges from 525,000 to 800,000 psi and maximum compressive stress ranges from 10,000 to 25,000 psi.

*Hi-Torque Engineering Company, Chicago, Illinois

Many of the components, such as the output shaft and the shifter mechanism, are not analyzed for the specific Le Mans loading, since their anticipated loading and usage does not greatly exceed those loadings applied in test to their production counterparts. PVT factors[5] for transmission gears, however, were computed and range from 4.4×10^5 to 2.6×10^6. With the use of an EP-type lubricant suitable for hypoid gearing, no scoring problems were encountered.

Aside from the use of magnesium cases to reduce weight of castings, and the use of bronze shifter forks to reduce wear, no materials, other than those used in production were considered.

Design Evaluation and Development

During the initial design evaluation phase, nearly all of the criteria for lubrication and cooling were established and then implemented from time to time as the program progressed. In addition, several design and process weaknesses were observed and corrected.

The lubrication system of the transaxle, as developed for competition, uses a common lubricant in the axle and transmission sections. A schematic of the system is shown in Figure 4; a single-lubricant pump is used to maintain flow to and from the external cooler.

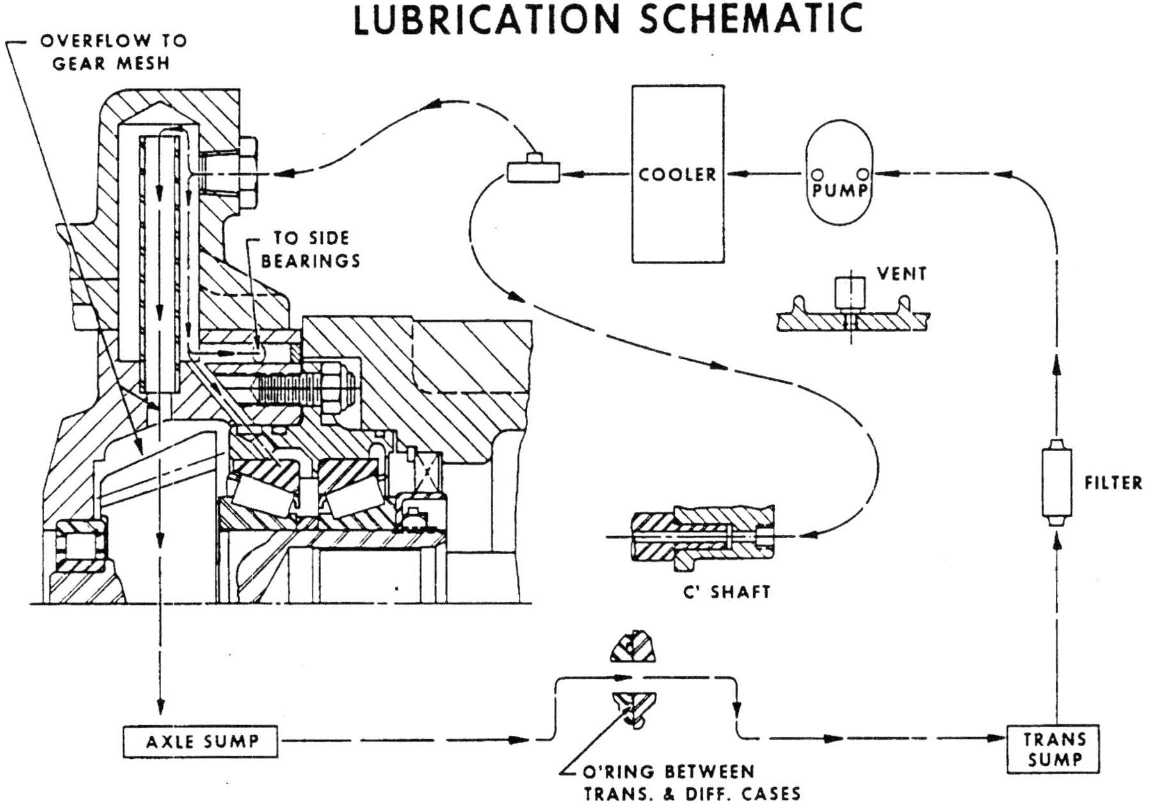

Figure 4

As indicated in this schematic, only the cluster gear needle bearings, pinion and differential tapered-roller bearings are pressure-lubricated. All oil not fed to the bearings is dumped on the pinion gear teeth as they emerge from ring gear engagement. The oil is pumped from a vented sump in the transmission section at a temperature of near 200°F through a 40 micron oil filter to an external cooler, and returned at approximately 180°F via bearing etc., to the axle sump. From the axle sump, the oil completes its circuit by flowing through mating port holes in each case wall to the transmission sump.

Prior to pressurizing the oil flow to the pinion bearings and also dumping oil on the pinion gear, the pinion gear and bearing supporting tapered bearings failed on test. Because of these failures and also because the bearing B_{10} life was low on the front tapered roller bearing, a larger size bearing (one with eight times longer rated life) was adopted. With the change to larger bearings, the oil flow to these bearings was pressurized, as well as, the oil flow to the differential side bearings. This latter modification not only provided ample lubrication and cooling for the side bearings but achieved satisfactory lubrication in the locking differential. Lack of adequate lubrication had been a problem in the differential. The side-tapered roller bearings splash-feed sufficient lubricant into the differential case through port holes in the end caps.

Deflection in the flywheel housing resulted in frictional heat being generated between the transmission input shaft and its tubular enclosure mounted in the axle case. This condition resulting in a failed clutch and throw-out bearing; it was remedied by increasing the stiffness of the flywheel housing and increasing the clearance between the two interfering components.

A further interference occurred between the transmission input shaft and transmission front ball bearing outer race. This specific problem was unique in that no functional deterioration was observed, and the fault was not found until teardown inspection; although, the bearing outer race had separated into several pieces. The problem was resolved by increasing the clearance between gear face and bearing race.

A third problem arising from deflection in the transmission occurred in the gear pump mounted on the face of the transmission rear cover and driven by a solid coupling to the transmission output shaft. Loading on the pump housing and gears, through the pump drive shaft, resulted in one pump failure and some other near failures. A special flex-coupling was designed and installed between the transmission output shaft and the pump shaft. This new arrangement proved satisfactory.

In vehicle tests, the outer race of one of the transfer drive rear bearings spun and damaged the housing cover. Although the inner race has only a finger-press fit, the spinning occurred at the outer race due to a differential in thermal expansion between the housing and bearing outer race. To avoid this problem occurring when competing in a race, a key was inserted in the cover with a flat ground on the outer race of the bearing.

A transmission gear failed in the 1964 Le Mans race; this particular input gear was one of the early prototypes, and was later found to be dimensionally incorrect in that the oil hole between clutch teeth and gear teeth was out of location. Another gear failure occurred in test at Sebring and was found to be due to faulty shot-peening after heat treat; this was a third-speed constant mesh gear.

The outer race of the locking differential became a problem early in the program. The outer race, in its original design, was too weak and split at the end opposite the ring gear mounting flange. This end of the race was reinforced by increasing its outer diameter to make the wall thickness comparable to that of the flanged end. Other problems, such as heat treat and finish grind, however, continued with this component throughout the program.

Initial set and wear of the drag springs in the locker assembly reduced its functional capability at high speeds. This condition was satisfactorily offset by reducing the amount of initial set in the spring, thus retaining more spring tension, and by controlling the character of the inner race surface on which the spring makes contact.

Static loading on the transaxle assembly to determine the maximum stresses in the housing components revealed the stresses to be well under their yield strength. Deflection studies[5] on the transaxle established the deflection in the axle section under full torque load at ring and pinion to be within acceptable limits.

Initial dynamometer experience paralleled the vehicle test failures closely; i.e., an outer race on the differential failed in dynamometer, and transfer gear bearing outer race damage to the cover was noted. On a subsequent dynamometer run after design changes, the transaxle completed the 48-hour test without incident.

In the clutch mechanism, a straight thrust-type throw-out bearing failed during the initial tests and was replaced by an angular contact bearing. The throw-out lever and fork, consisting of a welded and pinned assembly, encountered problems in obtaining a satisfactory weld; the condition was avoided by closer control of the welding technique.

Summary

The design analysis and test evaluations, after the incorporation of the development improvements, indicated the transaxles would be capable of meeting all operational requirements of the Le Mans race. The potential problem areas, which could never be fully covered, were quality of the components and their proper assembly. A degree of uncertainty, therefore, remained in the complete units throughout the race.

It is not expected that the present designs and stage of development will long retain superior competitive value; the weight and size of the unit are recognized handicaps and may soon require modifications to meet future requirements.

Acknowledgments

Individual contributions by personnel of the Ford Motor Company to the successful design, development and application of the Mark II GT transaxle have been significant, but are too numerous to list all of them individually. Messrs. E. I. Hull, R. D. Negstad and R. C. Lunn, Special Vehicle Activity, Ford Division, are acknowledged as the originators of the preliminary transaxle and clutch arrangements. The finalization of the design and build of the prototype transaxle was done by Standard Transmission Engineering, Experimental Engineering and Axle Engineering Departments, Product Engineering Office, Transmission & Chassis Division. Dynamometer tests were run by the Testing Department, Engine & Foundry Division; vehicle evaluation and track tests were coordinated by Special Vehicle Activity, Ford Division. The Timken Roller Bearing Company made the deflection tests on the ring and pinion axle gears. Thanks are again expressed to the many vendors who supplied component parts and assemblies, and generously supplied engineering data relative to their use in the subject application.

References

[1]H. Wohlers - Bending Stress of Spur and Helical Gearing - Progress Report AGMA 101.02, October, 1951, Section I, Pages 2 and 16.

[2]H. J. Grover, S. A. Gordon and L. R. Jackson - Fatigue of Metals and Structures - NAVAER 00-25-534, 1954, Page 45.

[3]Fafnir Bearing Catalogue No. 56 - Page 4.

[4]R. J. Roark - Formulas for Stress and Strain - Third Edition - McGraw-Hill, 1954, Pages 158 and 288.

[5]J. C. Straub, PVT Values for Gear Teeth Progress Report - AGMA 101.02, October, 1951, Section II.

670070

Mark II GT Sports Car Disc Brake System

Part I. Design and Development
Joseph J. Ihnacik, Jr.
FORD MOTOR COMPANY

Part II. Testing
Jerome F. Meek
FORD MOTOR COMPANY

INTRODUCTION

When the decision was made to build the first Ford GT sports car in 1964, many studies were conducted to select the ideal engine, drivetrain and suspension designs. One chassis system concept selected with virtually no investigation was the brakes. This does not imply that the brake system was subordinated to other vehicle components. On the contrary, good brakes were considered of paramount importance. It was believed, however, that disc brakes were the only logical choice. Over the years, disc brakes have created a very favorable impression in the world of sports cars, and there was no doubt that pound for pound the disc brake was more efficient than the conventional drum brake. Therefore, if disc brakes were not selected, an entirely new braking concept would have been required. This idea was discarded as not feasible within the available time. The selection of disc brakes was justified during the past racing season, as in part the success of the Mark II GT can be attributed to the brake system.

Since the Daytona and Sebring victories were crowned with an impressive Grand Prix win at Le Mans, it is appropriate to reflect briefly on the events that led to the universal

ABSTRACT

The design and development of Mark II GT brake system within the parameters dictated by the Mark I chassis presented many problems. The Mark II GT with its larger 427 cubic inch engine had more weight and much higher performance than the Mark I. Space limitations of the carryover wheels and suspension imposed a severe handicap on individual brake component design. This was compounded by shortening the normal one year development time to a three month period.

Part I of this paper is devoted to the consideration of factors which control the design of a brake. The concept of kinetic energy and its effects on brake performance is reviewed briefly. Use of the ventilated rotor design is explained for applications where severe heat is a problem, as in the case of the Mark II GT.

The development of the brake system from the 24 hour Daytona endurance race to the Le Mans Grand Prix race is reviewed. And unique rotor problems resulting from the various energy loads experienced at Daytona, Sebring and Le Mans are analyzed.

In Part II the brake dynamometer, its automatic programmer and the logic of race simulation duty cycle are described.

Use of Ford's new Reliability Laboratory brake dynamometer for screening of potential rotor designs is explained. In the screening process, dynamometer results proved that significant brake development work can be performed in a laboratory where a race can be simulated under carefully controlled conditions.

adoption of disc brakes for Grand Prix cars. Disc brakes were introduced at Le Mans during the 1952, 24 hour Grand Prix race, and with each succeeding year more and more entries favored disc brakes. In June of 1953 the race was won by a Jaguar equipped with disc brakes. During the 2540 miles of the gruelling race both lining life and overall performance of the brake system were highly impressive. This practical demonstration of improved braking, coupled with satisfactory durability, contributed to the general acceptance of the disc brake for sports cars as the ten year trend shown in Figure 1 indicates.

Durability was by no means the only motivation for the acceptance of disc brakes. It must be agreed that disc brakes fitted with competition brake linings also have a high resistance to fade, which in part accounts for the linearity of the disc brake torque. The linear response of disc brakes with smooth progressive action permits drivers to go deeper into turns before braking, thus gaining vital seconds on every lap. These characteristics of linear response and resistance to fade under repeated brake applications must be considered as prime factors in the switch to disc brakes. By 1963, just ten years after the first successful demonstration of the disc brake at Le Mans, all entries were so equipped.

This optimistic picture does not imply that all GT sports car brake problems were solved with the adoption of disc brakes. It is interesting to note that after 13 years of disc brake development on the Grand Prix circuits — since the 1953 Jaguar victory — brake engineers are still faced with unresolved problems. Rotor failures, fluid boil and short lining life are just a few of the problems that must be coped with today. One possible explanation for this dilemma is that the 1953 race was won by the Jaguar at an average speed of 105 mph. The 1966 Ford Mark II GT Le Mans winner averaged 125 mph, an increase of 19 per cent. To achieve the average speed of 125 mph, the engine, drive-train and other chassis components of the Mark II were increased substantially in size and weight over the C-type Jaguar. The combination of high speed and weight of the Mark II GT imposed an unprecedented heat load on the disc brakes, and there is no indication that the trend to the use of higher horsepower engines on the GT sports cars is leveling off. It can be safely speculated that demands on brakes will be even greater in years to come. This will be especially true of the cars equipped with automatic transmissions.

How the original Mark I sports car brake system was up-graded to dissipate the higher Mark II heat loads with minimum changes to other vehicle components is described in Part I of this paper. Highlights of the Daytona, Sebring and Le Mans Grand Prix races as well as some of the unique problems encountered will be discussed. Part II of the paper describes the accelerated development work made possible with the aid of Ford's new brake dynamometer.

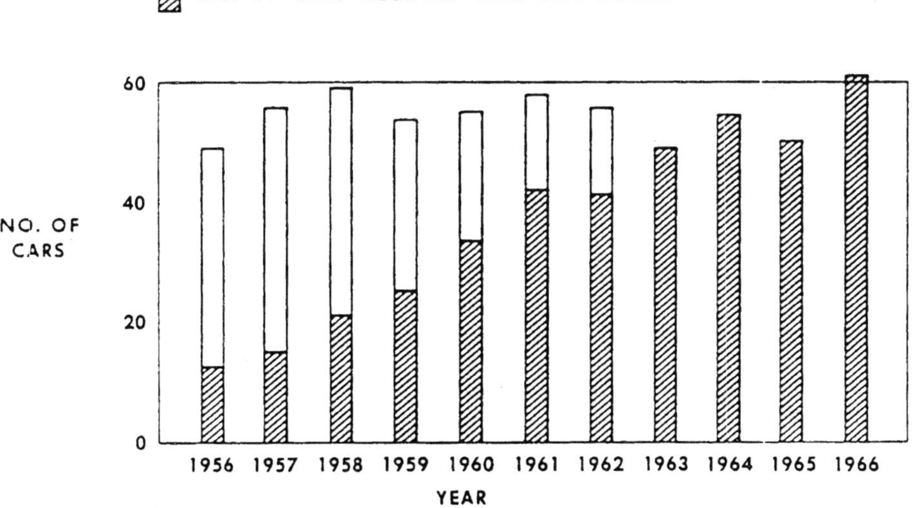

Figure 1 – Le Mans entries showing trend to disc brake usage

PART I
DESIGN AND DEVELOPMENT

Figure 2 – Mark I GT brake assembly showing the original 1/2 inch solid rotor

BRAKE DESIGN FEATURES

The design of the Mark II disc brake system was governed to a great extent by the parameters established for the 289 cu. in. Mark I GT brake shown in Figure 2. As frequently happens in the racing business, a last minute decision was made to increase the performance of the car. To achieve this, a new 427 cu. in. engine was installed, with the necessary chassis modifications to support the new engine. These changes increased the car weight by approximately 500 pounds to a gross weight of 2860 pounds (Figure 3). The change produced an extremely heavy high performance vehicle with relatively low performance brakes originally designed for the lighter and slower car. Any plans to up-grade the brakes at this point were limited to these guide-posts:

WEIGHT (GROSS), LB.	2860
FRONT (38%)	1087
REAR (62%)	1773
WHEEL BASE, IN.	95
TRACK, IN.	
FRONT	57
REAR	56
TIRE SIZE	
FRONT	9.75-15
REAR	12.80-15

Figure 3 – Mark II GT general specifications

- Four wheel disc brakes
- Ventilated rotors
- Heat sink capacity of disc limited only by the available space
- Fixed calipers with floating pistons modified to accommodate new rotors
- Carryover wheels and suspension
- Minimum change to existing vehicle design

To meet these objectives in the relatively short space of time, we initiated a joint development program with the Kelsey-Hayes Co.

MARK II GT BRAKE SYSTEM SPECIFICATIONS

	FRONT	REAR
CALIPER TYPE (GIRLING/KH)	CR - FIXED	BR - FIXED
CYLINDER DIAMETER – IN.	2.375	2.125
CYLINDER AREA – SQ. IN.	4.43	3.55
ROTOR TYPE	VENTILATED	VENTILATED
DIAMETER – IN.	11.56	11.56
THICKNESS – IN.	.775	.775
SWEPT AREA – SQ. IN.	133	133
LINING TYPE	MOLDED ASBESTOS	MOLDED ASBESTOS
AREA – SQ. IN. PER SHOE	10.9	7.0
VOLUME – CU. IN. PER SHOE	6.5	2.8
MASTER CYLINDER TYPE	DUPLEX	DUPLEX
DIAMETER – IN.	.625	.625
AREA – SQ. IN.	.307	.307
PISTON TRAVEL – IN.	1.25	1.25
VOLUME – CU. IN.	.384	.384
FOOT PEDAL TYPE	NON-POWER	NON-POWER
RATIO	3.3	3.3
TRAVEL – IN.	4.1	4.1
BRAKING DISTRIBUTION (DESIGN)	55	45

The solid 1/2 in. rotors originally designed for the Ford GT-40 sports car were replaced with 3/4 in. ventilated rotors shown in Figure 4. Girling type CR front calipers with opposed 2.375 in. diameter pistons were modified in the bridge area to accommodate the vented rotor. Similarly, the type BR fixed rear calipers with opposed 2.125 in. pistons were also modified to accommodate a common 3/4 in. rotor.

Additional system changes incorporated were new integrally molded RM-4528-19M shoe and lining assemblies and EPT (Ethylene-Propylene-Terpolymer) brake cylinder seals to cope with the 300° F (plus) fluid temperatures. The intense heat generated at the interface of the lining pad and rotor is conducted to the fluid by the high metallic content of the friction material, through the steel pistons. Additional heat is picked up directly by the caliper housing through the process of radiation. The total effect of the transient heat resulted in occasional vaporization of brake fluid during the development stage. Periodic flushing of the hydraulic brake lines with Dow HD-50-4 brake fluid to purge the system of moisture, combined with improved ram air ducting, have virtually eliminated the fluid boil problem. Although other brake components have undergone changes to improve their performance,

Figure 4 – Mark II GT sports car disc brake with 3/4 inch ventilated rotor

the lining, seal and fluid compounds continue to provide satisfactory service.

Two notable design improvements of a convenience nature prompted by the need for rotor and lining changes paid off handsomely during the 1966 racing season. Phil Remington of Shelby-American developed a quick-change brake pad retainer, shown in Figure 5, which facilitates removal and replacement of brake lining pads. During the same period John Holman of Holman-Moody redesigned the hub and rotor assembly to speed up rotor changes during the pit stops. The new design incorporates outboard mounting of the rotor and hat section assembly as shown in Figure 6. Retention of the rotor assembly is accomplished by the wheel; and braking torque is transmitted through the wheel drive lugs. New rotor installation with this arrangement can be accomplished in four to five minutes, while a skilled pit crew can replace worn lining pads in one to two minutes. Normally these changes are accomplished during a fueling stop to reduce pit time to a minimum. The merit of both components was amply demonstrated during the 1966 racing season. There is no doubt that the adoption of these features by other racing teams is imminent. The 1967 racing season will certainly see a greater use of quick-change brake components.

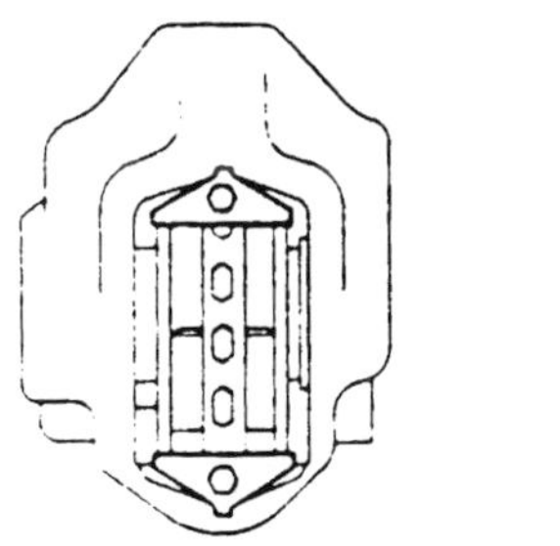

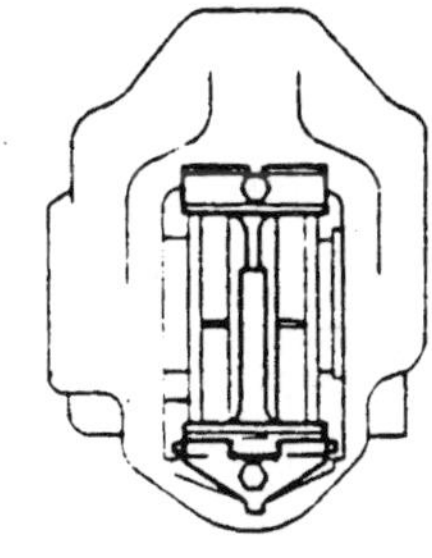

Figure 5 – GT brake lining pad retention brackets

Figure 6 – Mark II GT brake assembly with quick-change rotor

The final brake design feature which merits a brief explanation is the whipple tree master cylinder and brake pedal arrangement shown in Figure 7. This balancing beam and push-rod arrangement carried over from the Mark I sports car is extremely functional. It provides a wide range of adjustment in braking forces at the front and rear wheels by simply shifting the fulcrum point on the threaded beam in the brake pedal pushrod sleeve. In addition to the flexibility in braking distribution adjustment to suit individual drivers and track conditions, the dual brake system has served a useful purpose in providing partial braking on several occasions when hydraulic brake line failures occurred.

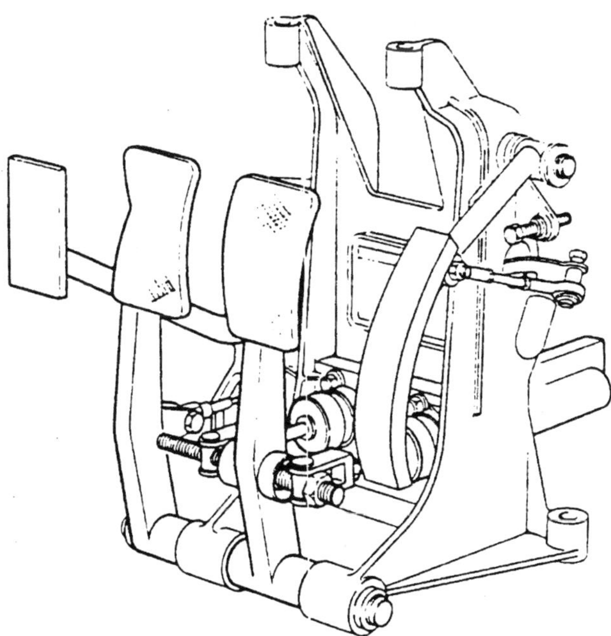

Figure 7 – Mark II pedal support bracket and brake master cylinder assembly

PERFORMANCE

Braking capacity, which determines the ability of a vehicle to slow down or stop, is without a doubt one of the principal factors that must be considered in the overall performance of a sports car. Unlike the Indianapolis

500 vehicles which use brakes primarily for pit stops, the sports cars rely heavily on the brakes to maneuver the hairpin turns of the Grand Prix circuits. Brakes are applied with punishing frequency for 12 to 24 hour periods.

While a detailed analysis of a braking vehicle is beyond the scope of this paper, a few significant factors affecting the vehicle's state of motion are pertinent and will be mentioned briefly.

The major forces affecting the deceleration of a vehicle are:

- Wheel braking force
- Engine braking
- Inertia of rotating parts
- Aerodynamic drag
- Rolling resistance

Influence of air resistance as a decelerating force is small at normal vehicle speeds, particularly for streamlined vehicles. For high speed vehicles the air resistance aids braking action and should be considered where accurate results of energy conversion rates are required.

In the case of a vehicle equipped with an automatic transmission, the rotating-mass effect of wheels and transmission must be added to the translatory mass of the vehicle. In the case of Mark II equipped with an automatic transmission it can be safely assumed that the effect of the aerodynamic drag cancels the inertia force of the wheels and transmission.

Braking potential of an engine can be utilized only with drive-trains that have a positive connection between engine and axle. This form of braking is particularly feasible for low values of deceleration. To achieve high rates of deceleration on hairpin turns, brakes are applied very suddenly, minimizing the engine braking effect. Frequently, in such cases of high deceleration the engine requires mechanical braking effort to synchronize the speeds of rotating parts for smooth shifting of gears. Depending on the inertia of the engine and the gear reduction ratio, a critical value of deceleration can be calculated above which the engine should be disengaged for optimum braking effect.

In stopping a moving vehicle, the brake performs an important function — that of converting energy of motion into heat. The temperature reached by the brake system during this process greatly affects the behavior of the rotor and the lining. Consequently, it is important to understand the concept of converting the kinetic energy of a moving vehicle into heat energy.

The prime factors to consider in determining the required braking performance are gross weight and the maximum speed of a vehicle. These factors are important because the kinetic energy of a moving vehicle varies as its weight and the square of its speed, or

$$E = \frac{Wv^2}{2g} \tag{1}$$

where

E = Kinetic energy (ft-lbs)
W = Gross weight of the vehicle (lbs)
v = Maximum velocity of the vehicle (ft per sec)
g = Gravitational conversion factor (32.2 ft per sec^2)

Figure 8 shows graphically the effect of the velocity factor on the instantaneous energy of a Mark II vehicle. The concept of kinetic energy points out that a high speed vehicle requires brakes having a greater energy conversion and heat dissipation capacity.

If the rate of deceleration of a vehicle is known and if engine and aerodynamic braking forces are neglected, the rate of heat production at the surface of the brake rotor by the conversion of kinetic energy may be developed from the basic energy equation

$$E = 1/2\, mv^2 \tag{2}$$

for

$$v = at \tag{3}$$

$$E = 1/2\, ma^2t^2 \tag{4}$$

where:

m = Mass in slugs
t = Time in seconds
a = Deceleration (ft/sec/sec)

taking the derivation of E with respect to time we obtain the rate of energy conversion to heat

$$\frac{dE}{dt} = 1/2\, ma^2 \times 2t \frac{dt}{dt}$$

$$\frac{dE}{dt} = ma^2t \tag{5}$$

and since a = v/t the equation may be reduced to

$$\frac{dE}{dt} = \frac{ma\,(v)\,t}{(t)}$$

$$\frac{dE}{dt} = ma\,v \tag{6}$$

It can be seen that the rate of heat production is proportional to the velocity. It is highest at the instant of brake application when the speed is maximum and gradually decreases until it reaches zero, when the wheel stops rotating.

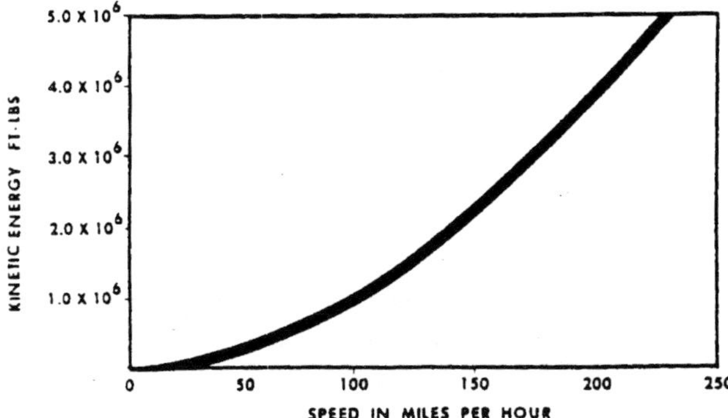

Figure 8 — Instantaneous kinetic energy curve of a Mark II GT plotted as a function of speed

THERMAL ASPECTS OF DISC BRAKE DESIGN

In the foregoing section, elementary equations of motion and relationships between forces acting on a vehicle in decelerating were discussed. However, braking performance limitations imposed by the energy conversion rate in the brake and the accompanying temperature rise in both the rotor and the lining were not considered. It is known that the brake performs an irreversible conversion of kinetic energy into heat. This conversion is a frictional process which takes place at the interface of the rotor and the lining. Heat created in the process raises the temperature of the friction surfaces. In the design of a brake system for a sports car severe brake applications of short duration are a rule rather than an exception. Under these conditions, equilibrium between the frictional heat generated in the brake and the heat dissipated by the brake to the surrounding air is of vital importance, as it affects the size and weight of the system. To provide a brake system capable of dissipating the heat generated during braking requires a knowledge of the thermal characteristics of disc brakes.

An investigation of the steady state heat characteristics of brake rotors conducted by Ford and Kelsey-Hayes engineers showed very graphically the advantage of a ventilated over a solid rotor of equal weight for energy conversion and heat dissipation under repeated brake applications. Figure 9 shows the results of a series of brake applications under controlled conditions for a solid and a ventilated rotor. For the steady state condition shown, the ventilated rotor reached a peak equilibrium temperature of 700°F after eight cycles. Equilibrium temperature is defined as the point at which the heat loss between brake applications

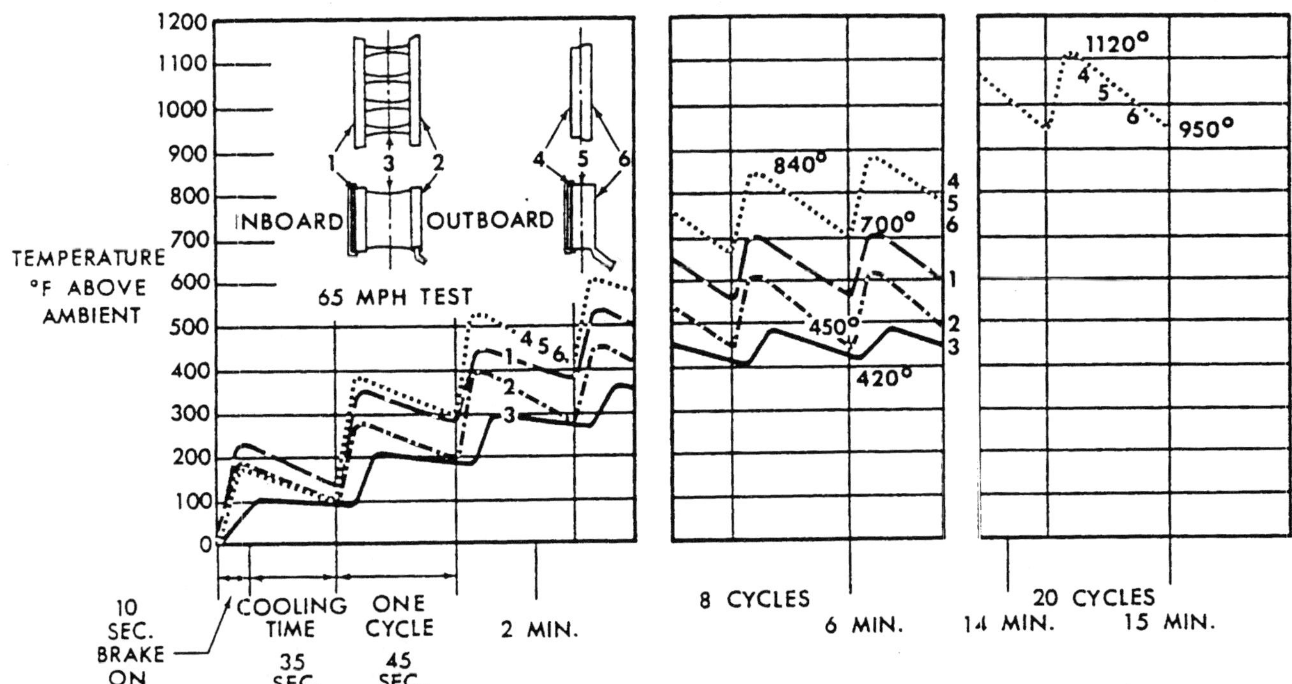

Figure 9 — Steady state temperature curves for solid and ventilated rotors of equal mass

is equal to the heat generated during braking. Eksergian (1)* Fazekas (2) and Petrof (3) have explored experimental and theoretical aspects of transient temperatures under these conditions. The solid rotor, because of its lower potential for dissipating heat attained a considerably higher peak equilibrium temperature of 1120°F. In the case of the solid rotor more of the kinetic energy converted to heat during each braking cycle is retained by the mass of the rotor due to less heat dissipation during each cooling cycle. As the curves in Figure 9 verify, subsequent brake applications drive the rotor to the higher equilibrium temperature.

When the brake is applied more heat is generated than dissipated, resulting in rotor temperature rise. This temperature rise during braking may be expressed in equation form as

$$T = \frac{E_h}{778\ wq} \qquad (7)$$

Where: E_h = Kinetic energy converted to heat (ft-lbs)

w = Weight of brake rotor (lbs)

q = Specific heat of the rotor material Btu per lb per degree F

Having established the temperature rise for a given brake stop, the measure of overall performance can be determined by equating the heat dissipation rate between brake application against the energy conversion rate.

HEAT LOSS

As we have stated earlier, in the case of the solid rotor, the rate of heat loss by radiation is relatively small in comparison to the rate of energy conversion to heat by the brake. However, radiation is one of two effective means for heat dissipation. The rate of heat loss by radiation can be determined as follows:

From Stefan-Boltzmann law of radiation

$$E_h = \sigma (T_2^4 - T_1^4) \qquad (8)$$

Where: T_1 = Ambient temperature (°R)

T_2 = Temperature of exposed rotor surface (°R)

σ = 0.174 x 10^{-8} Btu/Sq Ft/Hr/°R^4 (9)

Based on King's "Basic Laws and Data of Heat Transmission".

Combining Equation 8 and 9 a general equation for the rate of heat loss can be established

$$H = \frac{\sigma \xi A_S (T_2^4 - T_1^4)}{3600} \qquad (10)$$

Where: H = Rate of heat loss (Btu per sec.)

A_S = Area of exposed disc surface (sq. ft.)

ξ = Emissivity coefficient

Based on an emissivity coefficient ξ of .8 and a Stephan-Boltzmann constant of radiation of .174(10^{-8}) Btu per hour per square foot per degree Rankine to the fourth power the rate of heat loss becomes

$$H = .386\ (10^{-12})\ A_S\ (T_2^4 - T_1^4) \qquad (11)$$

In the case of the ventilated rotor there is still another effective way of heat transmission — that of forced convection made possible by the rapidly moving air stream directed into the vents of the rotor. Theoretical and experimental work by Petrof (3) Eksergian (4) Newcomb (5) and Koffman (6) support the generally accepted fact that ventilated rotors are superior to solid rotors for heat dissipation by convection. It was logical to conclude that a ventilated disc brake rotor offered the best short range solution for the Mark II GT brake design.

In the previous section it was stated that the largest part of the heat exchange in the brake is accomplished by radiation and convection from the rotor surface area to the ambient air. The amount of heat convected is a function of the size and quality of the contact surface and the velocity and turbulence of the air flow.

To take full advantage of the approximately 325 square inches of surface area provided by each of the ventilated rotors we installed air scoops and ducts as shown in Figure 10 to direct a steady stream of cooling air to the rotor surface.

The effects of the ram air were evaluated at the Riverside, California track with excellent results. The front brake rotor temperature, where the air stream has a relatively direct path, showed a drop in peak surface temperature from 1500° F. to 1250° F. Directing an

* *Numbers in parentheses designate references at the end of the paper.*

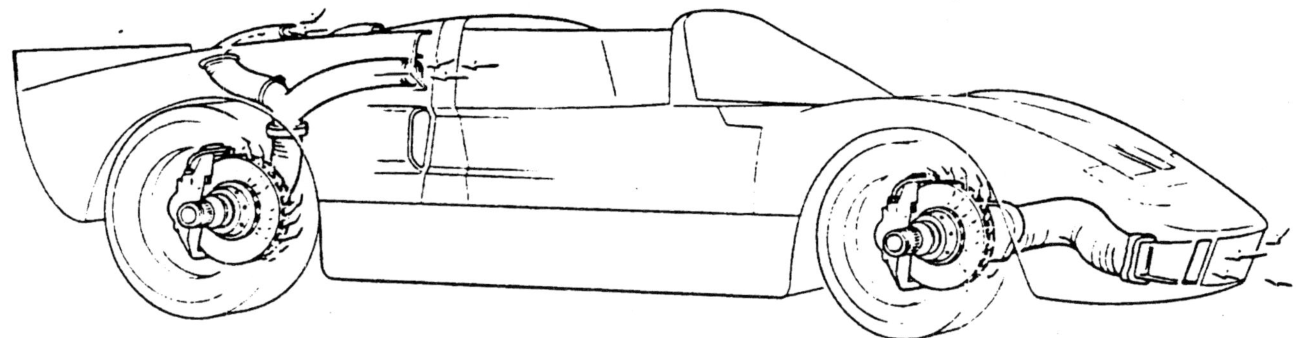

Figure 10 — Mark II GT ram-air brake cooling ducts

air stream to the rear brake presented a more serious problem. The air at the rear of the Mark II is generally at a lower pressure and is more turbulent than that at the front of the vehicle. But the most serious obstacle to an adequate air duct is the limitation imposed by the engine and transaxle package. One solution to this problem was used with some success at Sebring and Le Mans. A periscope-like air scoop mounted on the rear deck reached out into the high velocity air stream to supplement a weaker stream channeled from the rear quarter body scoop. The success of these air ducts can be measured in part by the reduction in rotor surface temperatures, but equally important is the side effect of lower brake fluid temperatures and improved lining life. Since the improved air ducts were incorporated, the problem of fluid vaporization has been virtually eliminated.

TWENTY-FOUR HOUR DAYTONA CONTINENTAL RACE

Because the sports car brakes must be capable of performing at high speeds under varying track and climate conditions, many factors must be taken into consideration in their design. The final brake design must be tested thoroughly under actual operating conditions to assure that the desired characteristics and performance have been obtained. This phase of development of the Mark II GT sports car brakes was spread over a three month period, culminating with a 72 hour "run in" at Daytona. Although all brake components were evaluated, the main purpose of the test was to develop a durable rotor.

After reviewing all considerations with respect to thermal shock, mechanical strength and resistance to wear, we specified a high grade of cast iron with randomly oriented graphite flakes. To provide assurance that the friction surface will not spall under the severe braking loads required to control the car on the Daytona circuit, we utilized a flame spraying technique to apply a thick coating of silicon carbide in a matrix of copper. The flame sprayed coating resists wear and has a high coefficient of friction and thermal conductivity. Test results obtained during the "run in" prior to the race confirmed the fact that the iron and spray coating specified were suitable materials.

The Daytona International Speedway shown in Figure 11 is a 3.81 mile course consisting of 2.5 miles of high banked speedway course and 1.31 miles of twisting infield roads. This course proved to be a formidable test for the Mark II brakes. In the skillful hands of drivers like Ken Miles, Lloyd Ruby and Bruce McLaren, the Mark II GT responded with speeds up to 196 mph on the banked speedway course. To enter the winding infield course, cars must decelerate violently to 45 or 50 mph at the No. 1 braking zone. The car is accelerated to 120 mph on the eastbound infield straight. Reaching the end of the straight, the car must decelerate to 45 mph at zone 2. Coming out of the southbound turn, the car accelerates to 125 mph and snubs down to 110 on the southbound leg. Once again the car is decelerated to 30 mph to navigate the hairpin turn. The final leg of the infield course is limited to approximately 107 mph before the 5th and final braking zone is reached.

In braking the car from the speeds just described, the brakes must convert the kinetic energy of a moving vehicle into heat energy. From Equation 1 the energy of a moving vehicle may be expressed as

$$E = \frac{WV^2}{2g}\frac{(88)^2}{(60)^2} \qquad (12)$$

Where: V = Instantaneous velocity (mph)

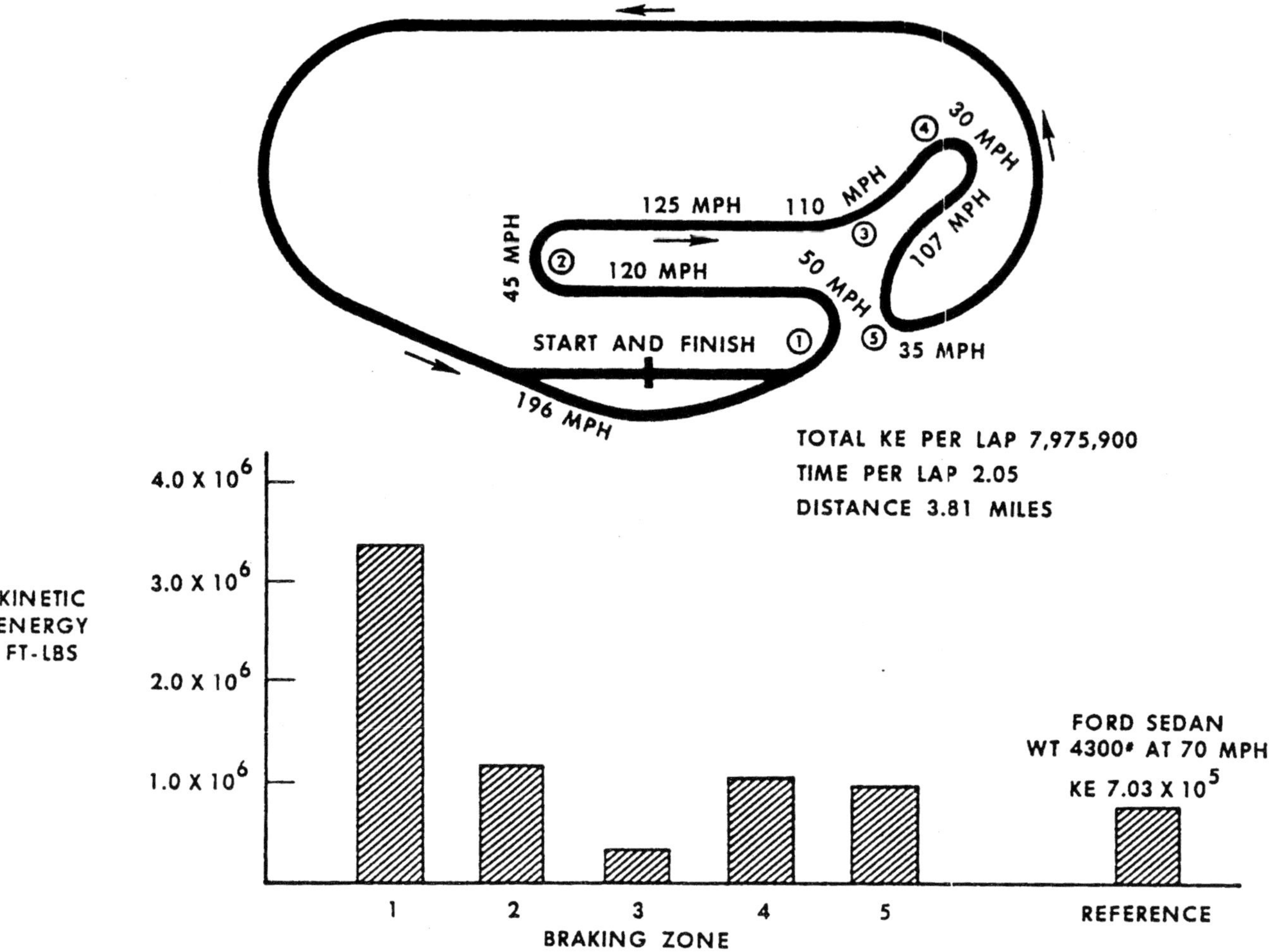

Figure 11 — Daytona International Speedway showing braking zones and the approximate energy converted to heat at each zone

But Equation 1 applies only to the case of a vehicle brought to rest. A vehicle decelerating from 196 mph to 50 mph at braking zone No. 1, the kinetic energy remaining at velocity V_2 is E_2 and the kinetic energy (E_h) converted to heat, if engine and aerodynamic braking forces are neglected, becomes

$$E_h = E_1 - E_2$$

$$= \frac{W(V_1^2 - V_2^2)}{2g} \qquad (13)$$

$$= \frac{2860\,(196^2 - 50^2)\,(88)^2}{2(32.2)\,(60)^2}$$

$$= 3,430,700 \text{ ft.-lbs.}$$

The values of the kinetic energy for stops 2, 3, 4 and 5 can be derived in the same manner. Plotting these values of the energy converted to heat against the braking zone produces a graphic profile of the heat energy that must be dissipated to the atmosphere on each lap of the course. Some appreciation of the amount of energy that is involved may be derived from Figure 11 by referring to the 703,000 ft-lbs of kinetic energy of a 4300 pound Ford sedan traveling at 70 mph (102.7 ft per sec). This energy is equivalent to 904 Btu of heat which must be dissipated in making a complete stop from 70 mph.

Neglecting the effect of aerodynamic drag, engine braking and rolling resistance, and assuming that the vehicle rate of deceleration was approximately 18 ft per sec^2 the time required to slow the Mark II GT from 196 mph to 50 mph is

$$t = \frac{(V_1 - V_2)\,(88)}{a\,(60)} \qquad (14)$$

$$= \frac{196 - 50\;(88)}{18\;(60)}$$

$$= 12 \text{ sec.}$$

And since power is defined at the time rate of doing work, the average power developed by the brakes in retarding the vehicle from 196 to 50 mph is

$$P_a \text{ (Average power)} = \frac{\text{work done}}{\text{time taken to do the work}}$$

$$= \frac{3{,}430{,}700}{550 \times 12} \qquad (15)$$

$$= 520 \text{ hp}$$

Some measure of the efficiency of the brake system may be derived from the fact that the 520 hp is dissipated by 60 pounds of rotor iron and 12 pounds of shoe and lining assemblies.

Frequently the load on a brake is judged by the power at initial application, that is, initial energy rate. This value is based on the fact that the brakes of a moving vehicle begin to convert kinetic energy to heat at the moment the brakes are applied. The rate of energy conversion is initially high, but decreases with speed until both speed and power are zero. The instantaneous power developed by the brakes in stopping from 196 mph may be expressed as follows:

$$P_i = \frac{WaV\,(88)}{550g\,(60)} \qquad (16)$$

$$= 8.29\,(10^{-5})\,WaV \qquad (17)$$

Where: P_i = Instantaneous power developed by brake (hp)

W = Weight of car (lbs).

a = Deceleration (ft/sec/sec)

V = Instantaneous velocity (mph)

Assuming the average rate of deceleration for the Mark II to be 18 ft/sec², the power developed by the brakes at zone No. 1 at initial velocity of 196 mph is

$$P_i = 8.29\,(10^{-5})\,(2860)\,(18)\,(196)$$

$$= 836 \text{ hp}$$

The ability of the brake to carry the power burden imposed upon it is related to the area of the brake surfaces. The ratio of the instantaneous power (P_i) and the effective area (A) of the friction contact surface in sq. in. is called the "power density". Thus, the Mark II with approximately 55 per cent of braking, at the front wheels generates a front brake power density of

$$P_i = .55\,(8.29)\,(10^{-5})\,\frac{WaV}{A} \qquad (18)$$

$$= \frac{.55\,(836)}{435}$$

$$= 10.7 \text{ hp per sq. in. of lining}$$

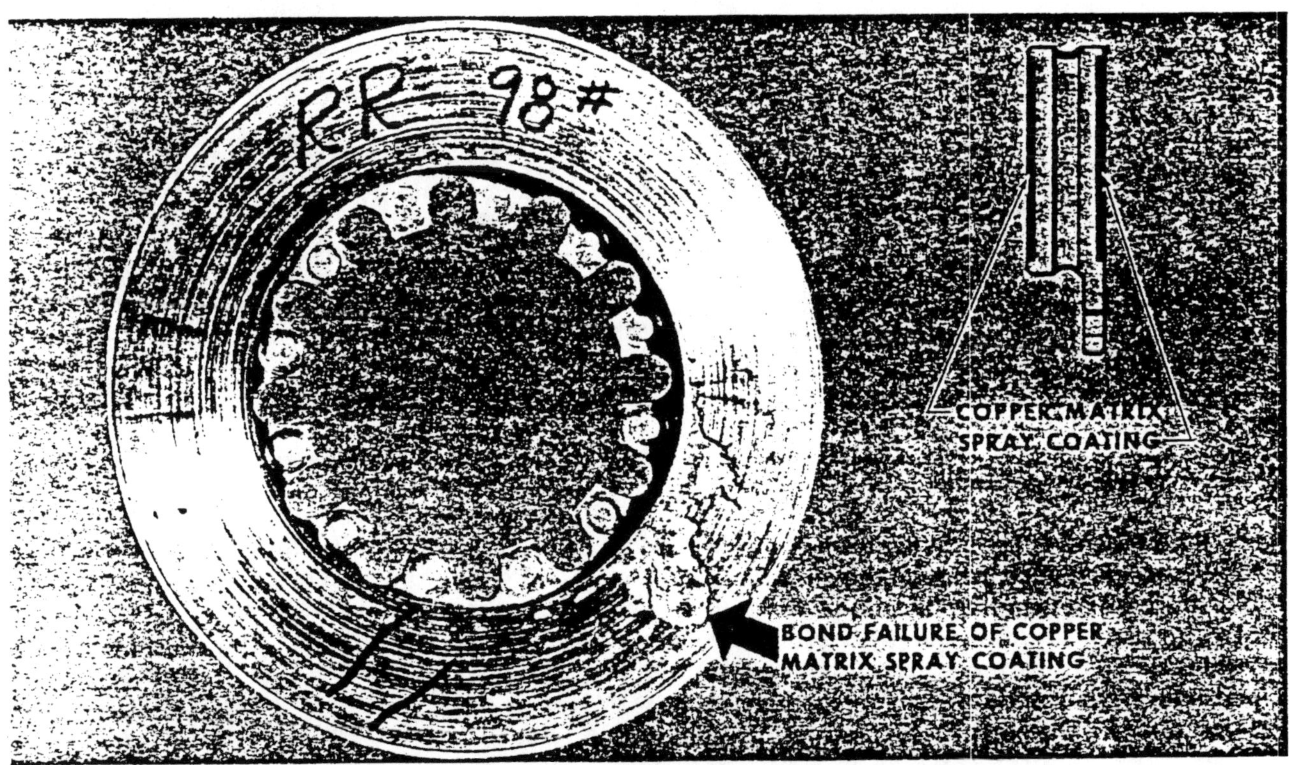

Figure 12 — Daytona Mark II GT brake rotor removed from car No. 98 after 18 hours

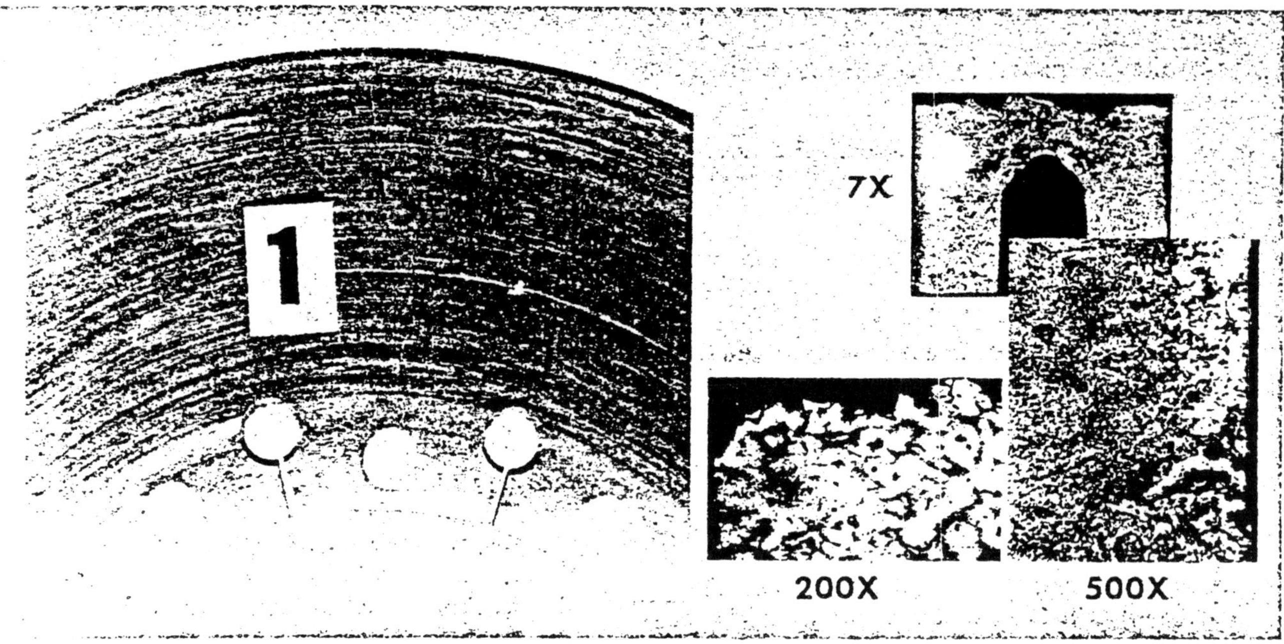

Figure 13 – Microstructural analysis of a development rotor showing heavy shrink and heat checks

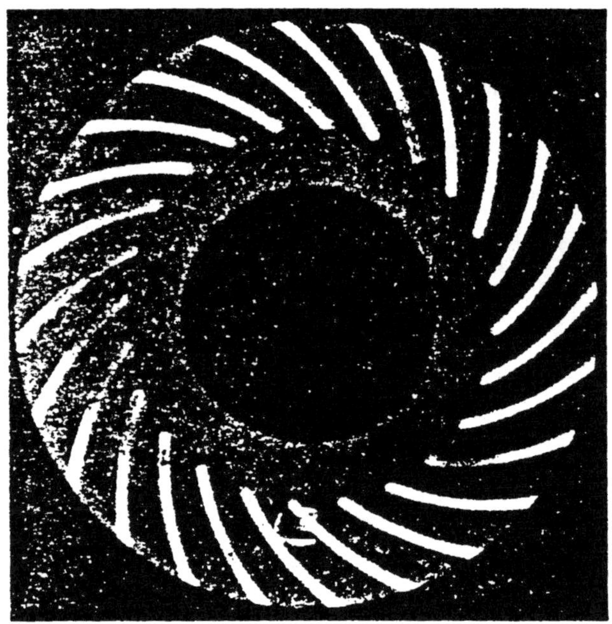

Figure 14 – X-ray view of a poor rotor casting showing dark areas of heavy shrinkage

Figure 15 – X-ray view of a sound casting shown by uniform color tone and sharp definition of rib structure

The area of friction material in contact with the rotor varies according to the particular application. For normal applications a maximum of 6 hp per sq. in. is usually recommended, but values of 10 hp have been used for certain installations where pad wear is a secondary consideration or where space limitations preclude the use of larger pads as in the case of the Mark II GT.

As the results show, the Mark II brakes were not only efficient but durable. Since the Daytona 3.81 mile course was traversed in roughly two minutes and the brakes were applied five times per lap, the heat was generated and dissipated with punishing frequency. During the race the brakes performed as expected for approximately 500 laps or 18 hours of the 24 hour endurance race before the thermal stresses weakened the base metal and caused some rotors to fail.

The failures in all cases were due to flaking of the flame sprayed coating, similar to the case shown in Figure 12. The failure was attributed to a combination of base metal breakdown due to heat checking and to heavy shrinking in the rib area, a condition which was encountered with a development rotor (Figure 13). To prevent failures of this nature in the future, we developed improved coring and casting techniques. In addition, as a final precaution, all castings were x-rayed prior to machining. Figure 14 shows a typical x-ray view of a poor casting with heavy shrinking, indicated by the dark area, in contrast a sound casting shown in Figure 15 presents a uniform color tone without the pockets of spongy metal. X-raying eliminated castings which otherwise may have caused structural brake failures.

SEBRING 12 HOUR GRAND PRIX

Long before the Daytona brake rotors had a chance to cool off, a new brake program was initiated to prepare the Mark II GT's for the next race — the Sebring 12 Hour Grand Prix. With less than two months remaining to complete system modification and to initiate procurement of hardware, the entire staff of the Manufacturing Research Cast Metals Section was assigned to the project. After weighing all alternatives, this group decided that the inability to check the soundness of spray metal coating ruled out the use of Daytona type rotors for Sebring. It was further decided that for the Sebring's twisting course nodular iron brake rotors could provide the necessary structural strength at the high temperatures recorded on a test run, provided the lower frictional characteristics of nodular iron could be tolerated. To determine this and other questionable

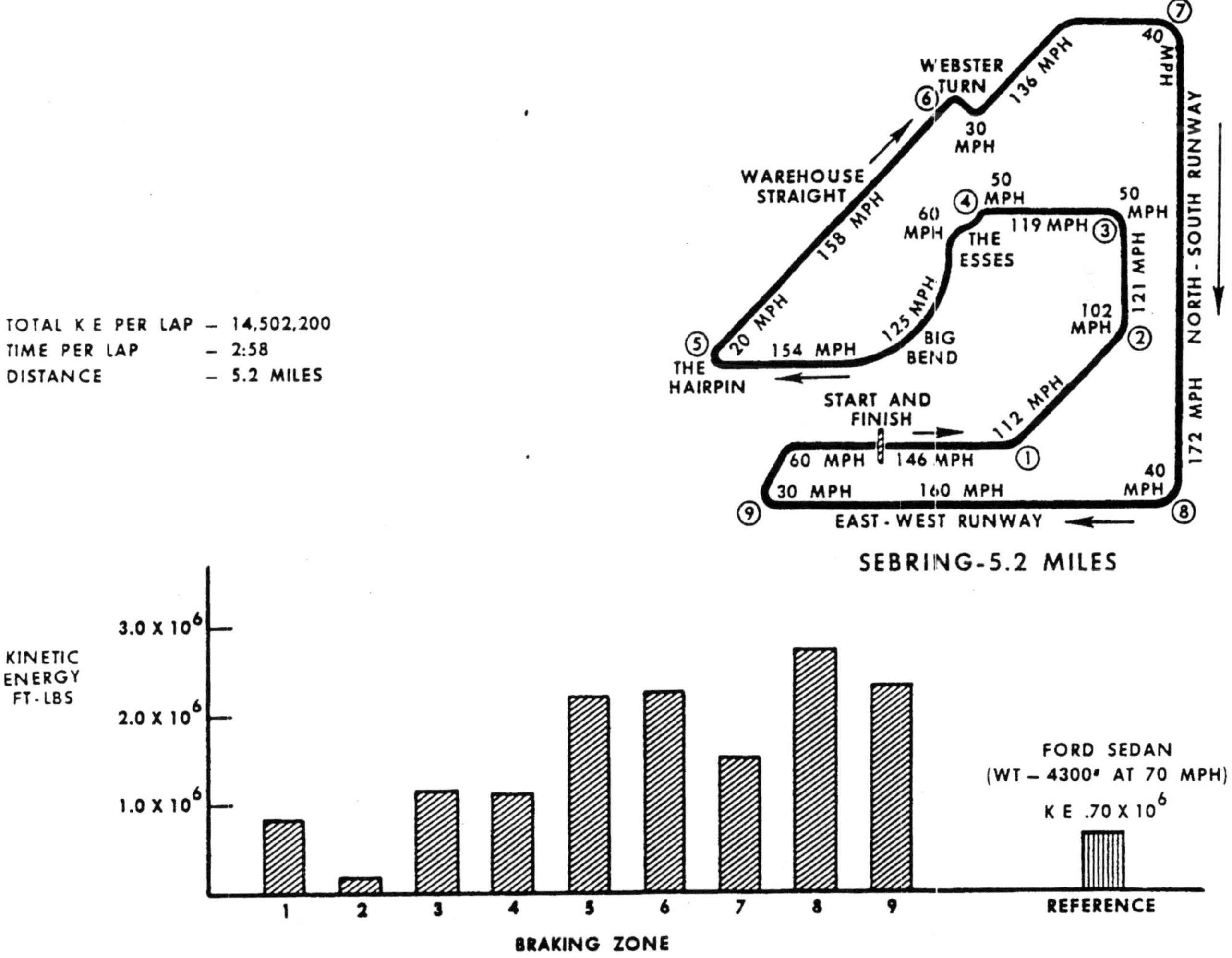

Figure 16 — Sebring 5.2 mile twisting course showing braking zones and energy loads

properties of the proposed high strength iron at elevated temperatures, we employed Ford's new brake dynamometer on a 24-hour basis. Nodular iron and other recommended rotor materials were evaluated on the basis of a race simulated from vehicle data recorded on the Sebring course. The dynamometer tests showed the nodular rotors were in fact superior to the Daytona type rotors. At this point, with only three weeks left before the race, a 24 hour durability test was conducted at the Riverside, California track. As expected, the pedal efforts for wheel lock-up were found to be high, but the lining and rotor combination survived the 24 hour test, correlating results with the dynamometer test. With the exception of a minor rotor dishing, .010 to .015 of an inch, which at this point was not recognized as a problem, the new brake package was considered ready for the Sebring Grand Prix.

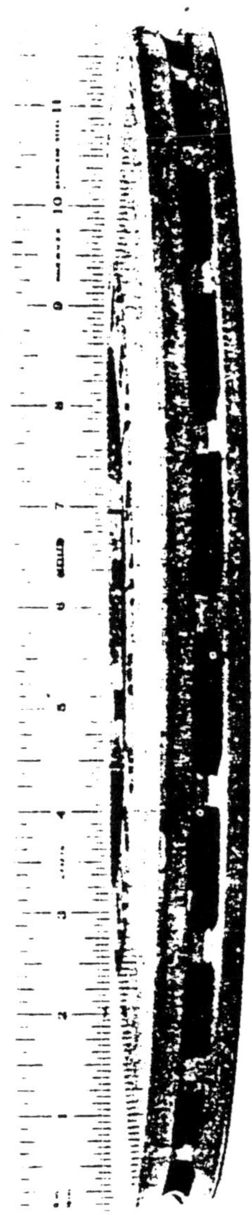

Figure 17 – Mark II Sebring rotor showing severe warping after 12 hours of racing

The 5.2 mile Sebring circuit shown in Figure 16 is unquestionably the most demanding in the country. The nine braking zones require a total of 14,502,200 ft.-lbs. of energy to be dissipated in a span of approximately 3 minutes. This punishing brake cycle was repeated 228 times in the 12-hour period.

The high rate of energy conversion once again took its toll of brake rotors. Although there were no structural failures, warped rotors such as the one shown in Figure 17 did present a problem. The saucer shaped friction face made lining pad replacement difficult, if not impossible. On several occasions new rotors were installed simply to facilitate pad replacement.

A post-mortem examination of microstructural evidence by the Ford Applied Research Laboratory showed that the rotor surface temperature exceeded 1500°F. Thermal stresses generated by alternately heating and cooling the iron produced some localized heat checking and general warping. This condition was duplicated by heating the brake rotor to 1480° for one hour followed by cooling in an air blast. It was not obtained when the rotor was heated to 1375°F and cooled in the same manner. This explains in part the discrepancies between the Riverside durability, where the rotor performed exceptionally well, and the Sebring race, where at the higher temperatures the rotors showed evidence of heat checking and warping.

LE MANS 24 HOUR GRAND PRIX

Sebring's twisting course is generally considered the most torturous test of brake performance encountered on the sports car circuits, and the amount of kinetic energy that is converted to heat energy by the brakes per lap certainly supports this claim. Figure 18 shows that there is, in addition to the total kinetic energy per lap, still another consideration in measuring brake requirements for a given circuit — the maximum instantaneous energy for any single brake application On the Le Mans circuit the Mulsanne corner after the high speed straight is the most challenging single brake retardation. It exceeds all of the Sebring or Daytona high speed stops.

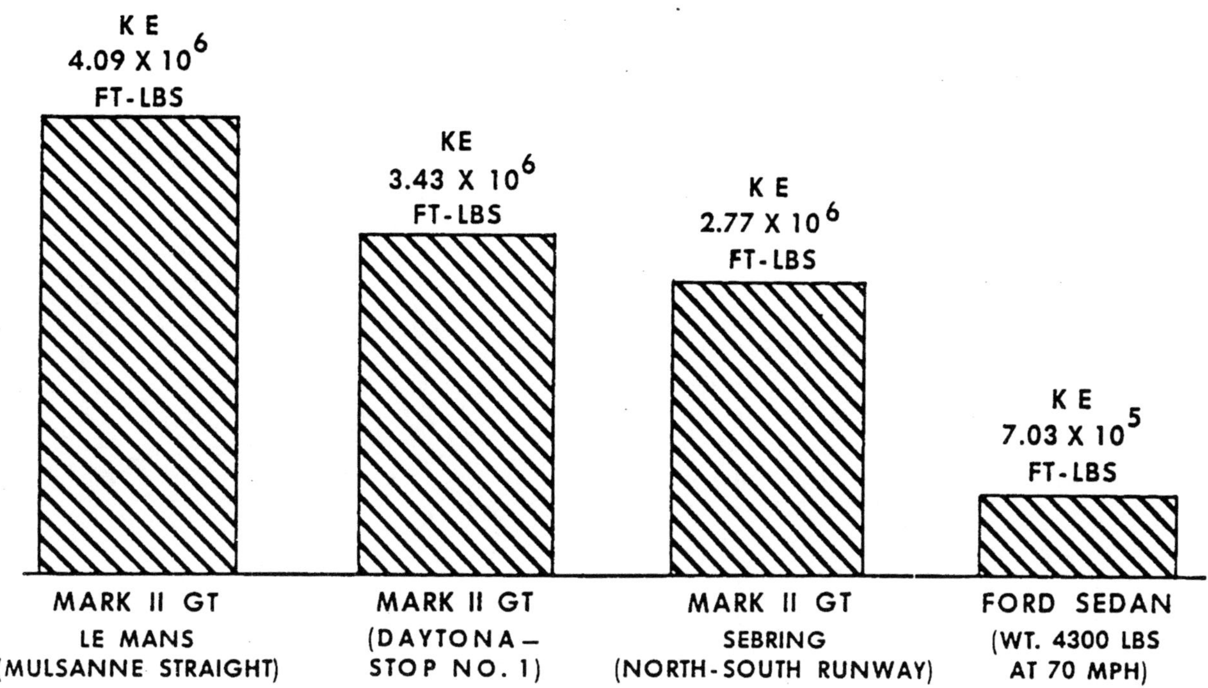

Figure 18 — Maximum energy converted to heat by the Mark II brakes during a single vehicle retardation

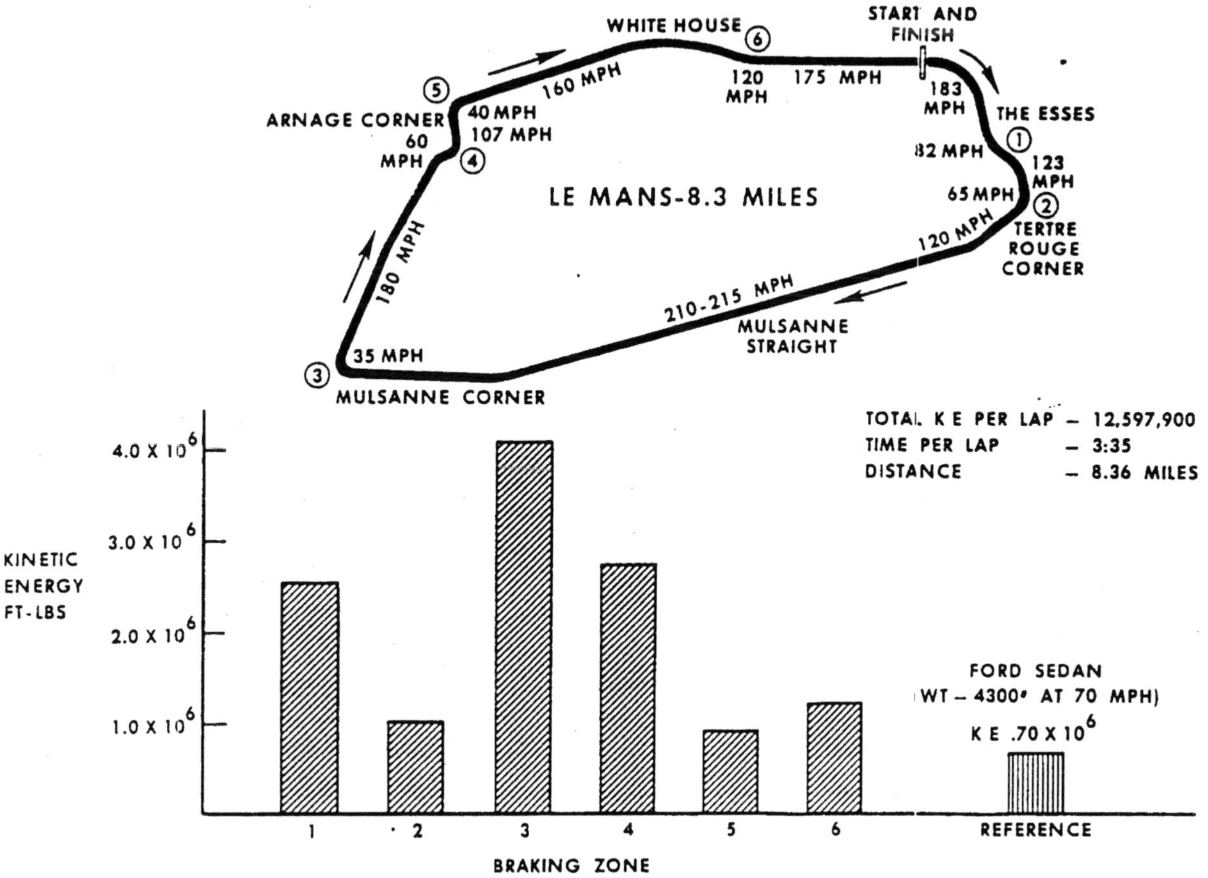

Figure 19 — Le Mans circuit showing the high speed Mulsanne straight and braking zones with related brake energy loads

After the experience at Sebring it was self evident that the energy conversion and cooling rates at Le Mans would require a material with considerably more thermal stability than the nodular iron offered. Once again a search for a suitable material was initiated. Experimental rotors were made from several cast iron compositions modified with alloy elements of chromium and molybdenum, primarily to harden the cast iron matrix but also to increase the thermal stability of iron carbide. These rotors were thoroughly evaluated on the dynamometer for susceptibility to cracking under thermal shock. Final test of the selected iron-moly rotor material was conducted on a vehicle at the Riverside circuit at a record braking pace with only slight surface heat checking.

The Le Mans course shown in Figure 19 is considered one of the fastest of the Grand Prix circuits. The 3.6 Mulsanne straight provides the Mark II with an opportunity to take advantage of its power as it reaches, with relative ease, speeds of 210 to 215 mph.

To negotiate the Mulsanne corner, the driver must reduce car speed to approximately 35 mph. In the process, 4,095,600 ft.-lbs. of kinetic energy are converted to heat. The instantaneous power developed by the brakes in com-

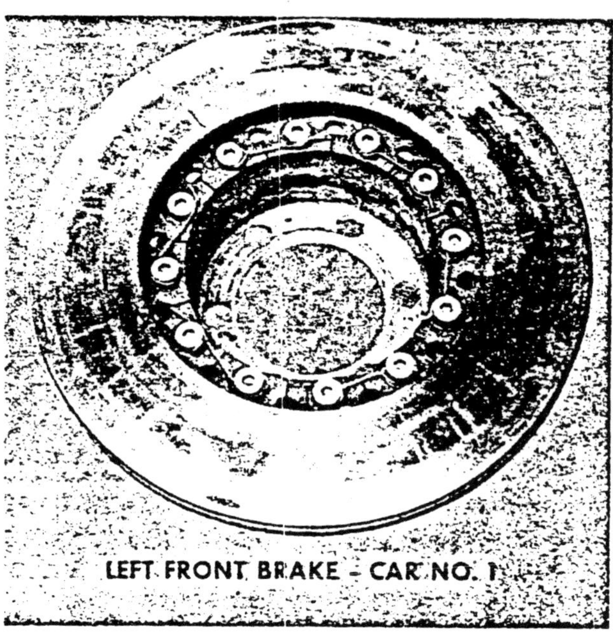

Figure 20 – Le Mans rotor showing extensive radial cracks after 65 laps

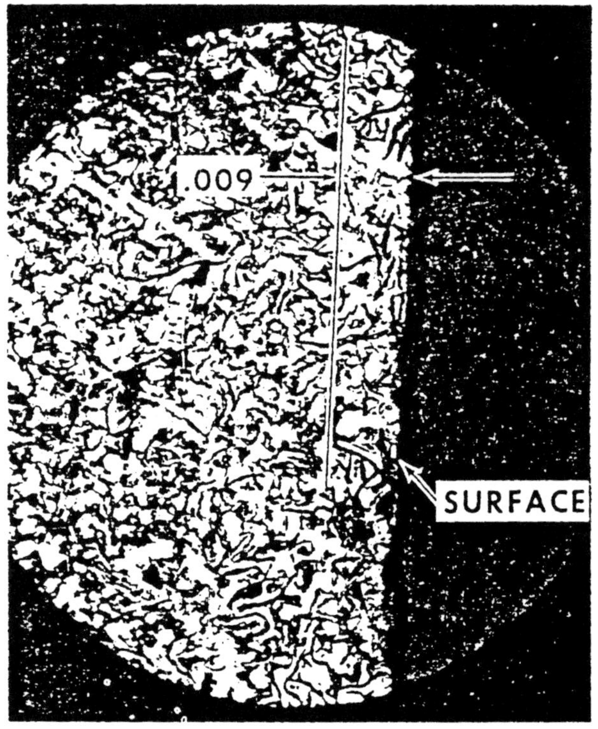

HEAT AFFECTED ZONE ON A RIVERSIDE DURABILITY ROTOR (100X)

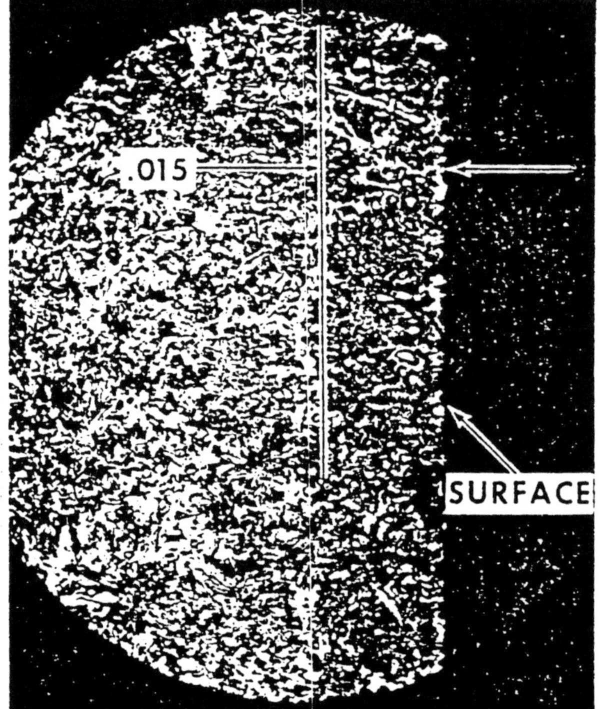

HEAT AFFECTED ZONE ON A LE MANS RACE ROTOR (100X)

Figure 21 – Photomicrographs of (a) Riverside rotor showing .009 in. heat affected area (b) Le Mans rotor heat affected zone of .015 in. shows effects of more severe brake usage

pleting this retardation can be determined from Equation 17, derived earlier in the paper. Assuming an initial speed of 210 mph, we can compute the instantaneous power developed by the brakes as follows

$$P_i = 8.29\,(10^{-5})\,(2860)\,(210)\,a$$

and for a deceleration rate of 18 ft/sec^2

$$P_i = 8.29\,(10^{-5})\,(2860)\,(210)\,(18)$$

$$= 896 \text{ hp}$$

The effects of this energy conversion were manifest on the rotors removed from the racing vehicles. One such rotor (Figure 20) removed from car No. 1 after only 5-1/2 hours of racing showed extensive radial cracks. A later metallographic examination by the Ford Manufacturing Development Office indicated that the heat affected zone at the braking surface shown in Figure 21 was approximately .015 in. deep while the 24 hour Riverside durability rotor shows a much lower heat penetration of .009 in. The microstructure in the heat affected zone of both rotors experienced temperatures in excess of 1500°F. With knowledge of the rotor soak temperature derived from the Riverside test data, an isochronic temperature gradient (Figure 22) was constructed to show the extreme temperature conditions experienced by the metal through the rotor cross section. This gradient explains in part the volume change at the surface as it goes through the transformation point at the critical temperature of the metal. Since the base metal is at a temperature substantially below the critical temperature, thermal stresses are developed at the interface. Thermal stresses generated by alternately heating and cooling the material in time produce localized heat checking, which if sustained generally produces radial cracks. Investigations of this phenomenon are discussed in detail by Fazekas (2) and Crosby (7).

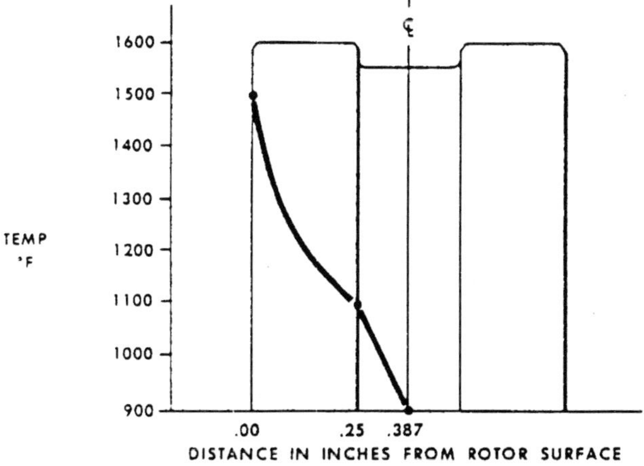

Figure 22 – Isocronic temperature gradient in a 3/4 in. ventilated rotor

Although the main part of this discussion appears to be preoccupied with the problems under racing conditions, there were other interesting observations recorded during the Le Mans race which should be mentioned. In the course of the 24 hour event, Car No. 2 driven by Bruce McLaren completed 360 laps of the circuit at an average speed of approximately 125 mph — including pit stops. In the process, the car required one change of rear lining pads and two changes of front brake lining pads. The general pattern for rotor changes indicates that the early pace of 135 to 140 mph resulted in a Mark II rotor life of 5-1/2 to 6-1/2 hours. After the competitors' vehicles were eliminated and the Mark II cars settled down to a cruising speed of 125 mph, no additional rotor changes were required. Both the No. 1 and No. 2 cars completed the race with the original rear brake rotors.

CONCLUSION

There is no doubt that as long as brakes are used for stopping from high speed, thermal stresses will occur. It is believed, however, that while these stresses experienced under racing conditions at Daytona, Le Mans and Sebring cannot be eliminated completely within the space limitations imposed by the Mark II package, they can be minimized by using an austenitic form of ductile iron. When properly heat treated, the ductile iron should provide greater dimensional stability, improved resistance to heat checking and better oxidation resistance. Certainly, these highly desirable thermal characteristics of ductile iron warrant a thorough investigation of the material for future high performance brake application.

Throughout the discussion a great deal of emphasis has been directed at the metallurgical aspects of the brake rotor design. Experience shows, however, that metallurgy alone cannot solve the problem. For a successful solution to the high performance brake problem, additional knowledge is needed in the area of the mechanics of heat transfer. Determining the proper wall thickness and the optimum amount of cooling air for ventilated rotors is of great importance. Rotor wall plays an important role in the braking cycle. It must store the energy of heat for dissipation to atmosphere at the completion of the braking cycle and must conduct the heat away from the lining interface to the vented surface area. A balance of these functions must be established to reduce thermal stresses by maintaining relatively low temperature gradients. Only with a working knowledge of these functions can the engineer make a notable contribution to future brake design.

While much of this work was aimed at sports cars, there is no doubt that the brake systems on all Ford Motor Company products will benefit from the experience. The design of efficient brake systems based on economy of space and weight is of paramount importance to the manufacturer. It is equally important to the customer who receives a superior product for his money.

REFERENCES

1. C. L. Eksergian, "Steady State Thermal Capacity of Ventilated Rotors", Kelsey-Hayes Company Report, July, 1962, Romulus, Michigan
2. G. A. G. Fazekas, "Temperature Gradients and Heat Stress In Brake Drums", paper presented at S.A.E. Summer Meeting, Atlantic City, New Jersey, June 6, 1952.
3. R. C. Petrof, "Transient Temperatures in Brakes", Ford Motor Company Report No. AR 65-14, May, 1965, Dearborn, Michigan
4. C. L. Eksergian, "Design Approach to the Automotive Disc Brake", Paper Presented at S.A.E. Meeting, Detroit, Michigan, March, 1956
5. T. P. Newcomb, "Temperatures Reached In Disc Brakes", Journal Mechanical Engineering Science, Vol. 2, No. 3 (1960), p. 167
6. J. L. Koffman,"Ventilated Disc Brakes", Automotive Egnineer, July, 1956, p. 277
7. V. A. Crosby, "Metallurgical Development In Brake Drums", paper presented at S.A.E. Meeting, Detroit, Michigan, January, 1959

PART II
TESTING

INTRODUCTION

Although the brake system had experienced rotor durability problems, it was not until after the 1966 Daytona Continental 24-hour race that the magnitude of these problems was recognized. It was obvious that the brake system would require major improvements to be competitive at Sebring or Le Mans. Furthermore, timing was critical. Only three weeks were available for rotor development for the Sebring race and eight weeks to the ultimate goal at Le Mans.

TEST EQUIPMENT SELECTION

Due to the critical timing, the test equipment and procedure selections were of primary concern. A test program that normally takes several months had to be compressed into three weeks. The key to fast, accurate rotor development proved to be the new brake dynamometer in the Ford Reliability Laboratory. The brake dynamometer was chosen for development testing for the following reasons.

1. The brake dynamometer would operate an equivalent to 10,000 miles of Sebring racing or eight Sebring races without wearing out a single vehicle. It would be very difficult to do this with an actual Mark II GT without major vehicle component replacements, which would delay the test program.
2. Brake dynamometer testing eliminated the need for skilled racing drivers and mechanics. A complete team of personnel, much the same for an actual race, is needed to conduct 10,000 miles of racing durability.
3. Testing with the brake dynamometer eliminated the variable of weather. When track tests are conducted, it is often difficult to tell whether changes in performance are due to weather or due to a design change.
4. The brake testing could be conducted using only one brake assembly, rather than a complete set of four prototype brakes which are required for a track test. The dynamometer can test a complete vehicle system of four brakes which is useful in brake distribution studies or in brake "pull" studies, but the Mark II GT had neither of these problems.

DETERMINING TEST REQUIREMENTS

To duplicate the conditions that a brake experiences, it was necessary to measure these conditions on a car at racing speeds. Since the initial goal was to develop a brake that would meet the grueling requirements of Sebring, accurate data for this track was required.

Although the brake dynamometer eliminated the need for a test car and a skilled driver, their use was required for a few hours to measure the brake operating conditions to set up the laboratory test. A Ford Mark II GT was instrumented and driven around the Sebring race course, recording the conditions the brakes were experiencing. The car was driven at average lap times of 2:58 minutes by several experienced drivers who would be driving the Ford Mark II GT's during the race. We utilized the instrumentation that had been used to obtain engine data to obtain the brake data. The compact test instrumentation in the car continuously recorded engine speed, throttle position, and brake rotor surface temperature. The oscillograph records obtained were analyzed to determine the following:

1. Car speed at brake application and brake release.
2. Car deceleration during brake application.
3. Maximum brake rotor surface temperature during brake application.

The car speed was calculated from engine speed, transmission gear ratios, rear axle ratio, and tire size. The brakes were assumed to be applied when the driver closed the throttle and assumed to be released when the driver opened the throttle. The vehicle deceleration was calculated from brake "on" time and the car speed change during this time. The resulting calculated deceleration was 18 ft/sec^2 for all brake applications.

The brake operating temperatures were measured by a radiation-type transducer which sensed surface temperature by the color of the glowing brake rotors. With this type of transducer, the brake had to be "red-hot" before the transducer would function. A thermocouple in the disc rotor would have been more desirable since temperatures over the complete operating range could be recorded. The radiation transducer installation, however, was relatively simple, while a thermocouple installation would have required a more complicated slip-ring assembly installation.

THE FORD BRAKE DYNAMOMETER

When the actual braking requirements for Sebring were known, the dynamometer phase of the testing program began. To better understand these tests, and how the brake development was accomplished so quickly, a description of the Ford brake dynamometer would be helpful.

The dynamometer consists of the following basic sections:

- Rotating inertia flywheels which simulate the translational inertia of the vehicle.
- Drive motor to bring rotating inertia flywheels to test speed.
- Test stations where brakes are mounted.
- Dynamometer controls and instrumentation.

As shown in Figure 23, the dynamometer can simultaneously test a complete set of brakes for a four-wheel vehicle. The dynamometer installation in the laboratory is shown in a photograph in Figure 24.

Dynamometer Inertia Section

The dynamometer inertia section, which simulates vehicle weight, is shown in Figure 25. Inertia loads can be varied between 15 and 453 $slug\text{-}ft^2$ by means of removable inertia flywheels piloted on the dynamometer shaft and bolted to an inertia mounting flange. The dynamometer inertia consists of:

- 15 $slug\text{-}ft^2$ of motor and dynamometer shaft inertia
- 21 — twenty $slug\text{-}ft^2$ flywheels
- 12 — one and one-half $slug\text{-}ft^2$ flywheels

The small and large inertia flywheels are stored in groups on either side of the inertia mounting flange. The flywheels in storage are moved over and bolted to the flange as required for a test. The inertia loading on the dynamometer is calculated as follows:

$$I = \frac{Wr^2}{4637}$$

Where: I = Dynamometer inertia, $slug\text{-}ft^2$

W = Vehicle weight, lb

r = Vehicle tire rolling radius, in.

The complete 453 $slug\text{-}ft^2$ of inertia represents the inertia imposed upon the four brakes of a 10,000 GVW truck, or one front brake of some of the largest Ford trucks. The Mark II GT tests were conducted with one front brake on the dynamometer. This one brake handled 870 lbs. of vehicle weight which is equivalent to 30 $slug\text{-}ft^2$ of dynamometer inertia.

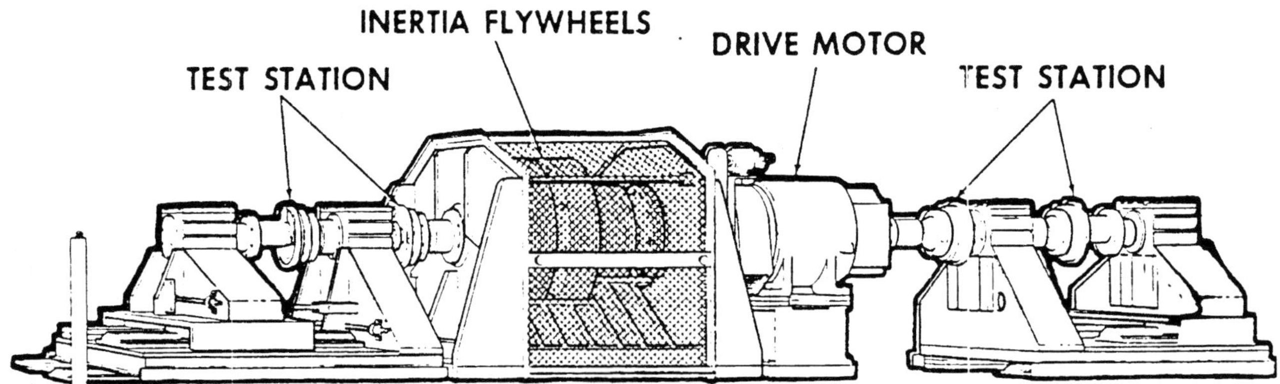

Figure 23 – Ford Reliability Laboratory Brake Dynamometer

Figure 24 – Reliability Laboratory Brake Dynamometer

Figure 25 – Dynamometer Inertia Section

Dynamometer Drive Motor

A DC electric motor drives the dynamometer up to test speed. As the test brakes are applied, the drive motor is normally electrically disconnected, immediately prior to and during the brake stop. For steady speed brake "drag" tests, the drive motor can be programmed to operate during the brake application. The motor can produce 125 hp from its 650 rpm base speed to the 2400 rpm maximum speed. Below the base speed, the motor can produce 1000 lb-ft torque on a continuous basis and up to 1500 lb-ft intermittently.

Dynamometer Test Stations

The four test stations on the dynamometer are positioned in an inboard and outboard arrangement. Normally, vehicle front brakes are tested at the outboard test stations while vehicle rear brakes are tested at the inboard stations. Figure 26 shows the test station arrangement at the end of the dynamometer. The front vehicle brake at the outboard test station is driven by the dynamometer shaft system which passes through the open center of the rear brake assembly at the inboard test station. Figure 27 shows the coupling required to drive a test brake drum or rotor at any test station. The coupling is normally fabricated from an automobile wheel "spider" welded inside a heavy circumferential steel ring. The test brake backing plate or caliper assembly is attached to the test station with an adapter. Figure 28 shows a front disc brake adapter and a rear drum brake adapter.

The brake backing plate adapter is mounted on a shaft which is restrained from rotation by a load cell. Each test station has a torque capacity up to 5000 ft-lb. The torque capacity of the two inboard test stations may be extended to 7500 ft-lb by relocating the load cells farther out on the torque reaction lever arms. The signal from each of these load cells is used to control the dynamometer and also is recorded as test data. More complete details of the controlling and recording aspects of the dynamometer are described later.

Each test station on the dynamometer is mounted on ways so that it can be moved back for easy removal and installation of the test brakes. Figure 29 shows both test stations moved back at one end of the dynamometer. The inboard test station moves on the machine base, and the outboard station moves on the base of the inboard station. This arrangement allows moving any individual station for access to a test brake without disturbing the adjacent station. Figure 30 shows both test stations in the normal running position.

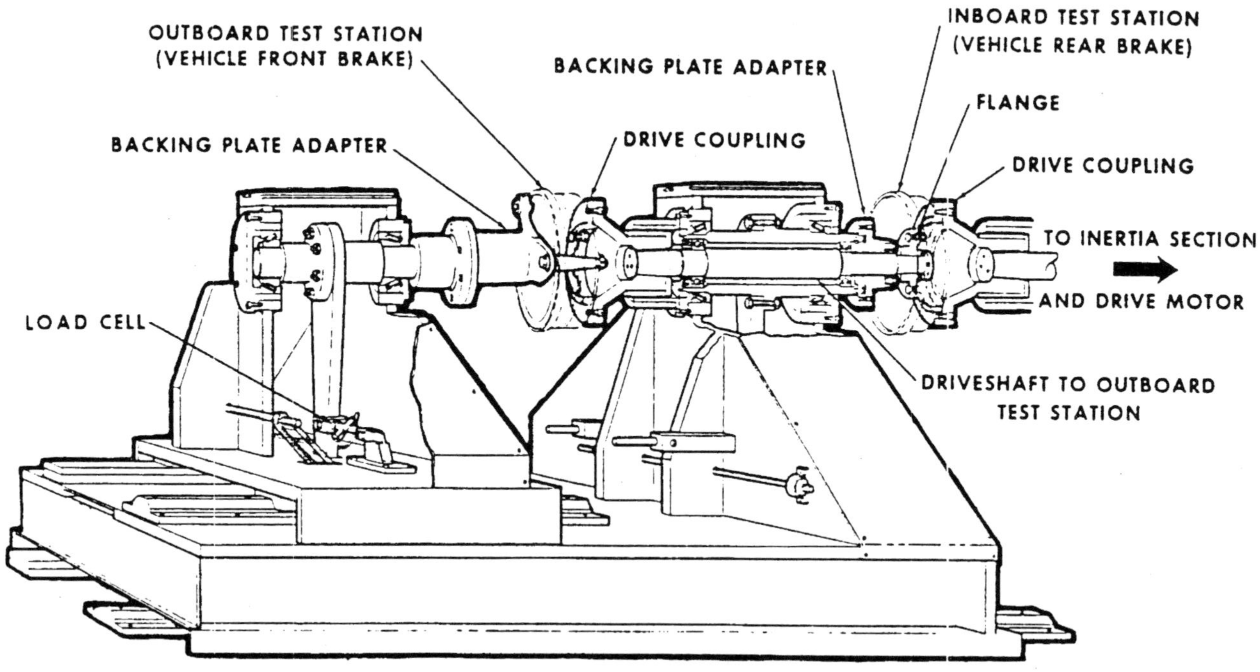

Figure 26 - Test Station Arrangement

Figure 27 — Coupling connecting test brake drum to dynamometer drive

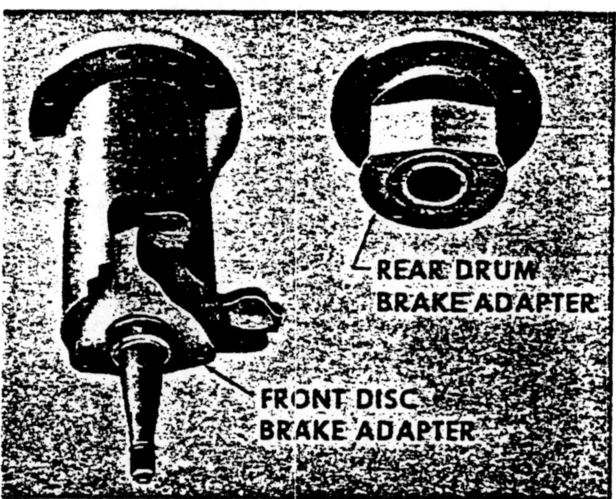

Figure 28 — Backing plate and disc brake caliper mounting adapter

Figure 29 — Test stations on dynamometer moved to provide access to test brakes

Installing the Mark II GT Brake

As was shown in Figure 28, a passenger car front brake adapter is normally fabricated from a front suspension spindle. Adapting a Mark II GT disc brake to the dynamometer presented problems not present in testing current passenger car and truck brake systems. The Mark II GT suspension components and wheels are made from thin-wall magnesium-aluminum alloy forgings which are difficult to attach solidly to the dynamometer. Also, the

Figure 30 – Test stations on dynamometer in test position

Figure 31 – Ford Mark II GT brake caliper mounting adapter on brake dynamometer

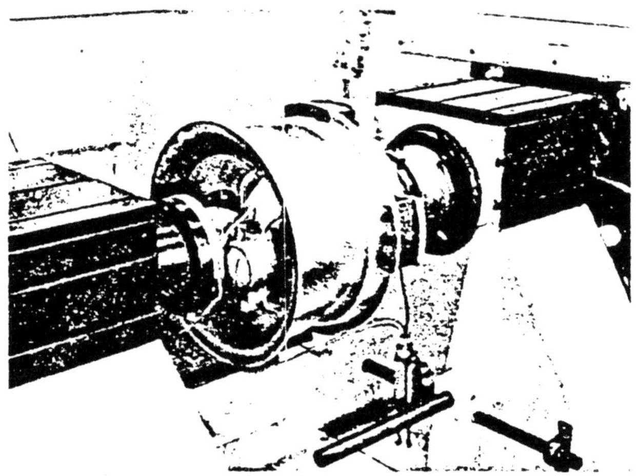

Figure 32 – Complete Ford Mark II GT brake and wheel mounted on brake dynamometer

critical time limitation forced all adapter and coupling construction to be of the utmost simplicity. The fabrication problem was solved by designing the adapter and coupling using aluminum. This permitted fast machining, and adequate strength was obtained without excessive size. Figure 31 shows the aluminum adapter, and Figure 32 shows the complete Mark II GT brake mounted on the dynamometer.

Test Brake Cooling

Brake cooling air is supplied to each test station on the dynamometer through 12 x 24 inch ducts. Air velocities up to 30 mph can be supplied to all four stations, and even higher air velocities can be provided for tests involving only one or two test stations. Figure 33 shows the air duct at one test station with the top half of the duct removed and the air return duct pivoted out of the way for access to the test brake. The cooling air is supplied from an under-floor plenum chamber, and exhausted to another under-floor plenum chamber on the opposite side of the dynamometer. The brake air cooling system has two variable speed fans. One fan pressurizes the supply plenum while the other evacuates the exhaust plenum. The air duct at the test station is maintained at slightly below atmospheric pressure so that no smoke or brake lining odor will escape into the laboratory.

The air velocity at each test station is controlled by varying the fan speeds and the individual test station damper position. During most tests, the brake cooling air velocity in the cooling duct is held constant to expedite the test cycle rather than duplicate the exact air velocity changes during braking. Heating coils are provided in the air system so that the cooling temperature can be controlled to any temperature between the outside ambient air temperature and 100°F.

Figure 33 — Dynamometer brake cooling ducts partially removed for access to test brakes

The Mark II GT tests were conducted with a complete vehicle wheel in place, causing the brake cooling air to flow through the ventilated rotor in the same manner as it would on the car. A nominal 50 mph cooling air velocity at 70°F. was used for all tests. Although this did not duplicate the variances in brake cooling during a lap at racing speeds, it did approximate the average cooling conditions.

Dynamometer Controls and Test Cycles

All brake tests are controlled and data recorded at the brake dynamometer operator's control console (Figure 34). The control system can be highly automatic and brake tests, such as completed on the Ford Mark II GT, can be conducted with the dynamometer virtually unattended. This allows test personnel to carefully observe all brake operations.

The dynamometer is capable of operating under four different control modes.

- Manual operation by the dynamometer operator.
- Constant speed operation with the brakes applied and released automatically at preset time intervals.
- A programmed test cycle that is automatically repeated up to 9999 times.
- A programmed brake test whereby up to sixty different test cycles follow each other automatically.

All programmed brake tests are based on a standard test cycle illustrated in Figure 35. The brake cycle consists of:

- Acceleration to stabilized braking speed.
- Brake application with deceleration to dwell speed.
- Dwell after brake release.
- Acceleration to brake cooling speed.
- Brake cooling period.

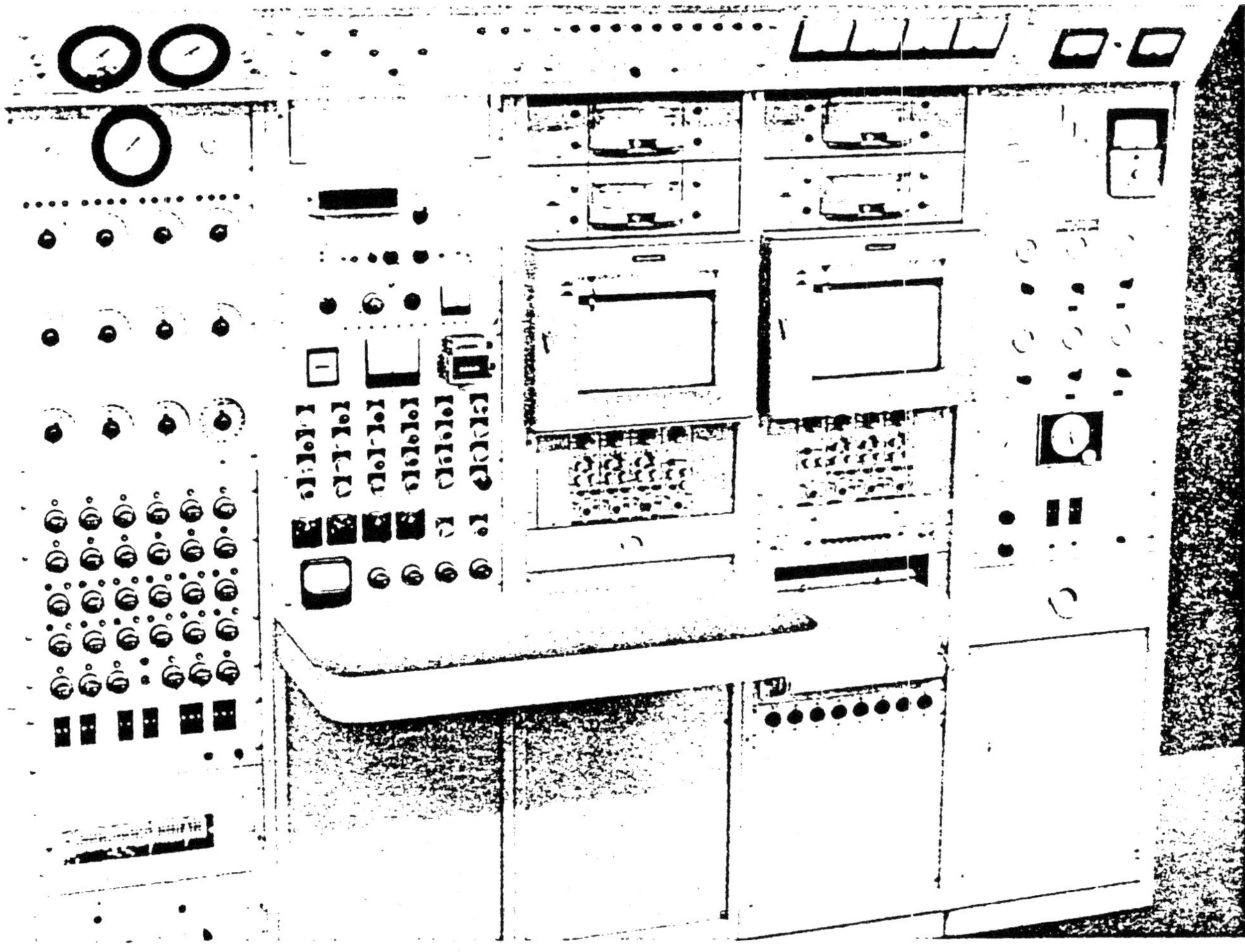

Figure 34 – Brake Dynamometer Operator's Control Console

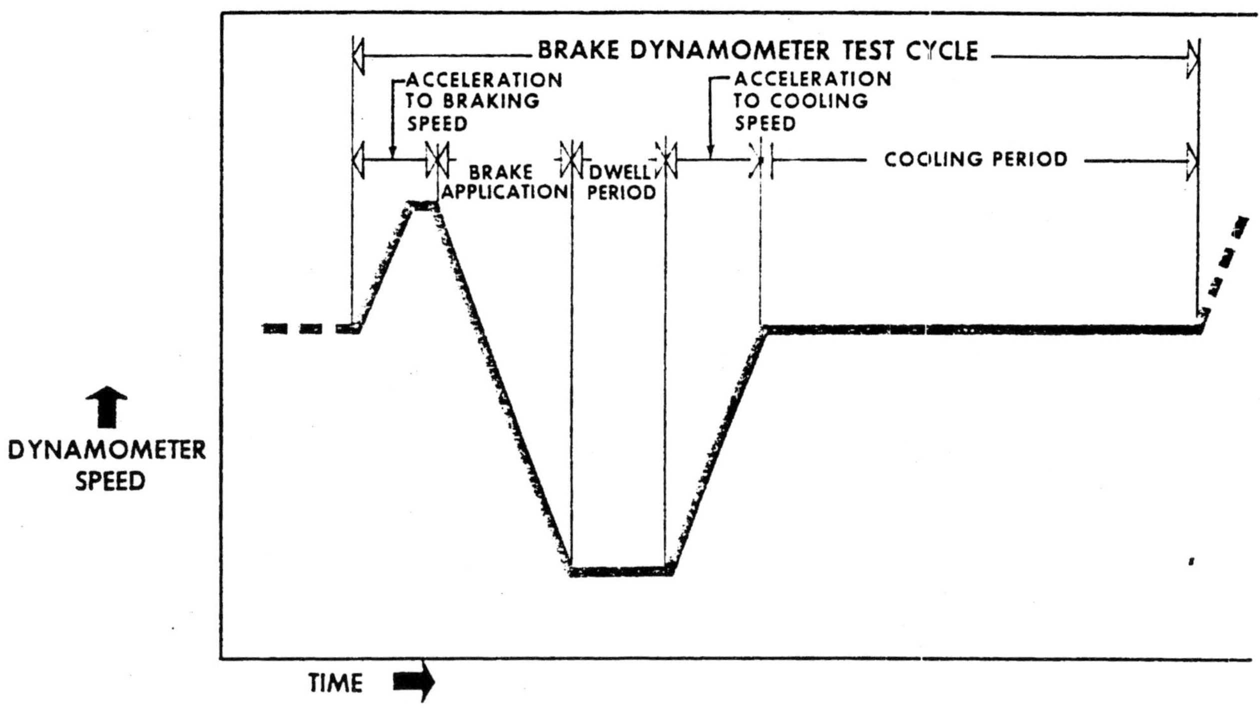

Figure 35 – Typical Brake Dynamometer Test Cycle

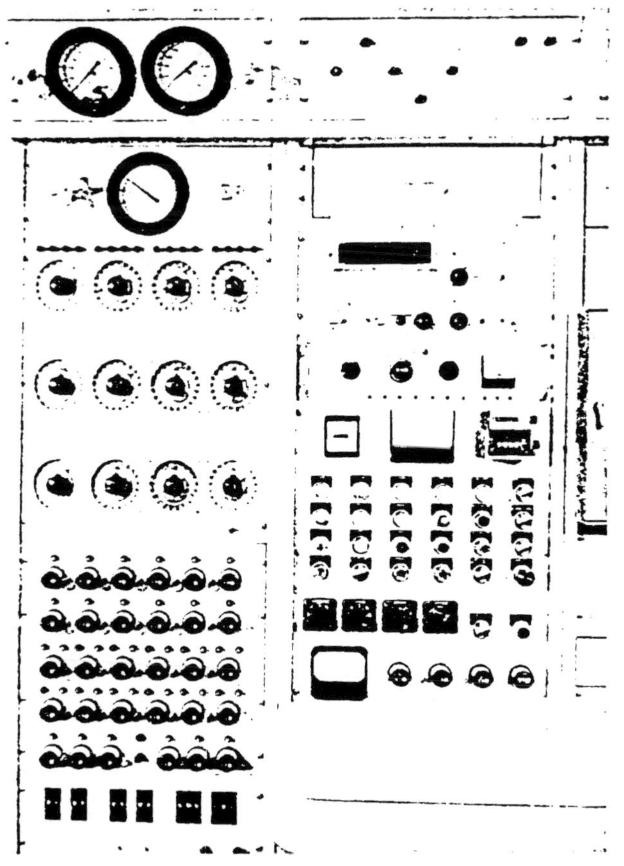

Figure 36 – Brake Dynamometer Programmer

To control the various parts of a programmed test cycle, a series of timers and potentiometers are used. The panel for the programmer controls is shown in Figure 36.

The timers normally program the time intervals for the different phases of the test cycle, while the potentiometers control dynamometer speed. For a repetitive programmed test cycle, only a limited number of the potentiometers and timers are used. For a typical brake cycle, the dynamometer speed at brake application, brake release, and brake cooling are preset on the potentiometer controls. Additional potentiometers are preset so a brake deceleration can be programmed at a constant brake line pressure or at a constant value of the sum of the brake torques. To perform these two functions, brake line pressures up to 2000 psi are supplied to the test brakes by a servo-controlled hydraulic power supply.

Additional capabilities of the dynamometer permit the length on one brake cycle to be controlled by either time or minimum brake temperature. Typical brake cycles with these two control methods are shown in Figure 37. A time-controlled cycle can be preset up to 1200 seconds. A brake temperature-controlled cycle can be preset for minimum brake temperatures up to 1000°F.

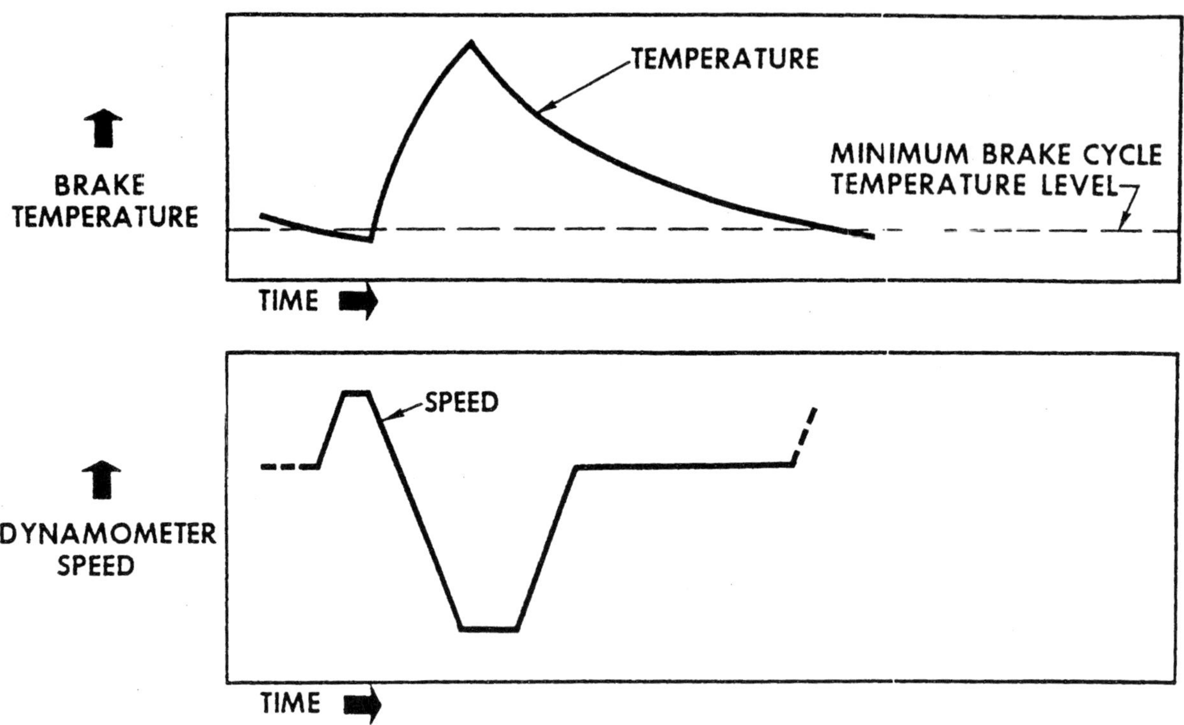

Figure 37 — Time or Temperature Controlled Brake Cycle

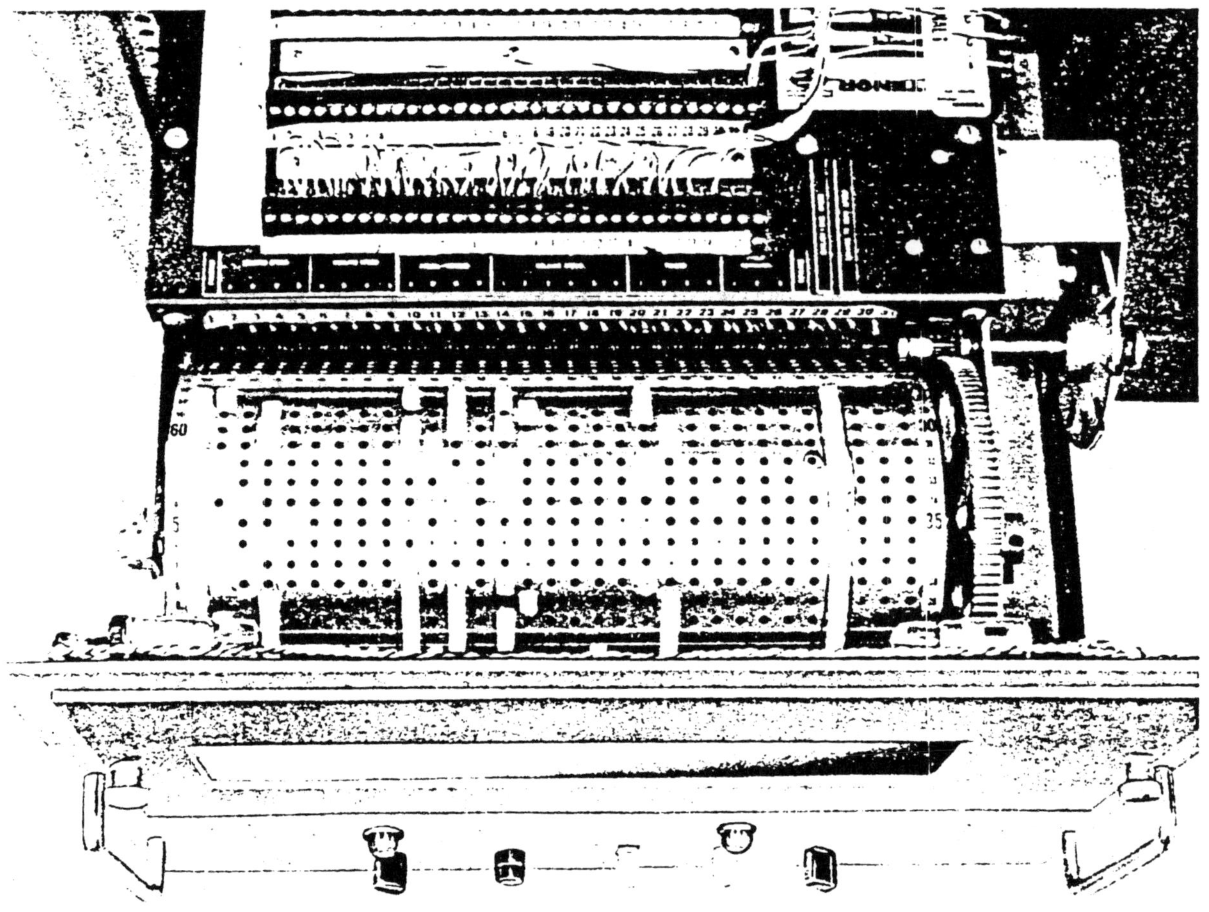

Figure 38 — Program Drum

As mentioned previously, as many as 60 different successive cycles can be controlled automatically by the programmer. This allows a variety of brake stops to be run, simulating an actual road test. The sequence of brake cycles is controlled by the program drum shown in Figure 38. The drum contains 60 rows to provide for programming the 60 different brake cycles. The drum is "pinned" to select the necessary preset dynamometer controllers required to produce the desired program cycle for each row. As each brake cycle is completed, the drum advances to the next row.

When operating under this mode, twelve potentiometers are provided to control dynamometer speed at brake actuation and brake cooling. For a given brake test, a particular speed control is pre-selected for use at brake actuation or brake cooling, depending on how the dynamometer patch board, shown in Figure 39, is connected. Also, if the brake cooling and actuation speeds in the following cycle are the same, the dynamometer may be programmed so that only one potentiometer control is required for these two points in a test. This is a very useful feature, since brakes are generally cooled at the speed at which they will be applied for the next stop. Also up to six different speeds for brake release and up to 12 different decelerations may be programmed into the dynamometer. Two basic timers are used in the control for brake cycle time and dwell time. Four additional timers can be pre-selected by means of the patch board and the program drum to perform any additional timing control for a programmed brake test.

For the Mark II GT tests, the braking schedule, as recorded at Sebring, was programmed into the dynamometer. Figure 40 illustrates the nine brake applications required for one lap at Sebring, along with a map of the course showing the location of the brake applications. The dynamometer speed at brake application and release was programmed to duplicate braking on the race circuit. The brake cycle time could be controlled by the minimum brake temperature, but this was not done due to the previously mentioned transducer limitations. Therefore, the test was controlled by adjusting the over-all brake cycle time to give the described maximum brake temperatures.

Dynamometer Instrumentation

Normally, the operating temperature of each brake is recorded on a two-channel strip chart recorder. A slip-ring assembly is provided at each test station so that temperatures can be conveniently measured with thermocouples installed in the brake drums or rotors. One inboard test station on the dynamometer is provided with a twenty-conductor slip-ring assembly which is used for special single brake tests where a larger number of temperatures may be required. The temperature recorders generally operate continuously during a brake test, since data on both temperature rise during a brake stop and temperature cool down rate between stops must be known.

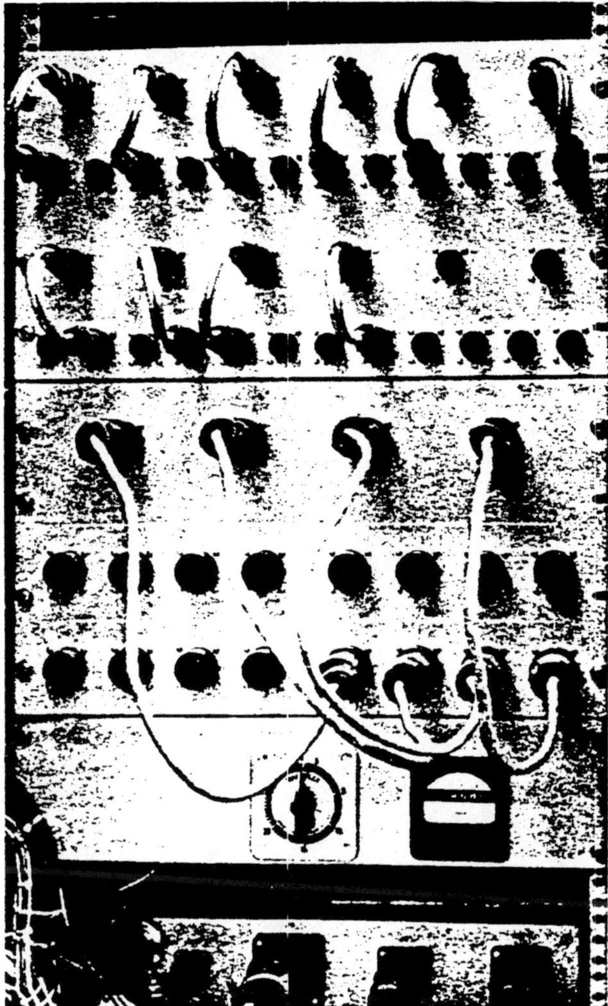

Figure 39 – Brake Dynamometer Programmer Patch Panel

In addition to the temperatures, the stopping torque from the individual test brakes, up to three brake line pressures, and the dynamometer speed are recorded on an eight-channel, multi-speed, direct-writing oscillograph. Figure 41 shows an oscillograph record from a typical four-station passenger car test.

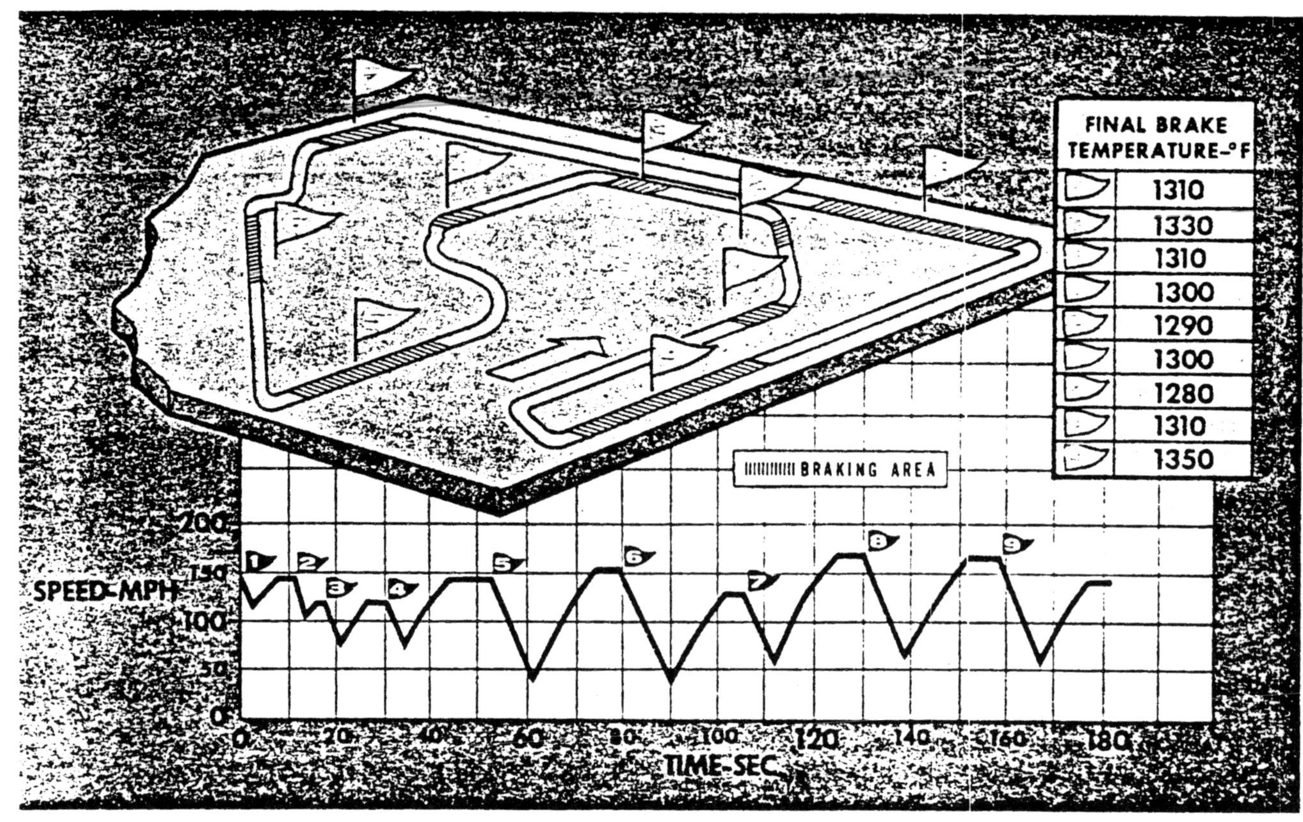

Figure 40 — Dynamometer Simulation of Sebring Race

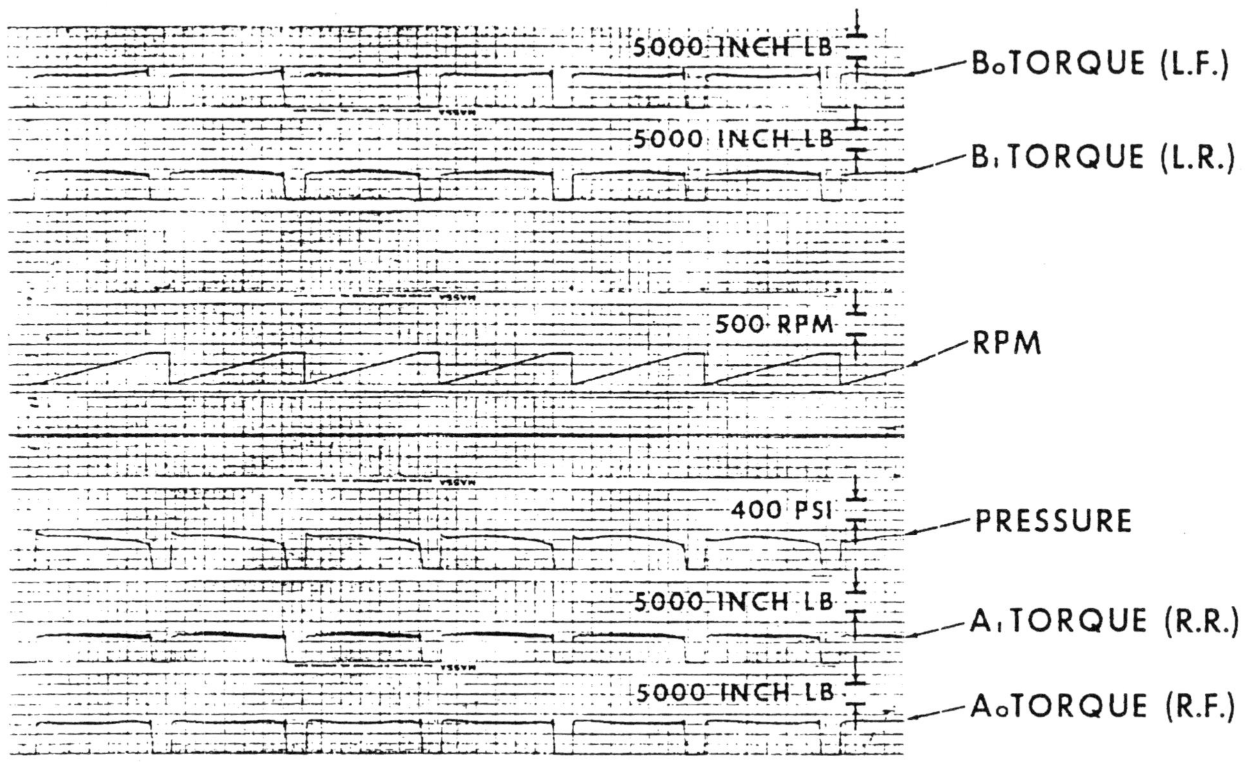

Figure 41 — Typical Oscillograph Record

The oscillograph recording torques, pressures, and dynamometer speed can be operated continuously, but generally, it is on an automatic operating mode so that it only records during the brake stop. The brake dynamometer operator pre-selects the oscillograph recording speed, and for long durability tests, he can preset the recorder to operate at a faster speed on certain pre-selected brake stops.

During the Mark II GT tests, the brake rotor surface temperature was monitored by an infrared radiation pyrometer to insure the same brake operating temperatures in the laboratory as on the track. The rotor interior temperature was measured with a chromel alumel thermocouple. The tests were conducted with only one front brake assembly, since this brake operated at the highest temperature as determined from track tests.

Dynamometer Automatic Shut Down System

Since the brake dynamometer is designed to operate virtually unattended, an automatic shut down system is incorporated in the dynamometer in case of machine or brake failure when the operator is not present. The dynamometer will either coast to a stop or be electrically braked by regenerative braking of the drive motor if any of the following system failures occur:

- Test brake fails to stop dynamometer in the allotted preset time.
- A test brake exceeds a preset temperature.
- Brake cooling air system malfunction.
- Dynamometer overheat, overspeed, or lubrication failure.
- Dynamometer motor-generator set overheat or overspeed.

MARK II GT TESTS

The brake rotor design which was used for the Daytona race and for some practice sessions on the Sebring circuit was the "baseline" rotor for the dynamometer tests. It was hoped that the problems experienced with this rotor could be duplicated in the laboratory on the brake dynamometer. Three of these "baseline" rotors were tested. Each ran an equivalent to 100 simulated Sebring miles on the dynamometer before surface pitting and cracks began to appear. These tests produced brake rotor failures in the laboratory which were identical to the failures experienced on the race track. Thus, the dynamometer proved to be the tool required to attain the ultimate goal of developing a reliable brake system for the Sebring and Le Mans races.

After the dynamometer tests proved duplication of actual race conditions, it was a matter of evaluation of the various proposed disc brake rotors to determine optimum design. During the tests, the brake rotors were inspected every 90 brake stops (10 Sebring laps) for failures. Brake lining wear measurements were made so that the effect of rotor material and rotor condition on the brake lining would be known.

Seven disc brake rotor designs were tested before the Sebring race. Also, up to three samples of the more promising designs were tested. In addition to helping to develop a brake rotor with a longer life, the rotor warping or "dishing" problem discussed earlier in this paper was discovered during dynamometer tests. Although this problem was not solved in time for the Sebring race, it was learned that the problem could be minimized by accurately centering the rotor between the two sides of the caliper assembly.

After the Sebring race, more testing was conducted to further refine the brake system for Le Mans. Between the Sebring and the Le Mans races, 22 different brake rotor designs, including up to four samples of the more promising designs, were evaluated on the dynamometer. In addition to subjecting the brakes to normal racing-operating temperatures, the best brake designs were tested at higher temperatures such as might be experienced if the driver was forced by competition to run a faster lap time. The higher temperature testing was accomplished by shortening the brake cycle times enough to get the desired temperatures.

Although the brake dynamometer was used for the major portion of the brake development testing, brake tests were also conducted using actual vehicles on the Riverside, California racing circuit. In this phase of testing, the more promising disc brake rotor designs from the dynamometer tests were evaluated on a Ford Mark II GT operating on a race track. The brake ultimately selected for the Le Mans race was developed in the laboratory and was verified on an actual car. This required only a few days of brake durability testing with a car.

The dynamometer testing was still not completed after the final brake design was selected for the Le Mans race. The necessary brake parts required for the race had to be manufactured and then quality-checked. During the development tests, it was also learned that some of the brake problems were a result of manufacturing defects rather than brake design defects. A number of the manufactured "production" rotors were tested on the brake dynamometer to insure that these parts had the expected durability and performance before the production batch was shipped to Le Mans.

CONCLUSION

Although the primary task of the Mark II GT test was to develop a winning brake system for the Sebring and Le Mans races, a major benefit was also derived. This was the valuable experience gained in the laboratory testing of brakes. The Mark II GT test did not use the full capabilities of the brake dynamometer, but it was a beginning for many other tests which will come in the future. Using the principles learned in this and succeeding tests, the dynamometer has proved to be a valuable tool for developing better brake systems.

670071

Laboratory Simulation, Mark II–GT Powertrain

B. F. Brender
C. J. Canever
I. J. Monti
J. R. Johnson

Ford Motor Company

In the latter part of 1965, the Testing Department, Engine and Foundry Product Engineering, was assigned the task of providing an "Indoor Laboratory Le Mans." As a laboratory test activity, we were quite aware of certain limitations, both from an equipment and from a control standpoint. Objectively, any laboratory approach must recognize its limitations in attempting to "duplicate" any part of the outside, or customer world.

Several obstacles were involved in this particular "duplication." First, we stand still. Second, we did not have the "educated" toe and associated "anatomical" sensor and servos that any competent high performance car driver has built into his ability to handle a vehicle such as the Mark II. Third, there were certain basic portions of the "cycle" involved that we did not feel we could reasonably duplicate, primarily the braking portion. As a result of these limitations, we were very careful, as most laboratory people are, to avoid the implication of "duplication." Within these limitations, we "simulated" the Le Mans circuit, and subjected our test objects to that simulation.

The objective was to simulate the same speeds, loads, and times experienced during the race, for the proposed powertrains to be used. These were:

1. A Mark II car with a 427 cubic inch displacement engine and a Ford designed four-speed standard transmission and axle.
2. A "J" car with a 427 cu. in. engine and a Ford-designed two-speed automatic transmission and axle.

The engines were supplied by the Engine and Foundry Division, and the transmissions and axle were supplied by the Transmission and Chassis Division.

The primary requirement to begin the simulation was dynamometer equipment containing regulation to produce inertia, grades, roadloads, and braking capability. Such a capability existed in our dynamometer building in the equipment installed in test room 17D (Figure 1). This room contained two-250 horsepower absorption 2500 rpm General Electric dynamometers with the following control circuits:

1. Multiple load — rolling resistance, grades, and braking.
2. Load proportional to speed squared — windage.
3. Load proportional to speed — speed boost.
4. Load proportional to rate of change of speed — inertia.
5. Speed separation — cornering.

The usual requirement of fixtures, brackets, tubes, tanks, and similar items were supplied as required to mount the engine and transaxle assembly between the two dynamometers.

We did not profess to be experienced drivers, in the relationship required. But we did not expect any difficulty in providing the necessary programming control for the Le Mans circuit. For some time, we had been bringing various vehicle tests into the laboratory for better environmental control and closer repeatability for comparative evaluation purposes. The programming of circular tracks, such as Indianapolis and Daytona, do not require a high degree of sophistication, because operation is primarily in one gear with repeatable timed sequences of full-throttle acceleration or closed-throttle decelerations. The Le Mans circuit was quite different and much more complicated.

Figure 1

LE MANS CIRCUIT

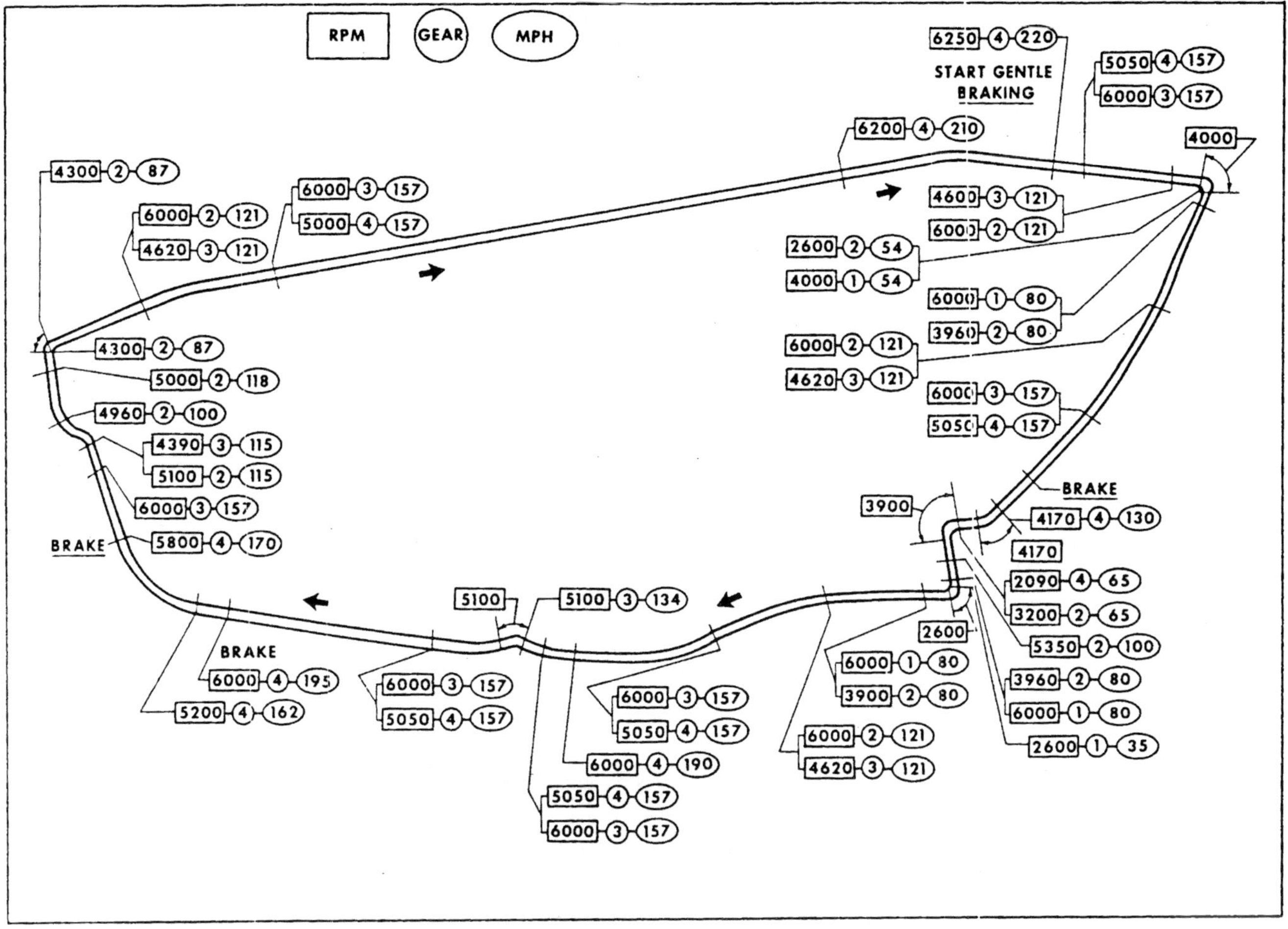

Figure 2

Three sources of information were employed to define the control and programming requirements that were the next "corner" to turn:

1. Vehicle data and characteristics were obtained from computer calculation, and verified by actual vehicle tests.
2. Circuit pattern records were made at Le Mans and other road test locations, on recording oscillograph equipment.
3. We found two drivers who had performed at Le Mans, and despite the fact that these drivers had other worries than remembering their toe positions, the tachometer reading, and gear position, they established a road map (Figure 2) of the circuit indicating vehicle velocity, engine speed, and gear selection patterns at the various locations around the 8.3 mile course. Mr. Phil Hill and Mr. Ken Miles were most helpful.

Our experience in this type of simulation, in respect to correlation to the expected environment, has shown that the validity of results from cycle reproductions is never better than the accuracy of the input information. These three sources in the required combination gave us our accurate input.

The control devices used were of two types. One, identified as the "Mark II Programmer," was used for the standard transmission. The other was a unit in use for general economy and emission cycle testing, identified as the "D.R.I.V.E.R." (Dynamometer Road Imitator Vehicle Economy Rater); this was used for the automatic transmission.

The basic approach used in this simulation is the same as employed for standard and conventional tests and considers three factors. The road is provided through the use of dynamometer-controlled circuitry which, in effect, is an analog computer. The required throttle action is provided by the appropriate cycle programmers. The shift pattern requirements, integrated with the required throttle action, is provided through the use of peripheral devices which afford the forces and direction necessary to shift the transmission in use.

In mounting the engine and transaxle, car conditions were adhered to as closely as possible (Figure 3). The final test set-up included the tuned exhaust manifolds as designed for the vehicle, and the electric fuel pump and engine oil reservoir were mounted in their relative positions to the engine. The fuel pump had to draw from a small fuel tank that maintained a constant fuel level. External oil coolers kept the oil temperatures of the engine and transaxle within prescribed limits.

MARK II PROGRAMMER

Considering the forces involved in operating the clutch, performing a shift, regulating speed of motion, etc., it was apparent that pneumatic mechanisms would be required at the transmission, clutch, and throttle units. Because of the number of air lines, solenoid valves and pressure regulators necessary for the test, it was decided to mount these on a control tray near the engine and transaxle. The programmer cabinet, housing the sequencing relays and switches, was located near the operator at the dynamometer console.

A basic understanding of the shift sequence in the car determined what was needed at the mechanisms and control tray. The procedure employed was to first, design the shift, clutch and throttle units; second, establish the necessary solenoid valves and regulators for the control tray; and finally, design the programmer to operate the solenoid valves to simulate the Le Mans circuit as near as practical.

SHIFTER MECHANISM

The standard transmission tested was a 4-speed transaxle unit with a shift rod at the transmission that required a combination of

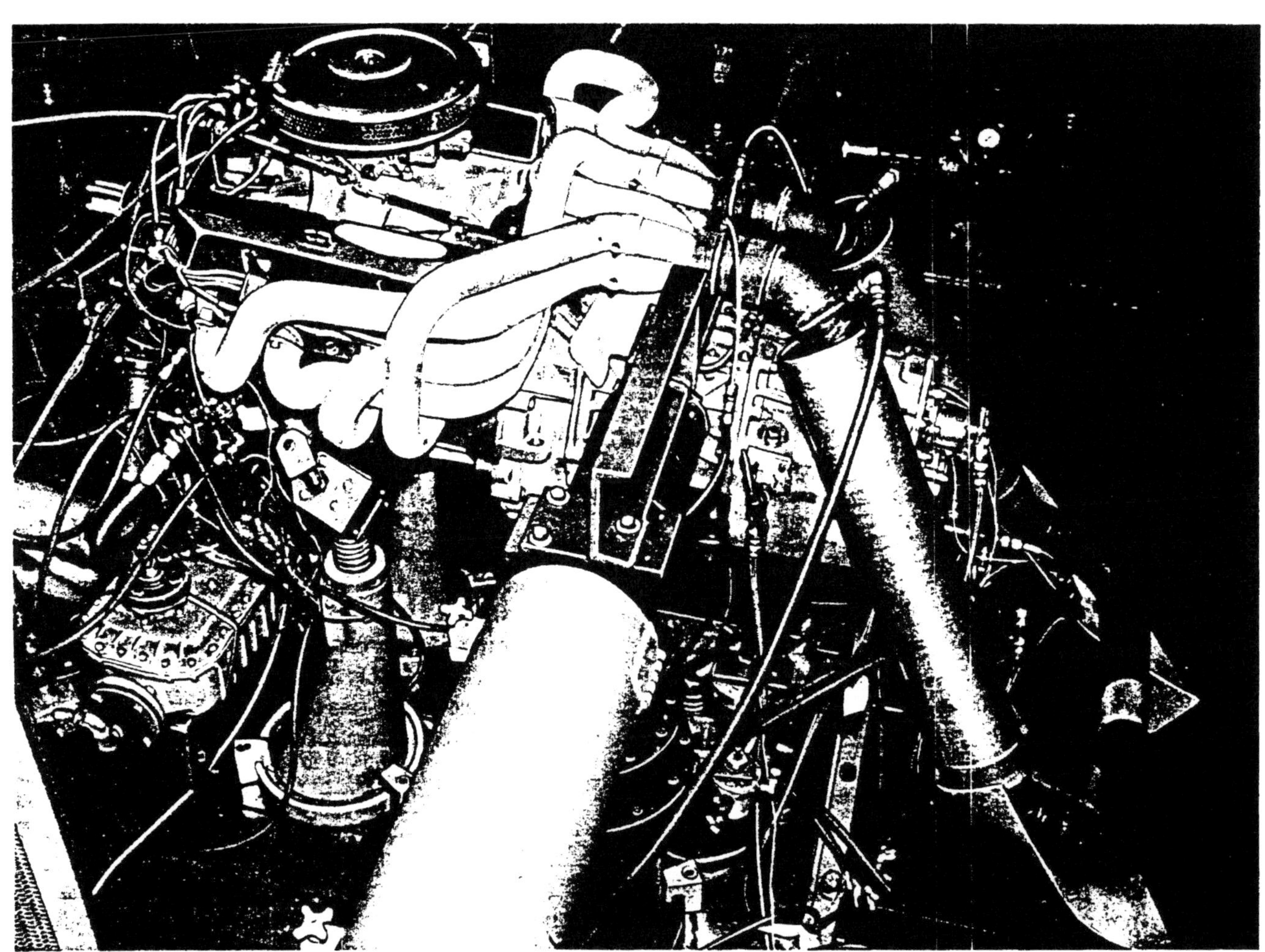

Figure 3

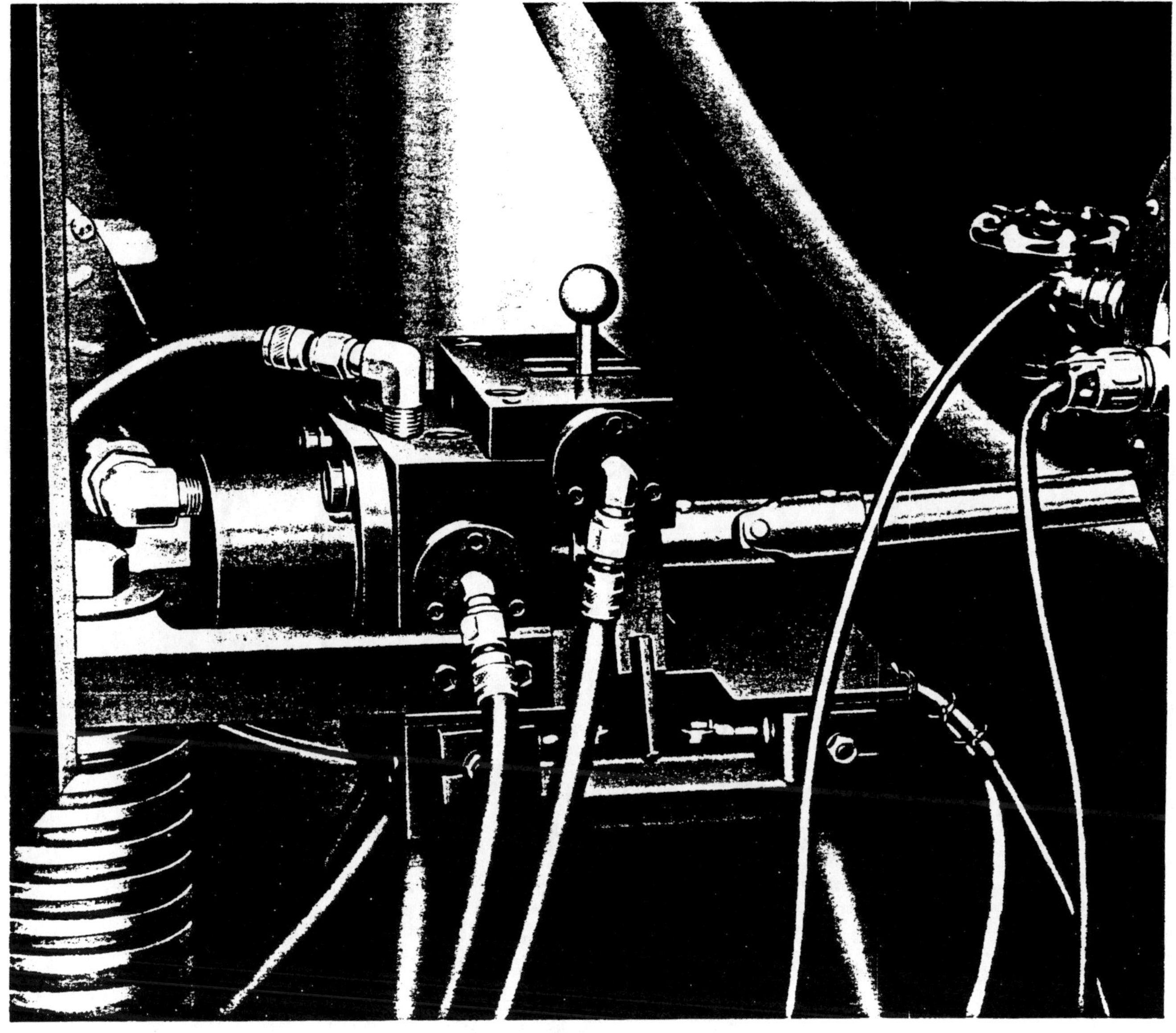

Figure 4

rotation and push-pull to give the "H" shift pattern. Reverse gear was not included in the test cycle. The test shift mechanism was mounted at the same location as the shift lever in the race car (Figure 4), and the interconnecting tube and U-joints were the same as those used on the vehicle. The unit incorporated an H-pattern guide similar to that on the car. The main actuating cylinder accomplished the push-pull necessary to shift into each gear. The rotational motion, needed at neutral position, was performed by several lateral pistons acting on a lever connected to the main piston rod. An additional set of transverse pistons with stop pins served to stop the main piston at neutral position when required. On this test, one stop-pin was used for the 4-2 downshift and the other for the 1-N shift at a stop. Microswitches were mounted on the mechanism at each gear position and at neutral.

Shift speeds were performed within 0.3 to 0.7 seconds. To help reduce the shift speed within these reasonable limits, an air-over-hydraulic system was used on the push side of the main shift cylinder. Provision was made to have a pressure regulator for each of the shifts, since shift effort and shift time were interrelated. These regulators and necessary solenoid valves were mounted on the control tray. Figure 5 provides a drawing of the shift mechanism and its pneumatic circuitry.

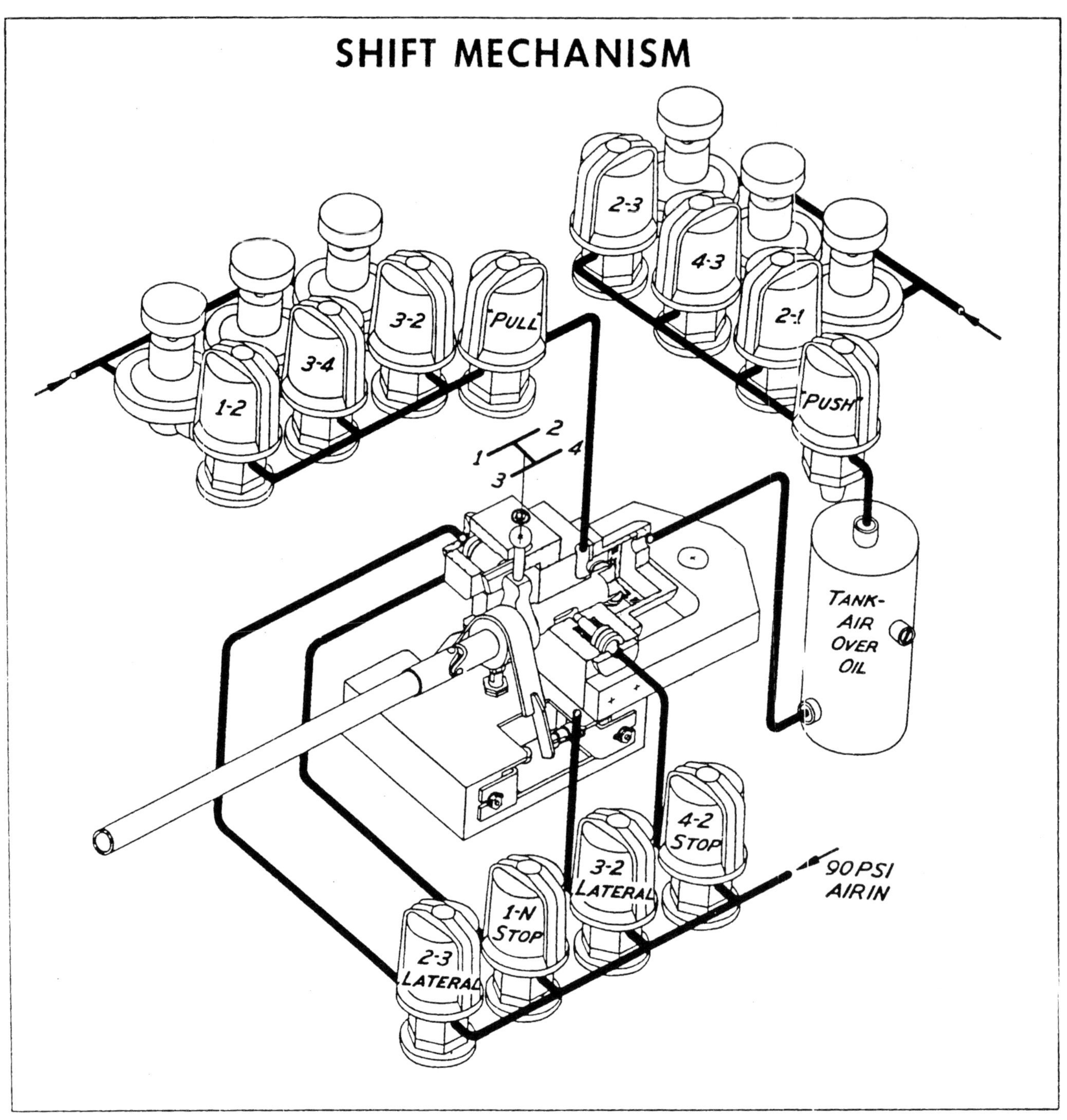

Figure 5

CLUTCH MECHANISM

An air cylinder and level arrangement were used to operate the clutch. Adjustable stop nuts limited the lever travel as required. A spring served to retract the lever and air cylinder when re-clutching. Air was used to actuate the lever 1/4" to 1/2" past the de-clutch point.

The rate at which this air was released or exhausted regulated the speed of re-clutching. Needle valves were used to regulate the re-clutch speed. Five needle valves proved sufficient; one for each of the four gears on acceleration, and one for all the gears on deceleration. Figure 6 illustrates the linkage and pneumatic circuitry of the clutch mechanism. Also included was a microswitch arrangement to permit shifting the transmission only if the engine was de-clutched. A ramp sleeve was adjusted to actuate the microswitch at a position slightly past the de-clutch point.

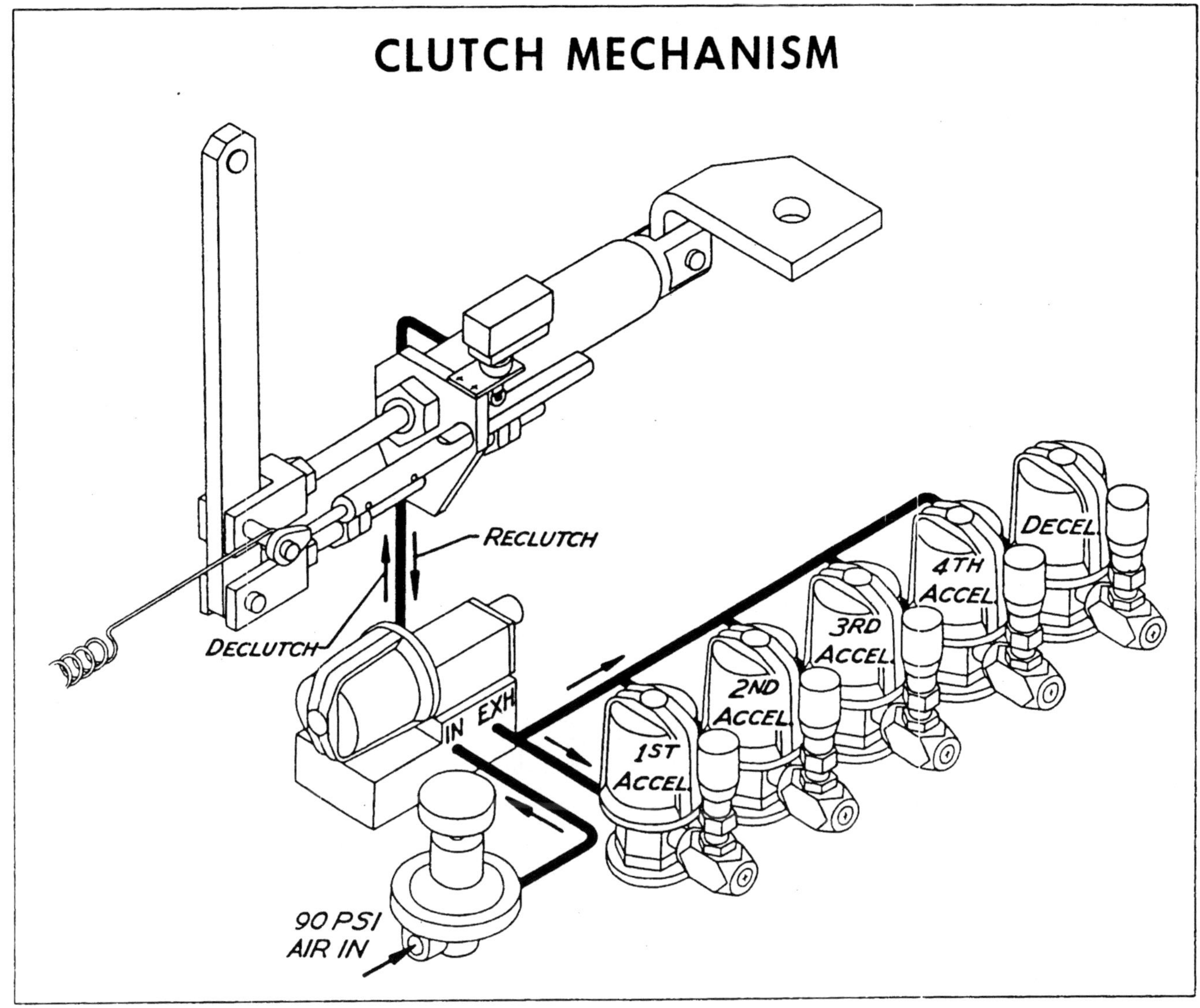

Figure 6

THROTTLE MECHANISM

A pneumatic, direct-action positioning actuator was used to operate the throttle. This provided the ability to set to any throttle position by simply varying the signal pressure. The unit was installed so that an air pressure signal would open the throttle — no pressure would return the throttle to idle. An adjustable W.O.T. stop was used to prevent damage to the carburetor, and a guide plate kept the piston rod from rotating and jamming the linkage to the carburetor. A throttle pressure regulator for each gear was provided at the control tray for acceleration. On this test, these regulators were adjusted to give W.O.T. To aid in synchronizing the throttle movement with the clutch for a smooth acceleration, needle valves were used to control the opening speed of the throttle. A "dump" solenoid valve served to cut-off and dump the throttle signal pressure when the unit was ready to shift. The complete pneumatic circuit for the throttle is shown in Figure 7.

When decelerating, it is customary in racing to "jog" or "feather" the throttle when downshifting. This was also provided for with one regulator sufficing for all the gear downshifts. An extra 4th gear throttle regulator was installed for the three mile Mulsanne straight at Le Mans.

CONTROL TRAY

The solenoid valves on the control tray are electrically operated by the programmer. The tray, in turn, pneumatically operates the throttle, clutch, and shift mechanisms. The tray was supplied with 90 psi air from our building

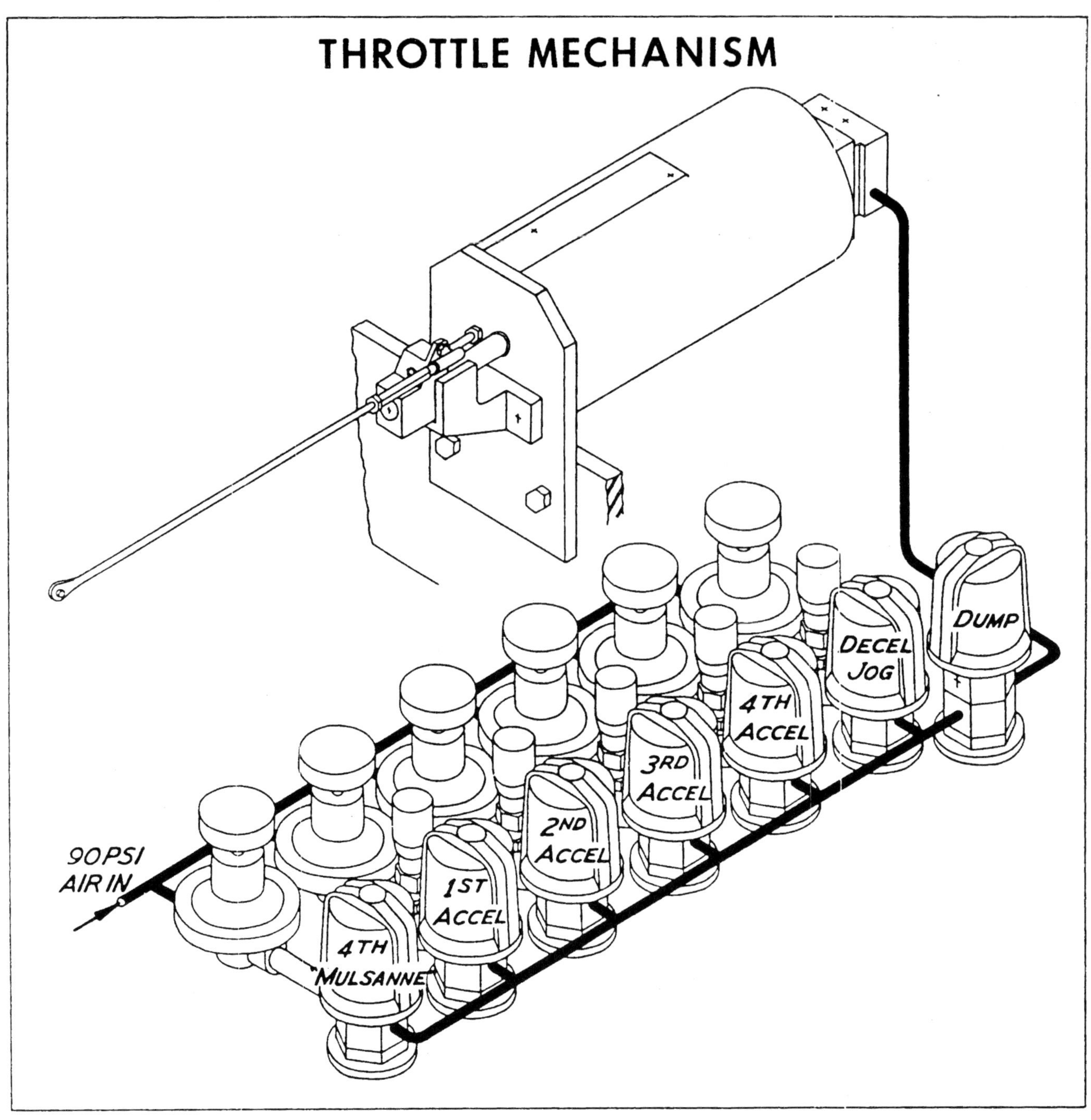

Figure 7

facilities. Adjustment of the pressure regulators and needle valves were made to give a smooth acceleration and deceleration pattern. Accelerations were at W.O.T. and decelerations were at "no-throttle", except for "feathering" the throttle when downshifting. The needle valves for the re-clutch rates and throttle opening rates were adjusted to synchronize the two correctly. The shift pressure regulators were adjusted to give shift times of 0.3 to 0.7 seconds for each gear.

PROGRAMMER

The programmer itself was modified from a shift-o-matic unit that is used for transmission durability and development work at our facilities. The unit was revised somewhat to provide shift patterns suited for simulating the Le Mans circuit. Figure 8 illustrates the pattern established for dynamometer testing and represents a practical solution to this particular problem.

An exact cycle pattern was considered. However, this would have necessitated construction of a new programmer with a resulting time delay. Figure 9 gives an actual acceleration pattern of a portion of the cycle in terms of engine and dynamometer rpm.

The cabinet contains the electrical sequencing relays and related equipment necessary to use the signals from a tach generator at one of the dynamometers, and the microswitches at the clutch and shift mechanisms to operate the shifter control tray and dynamometer to simulate the Le Mans circuit. Although shifts were made on the basis of engine speeds, a tach generator, directly connected to one of the dynamometers, was used to effect the shifts. The cycling pattern was derived from engine rpm, transmission gear, and vehicle speed combinations indicated on the Hill-Miles Le Mans circuit map. Analysis of this information showed that shift speeds were fairly consistent over the

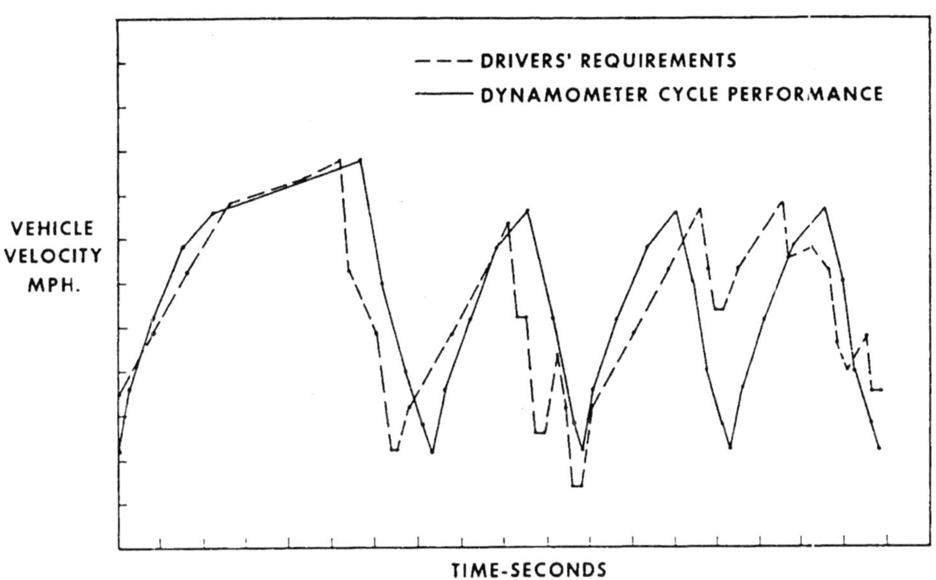

Figure 8

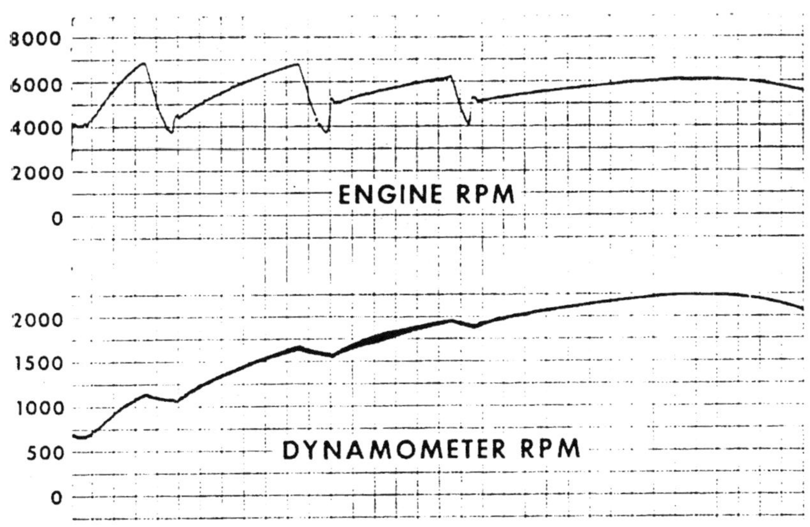

Figure 9

course. From this, it was decided that all shifting would be at the following engine rpm.

1 — 2 at 6800 rpm	4 — 2 at 3900 rpm
2 — 3 at 6800 rpm	3 — 2 at 3900 rpm
3 — 4 at 6200 rpm	2 — 1 at 2600 rpm
4 — 3 at 5040 rpm	

The cabinet centers around a stepping switch with a choice of several shift patterns available, any one of which can be selected at each step on the switch. The patterns available on this test were:

1 - 2 - 3 - 4 - 3 - 2 - 1,
1 - 2 - 3 - 4 - 2 - 1 where 3rd gear is skipped,
1 - 2 - 3 - 4 - 3 - 2 - 1 with a 3-mile run in 4th,
N - 1 when starting, and
1 - N when stopping.

Shift speed dials, coupled to potentiometers mounted on the cabinet, were adjusted to take the dynamometer tach generator output and trigger one of two voltage-sensitive relays to give the selected shift speeds listed above. Starting the "race" and stopping for the "pit stops" every two hours was performed through a "cycle" switch on the cabinet. A feature of the switch was that even though the operator would turn the switch off prematurely for a pit stop, the programmer would continue to complete a normal shift pattern back to neutral gear.

The electrical system contained additional features to implement a rapid and safe shift. Throttle pressure was automatically dumped whenever the engine was de-clutched, and to synchronize the two, delay action on the de-clutch was necessary. This was effected through an adjustable time-delay relay. The shift microswitch at the clutch mechanism had to be actuated before a shift force could be exerted. On acceleration, throttle pressure and re-clutching the engine were allowed only if the transmission was in gear; i.e., a gear position microswitch was actuated. Once the shift was accomplished, the shifting force was removed. The cabinet also provided visual indications of the gear position of the transmission, the part of the program being performed, and the number of laps completed.

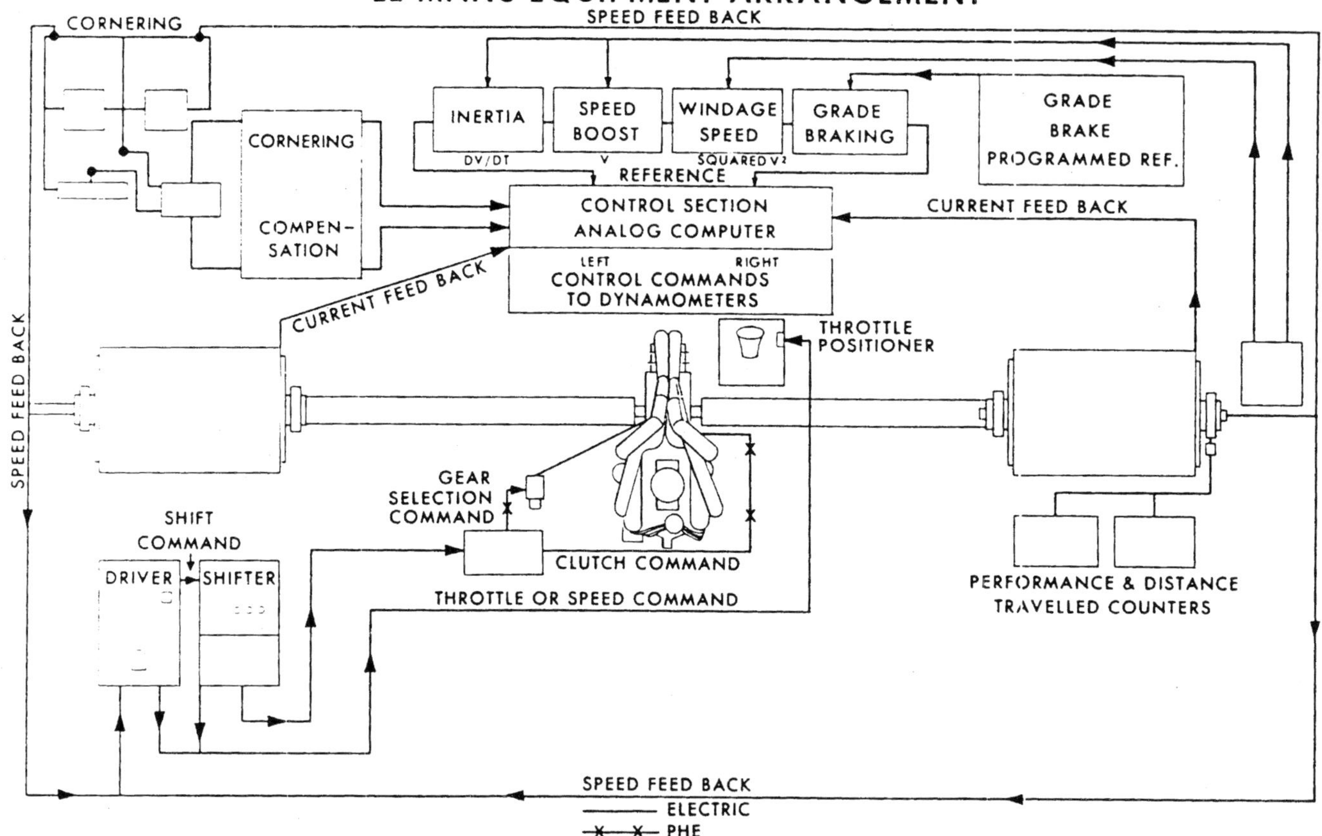

Figure 10

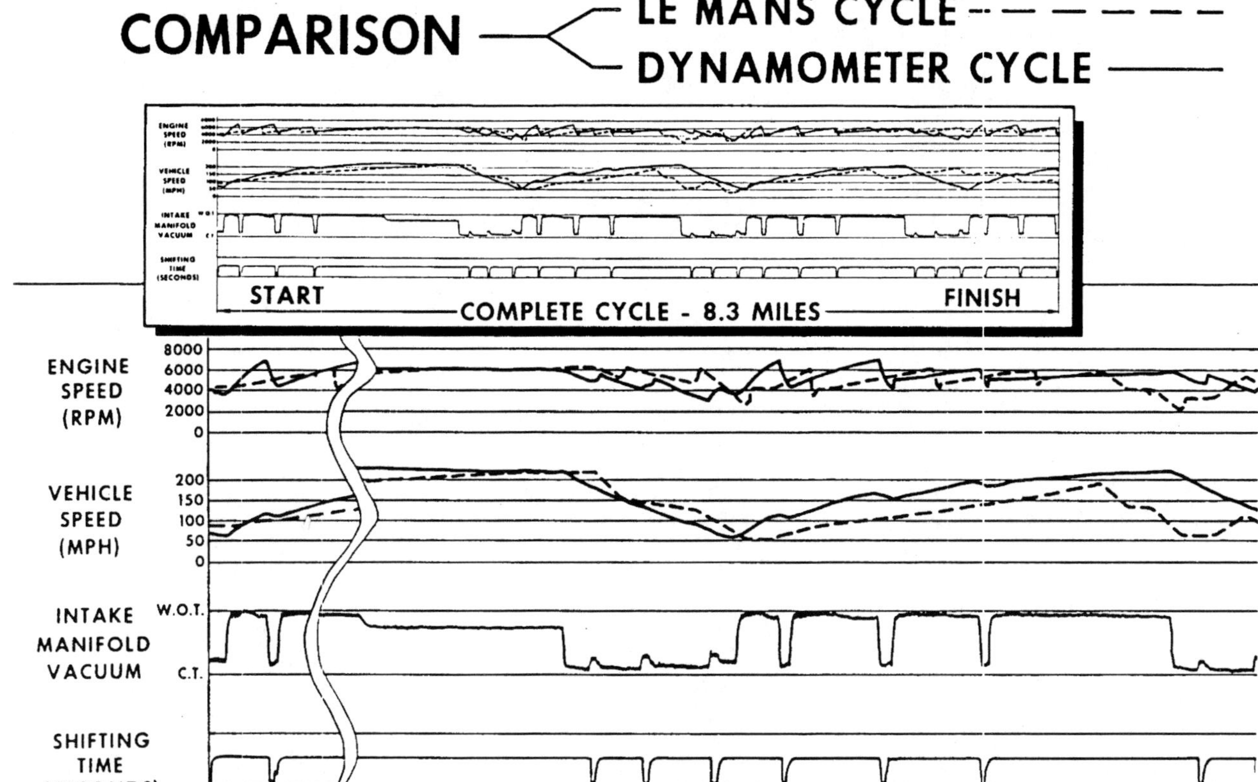

Figure 11

J-CAR PROGRAMMER

The 427 cu. in. engine with an automatic transaxle was cycled by an entirely different unit. This unit, which is called the "D.R.I.V.E.R.," was built specifically for economy test development work. It employs a punched-tape electrical system and was easily adapted to simulate the Le Mans circuit. Here again, pneumatic mechanisms were used to actuate the throttle and select the transmission gear. An air cylinder was used to select the HI or LO position on the transmission. A diaphragm-type positioner was installed to actuate the throttle.

An over-all schematic is shown in Figure 10, indicating general relationships of the equipment as described.

Figure 11 is an actual recording oscillograph trace of the dynamometer cycle showing engine rpm, axle rpm, and manifold pressure. Superimposed on this trace is the pattern established by actual vehicle-generated oscillograph traces.

By direction, all powertrain components were required to complete forty-eight hours of "simulated Le Mans dynamometer cycle" testing. (The actual event requires 24 hours.) Failures did occur during these tests. Investigations and design or manufacturing changes resulted, and these then were again subjected to the test.

Actual results during the 1966 Le Mans race correlated with the dynamometer results. The simulated dynamometer cycle testing precluded the recurrence of component failures in the race that were experienced during the 48-hour endurance run in the laboratory.

SUMMARY

The Testing Department personnel directly involved — equipment designers, equipment technicians, instrument technicians, dynamometer operators, and test engineers -- are confident that the successful "Laboratory Simulation — Mark II — Powertrain" approach and completion of the required tests through its use were a significant factor in the outcome of the 1966 Le Mans race. Due credit is extended to the engine designers, transaxle designers, manufacturing personnel, vehicle designers, vehicle builders, and the drivers and their pit crews. "What made the wheels go around" had to pass the laboratory test first. This approach did not "duplicate," but it did "simulate" what the components had to do. The result was a double victory for the Mark II Ford GT — first in the laboratory and then at Le Mans.

CPSIA information can be obtained at www.ICGtesting.com
Printed in the USA
LVOW092302220413

330358LV00004B/16/P